Molecular and Integrative Toxicology

Series Editor
Rodney R. Dietert

For further volumes:
http://www.springer.com/series/8792

Kathleen M. Gilbert • Sarah J. Blossom
Editors

Trichloroethylene: Toxicity and Health Risks

Humana Press

Editors
Kathleen M. Gilbert
Department of Microbiology and Immunology
Arkansas Children's Hospital Research Institute
University of Arkansas for Medical Sciences
Little Rock, AR
USA

Sarah J. Blossom
Department of Pediatrics
Arkansas Children's Hospital Research Institute
University of Arkansas for Medical Sciences
Little Rock, AR
USA

ISBN 978-1-4471-6310-7 ISBN 978-1-4471-6311-4 (eBook)
DOI 10.1007/978-1-4471-6311-4
Springer London Heidelberg New York Dordrecht

Library of Congress Control Number: 2014931978

Printed on acid-free paper

Humana Press is a brand of Springer

Springer is part of Springer Science+Business Media (www.springer.com)

Preface

This book is designed to highlight the best-characterized aspects of trichloroethylene (TCE) toxicity. These include both cancer and non-cancer endpoints. Epidemiologic data concerning the effects of TCE on human health are presented, as well as results obtained in animal models. When available, mechanistic information was provided, and future research directions were outlined.

As noted in recent toxicological reviews of trichloroethylene by the National Research Council and the US Environmental Protection Agency, the evidence on human health hazards from TCE exposure has strengthened in recent years. Among other things, this has led to TCE being upgraded as a carcinogen. Based on the likelihood of exposure together with likely negative health impact, TCE is ranked 16th out of 275 chemicals on the CERCLA (Comprehensive Environmental Response, Compensation, and Liability Act) list of hazardous chemicals. TCE use in the workplace has declined in recent years. However, due to improper disposal methods TCE is a common water contaminant. Although TCE levels in community water systems are generally monitored, TCE levels in private wells (a common source for irrigation, and the source of drinking water for about 10 % of Americans) are often unknown. Contact with TCE may be elevated for people living near waste facilities where TCE is released, residents of some urban or industrialized areas, or individuals using TCE-containing products.

The book begins with a history of TCE use in the US. Chapter 1 highlights the many commercial uses of TCE and explains why its use as a solvent, spot remover, defatting agent in food processing, and even anesthetic, was so wide-spread. Chapters 2, 3, and 4 characterize TCE immunotoxicity. Chapter 2 discusses the autoimmunity associated with adult TCE exposure in both human and mouse models. Chapter 3 focuses on dermal and systemic hypersensitivity caused by occupational exposure to TCE. Chapter 4 discusses TCE-induced protein oxidation and its role in immunotoxicity. Chapters 5 and 6 characterize neurotoxicity linked to TCE exposure, namely Parkinson's disease and other persistent and neurological changes. Most epidemiological and experimental studies of TCE toxicity have focused on adult exposure. However, the developing organ systems appear to be especially sensitive to environmental perturbation. Along these lines, Chaps. 7 and 8 discuss developmental toxicity associated with TCE exposure during gestation and/or early life. Chapter 7 focuses on the effects of TCE on the brain and immune system, while Chap. 8 centers on how TCE impacts the developing heart. Chapter 9 discusses the connection between TCE and kidney cancer, and highlights the regulatory pathway taken to have the chemical classified as a carcinogen. The last two chapters in the book discuss aspects of TCE toxicity that represent important future directions in TCE research. Chapter 10 discusses the possible role that TCE-induced epigenetic changes, specifically DNA methylation, plays in the toxicity of the chemical. Lastly, Chap. 11 focuses on mathematical modeling of TCE toxicity, and goes beyond pharmacokinetics to discuss use of pharmacodynamic modeling to tease out contributions to etiology.

We would like to acknowledge the chapter authors for their efforts in creating this book. Even though they all have many demands on their time, they generously contributed toward this effort to publicize the toxicity of this threat to human health.

Little Rock, AR, USA Kathleen M. Gilbert
Little Rock, AR, USA Sarah J. Blossom

Acknowledgements

We wish to thank Bev and Joe Gilbert and Clay Fendley for all their help and support over the years.

Contents

Chapter 1
History of TCE

Richard E. Doherty

Abstract The use of trichloroethylene (TCE) spans a period beginning in the early twentieth century and continuing to the present day. Although the largest use of TCE in terms of volume was in the degreasing of metals, it was also used in dry cleaning, textile processing, food processing, medical applications, chemical production, and a variety of consumer products. The use and production of TCE evolved significantly over time in response to market conditions, historical events, economic climate, technology development, environmental regulations, toxicity concerns, and the availability of competing products. The spillage and disposal of TCE resulted in the contamination of countless groundwater supply wells, the cleanup of which will likely continue for decades to come.

Keywords Trichloroethylene • History • Trichloroethene • TCE

1.1 Introduction

Trichloroethylene (TCE, C_2HCl_3), also known by its IUPAC[1] name of trichloroethene and a host of other chemical and trade names (see Table 1.1), has been used for a variety of industrial, commercial, medical, and consumer applications. Its widespread use, particularly in the mid-twentieth century, stemmed from its powerful solvent action on fats, greases, oils, resins, waxes, and a variety of other natural and synthetic substances. The use of TCE as an industrial metal cleaning agent,

[1] IUPAC, the International Union of Pure and Applied Chemistry, is recognized as the world authority on chemical nomenclature.

R.E. Doherty, PE, LSP
Engineering & Consulting Resources, Inc., 966, Acton, MA 01720, USA
e-mail: rdoherty@alum.mit.edu

K.M. Gilbert, S.J. Blossom (eds.), *Trichloroethylene: Toxicity and Health Risks*, Molecular and Integrative Toxicology, DOI 10.1007/978-1-4471-6311-4_1,

Table 1.1 Synonyms for trichloroethylene (Doherty 2000; IARC 1997; Barbalace 2013)

Chemical names:
1,1,2-trichloroethylene
1,2,2-trichloroethylene
1,1-dichloro-2-chloroethylene
1-chloro-2,2-dichloroethylene
Acetylene trichloride
Ethinyl trichloride
Trichloroethene
Ethylene trichloride
Trade names (manufacturer names in italics):
Algylen
Alk-Tri *(Dow Chemical)*
Anamenth
Benzinol
Blacosolv (industrial grade)*(G.S. Blakeslee)*, Blancosolv
Cecolene
Chlorylen, Chlorilen, Chlorylea
Circolsolv
Crawhaspol
Densinfluat
Dow-Tri *(Dow Chemical)*
Dukeron
Ethyl Trichloroethylene (industrial grade)
Ex-Tri *(Dow Chemical)*
Fleck-Flip, Flock Flip
Fluate
Gemalgene, Germalgene
Hi-Tri *(Dow Chemical)*
Lanadin
Lethurin
Narcogen, Narkogen
Narcosoid, Narkosoid
NCI-C04546
Neu-Tri *(Dow Chemical)*
Nialk, Nialk Trichlor (*Hooker Chemical)*
Perm-A-Clor (*Hooker-Detrex, Inc.*)
Petzinol
Philex, Phillex (industrial grade)
Stauffer Trichloroethylene *(Stauffer Chemical)*
Threthylen, Threthylene, Trethylene
Triad, Triad-E *(Hooker-Detrex, Inc.)*
Tri, Trial, Triasol
Trichlooretheen, Trichloorethyleen
Trichloraethen, Trichloraethylen
Trichlor Type 113/114/115/122 (industrial grade)
Trichloran, Trichloren
Tricloretene, Tricloroetilene
Triclene, Tri-Clene (*DuPont, Diamond Shamrock*)

Table 1.1 (continued)

Trielina, Trielene, Trielin, Trieline
Triklone (industrial grade)
Trilene, Trilen, Triline (anesthetic grade)
Trimar
Triol
Tri-Paint Grade (industrial grade)
Tri-Plus, Tri-Plus M
Trisan
Trivec
Tromex
Un 1710
Vapoclean
Vapoclor
Vestrol
Vitran
Westrosol

primarily in the vapor degreasing process, was undoubtedly its largest use in terms of quantity. However, other important uses existed. This chapter explores major uses of TCE and how they evolved over time, and briefly discusses TCE's role as an environmental contaminant.

TCE was first prepared in 1864 by Dr. E. Fischer of Neustrelitz, Germany (Fischer 1864). However, TCE was not utilized for commercial or industrial purposes until over 40 years after Dr. Fischer's work. Potential commercial applications in dry cleaning, textiles, varnishes, and as an extraction agent for fats led to the construction of production facilities in Yugoslavia in 1908 (Gerhartz 1986) and Germany in 1910 (Mellan 1957). In the United States, limited production by Dow Chemical and the Carbide and Carbon Chemicals Corporation (a predecessor of Union Carbide) began in 1921. The Roessler & Hasslacher Company (a predecessor of DuPont) began production in Niagara Falls, New York in 1925 (Doherty 2000).

TCE was produced by a variety of methods that evolved over time. The primary production process, developed in Austria in approximately 1905, involved the chlorination of acetylene to produce 1,1,2,2-tetrachloroethane, which was then dehydrochlorinated to produce TCE (Hardie 1964). During the 1970s, the rising price of acetylene gradually rendered this method uncompetitive, and the last major manufacturing plant in the United States (US) that utilized the acetylene process was closed by Hooker Chemical in 1978. More recent widely-used production processes include high-temperature chlorination of ethylene or 1,2-dichloroethane, and the oxychlorination of ethylene or C_2 chlorinated hydrocarbons (Mertens 1991). All three of these major production methods yield tetrachloroethylene (PCE) in addition to TCE.

Pure TCE slowly auto-oxidizes in the presence of air, and rapidly degrades when in contact with aluminum. Beginning in approximately the mid-1930s, commercial grades of TCE included low concentrations (typically 0.1–0.5 %) of stabilizing chemicals to counteract the deleterious effects of acids, certain metals, oxygen, heat, and/or light. Stabilizers used for TCE generally fall into one of three classes:

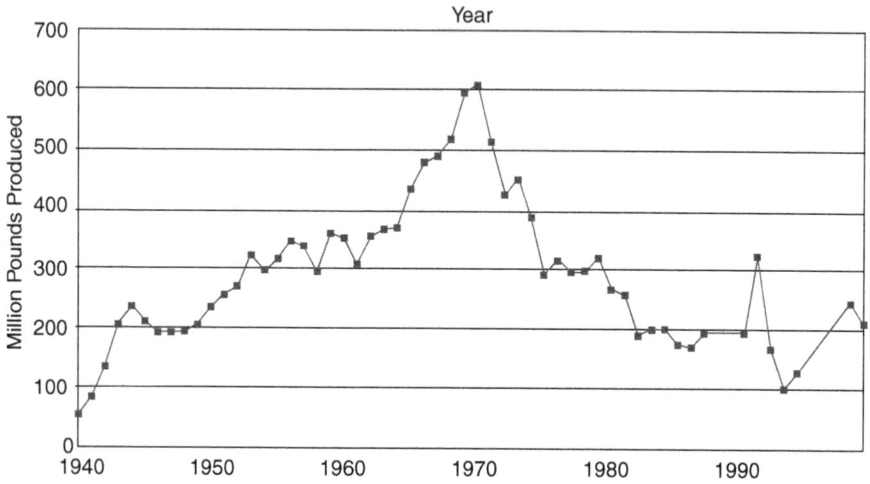

Fig. 1.1 US production of TCE (Doherty 2000; Lee et al. 2003; Leppart 1945)

acid acceptors to neutralize hydrogen chloride, metal stabilizers to form complexes with metal salts, and antioxidants to retard formation of oxidation products. Acid acceptors used prior to 1954 commonly included trimethylamine, triethylamine, and diisopropylamine, which were of limited effectiveness because they were consumed in the neutralization process. The introduction of a neutral, pyrrole-based stabilizer formulation by DuPont in the mid-1950s began the transition away from amine-based formulations. Other chemicals used as TCE stabilizers included alcohols, esters, ethers, substituted phenols and epoxides (primarily epichlorohydrin and butylene oxide). TCE used as an analgesic contained 0.008–0.012 % thymol as an anti-microbial agent. Dyes such as waxoline blue were sometimes added to help distinguish TCE from chloroform (Aviado et al 1976; Doherty 2000).

Production volumes of TCE in the United States are shown on Fig. 1.1. TCE production reached a short-term peak in the mid-1940s to meet World War II demands for production and cleaning of military equipment. Similar to nearly all industrial materials, TCE's production and distribution were strictly controlled by the United States government during the World War II years to allow military and, to a lesser extent, essential civilian demands to be met. The evolution of the various US government orders that restricted the use of TCE and allocated available supplies are described by Doherty (2012).

After World War II, TCE production continued to increase, primarily in response to its use in metal degreasing. Civilian and Korean War-related demand led to a December 1950 US ban on hoarding of TCE and 54 other scarce materials (J City Post Tribune 1950). Although TCE producers significantly expanded production capacity in the post-war years, supplies remained relatively scarce until the economic slowdown of the late 1950s. During the 1960s, decreased imports, high demand for vapor degreasing solvent, and military demand associated with the Vietnam War caused TCE's production in the US to increase significantly, reaching a peak in 1970 when approximately 600 million pounds were produced (US Tariff Comm 1971).

One of the first environmental regulations to significantly affect demand for TCE was California's Rule 66, a law intended to control the use of smog-promoting chemicals. Enacted in August 1966, Rule 66 caused many TCE users in California to switch to alternative solvents such as PCE or 1,1,1-trichloroethane (C&EN 1969). On a national level, the 1970 Clean Air Act regulated TCE as an air pollutant and set emission limits on users in ozone non-attainment areas. In March 1975, the National Cancer Institute (NCI) released its finding that TCE caused cancerous tumor growth in mice livers (NCI 1976). In 1976, the US Environmental Protection Agency (EPA) added TCE to its Hazardous Substances List. These and other factors (including the shutdown of TCE plants utilizing the acetylene process) led to significant reductions in production and use of TCE throughout the 1970s.

While environmental regulations contributed to a continued decline in TCE production and use in the 1980s, they resulted in a minor resurgence in use in the 1990s. The 1990 Clean Air Act Amendments and the 1990 amendments to the Montreal Protocol set dates by which the use of 1,1,1-trichloroethane, 1,1,2-trichloro-1,2,2-trifluoroethane, and carbon tetrachloride (among other chemicals) would be severely restricted. Many users of these chemicals turned to TCE as a substitute, contributing to a reported 10 % annual increase in TCE consumption in the US between 1993 and 1996 (Leder 1999).

Between 1996 and 2012, annual US exports of TCE ranged from a low of 36.8 million pounds in 2005 to a high of 84.4 million pounds in 1998. In 2011 and 2012, over 50 million pounds were exported each year. Major importers were Korea, the Netherlands, and China, each of whom imported over 100 million pounds from the US over the 17-year period (USITC 2013). Global consumption of TCE in 2011 was estimated at 945 million pounds, of which 255 million pounds was in the US (Glauser and Funda 2012).

1.2 TCE Uses

1.2.1 Dry Cleaning and Textile Processing

TCE, along with carbon tetrachloride, was one of the first chlorinated solvents to be used in dry cleaning as a substitute for petroleum-based cleaners. While the chlorinated solvents were historically more expensive than petroleum cleaners, they offered two major advantages: they had essentially no risk of fire or explosion, and they did not leave a residual odor on dry-cleaned clothing. TCE's use in dry cleaning began in approximately 1930, and increased throughout the decade. However, dry cleaning use decreased after it was found to cause bleeding of dyes on cellulose acetate fibers (Chem Wk 1953).

TCE formulations prepared for use in the textile industry were available in England as of 1912 (Chem Tr J Chem Eng 1912). TCE was used to scour wool, cotton, and other fabrics, and as a solvent in dyeing and finishing operations.

In the years following World War II, TCE was used to remove waxes and oils from natural and artificial fabrics in preparation for desizing, bleaching, dyeing, printing and finishing (Hardie 1964). As of the early 1990s, it was in use as a carrier solvent for spotting fluids, and in waterless dyeing and finishing operations (Mertens 1991). The 1990 restrictions on the use of 1,1,1-trichloroethane prompted the textile industry to use TCE to remove oil and grease stains from fabrics (Mirza et al. 2000).

1.2.2 Metal Cleaning and Degreasing

TCE's predominant use was as a degreasing agent, an application for which the chemical was remarkably well-suited due to its solvent action, noncorrosivity, rapid evaporation, lack of flammability, and ease of recycling. TCE was used for this purpose by a variety of industries, including electronics, defense, aerospace, aviation, rail, shipbuilding, and automotive, among others. Among the many hundreds of TCE's cleaning applications was the flushing of liquid oxygen tanks, liquid hydrogen tanks, and associated piping systems by the aerospace industry (PPG Ind 1999).

Although TCE was utilized for so-called "cold cleaning" (i.e., cleaning using room-temperature TCE), its major metal cleaning use was in vapor degreasing machines, the use of which dates back to the 1920s. By the mid-1930s, vapor degreasing technology had evolved to the point where these machines were used by many leading manufacturers of metal products (Davidson 1938). By the 1940s, the largest use of TCE was in the vapor degreasing of metals (Kirk and Othmer 1949). As of 1952, 92 % of TCE produced in the US was used in vapor degreasing (Chem Wk 1953). This percentage remained relatively constant until the mid-1970s, when it dropped to approximately 80 % due to the increased use of the less toxic 1,1,1-trichloroethane in vapor degreasers (Doherty 2000).

In its simplest form, a vapor degreaser consists of a metal rectangular container with an open top. A shallow layer of liquid TCE[2] at the bottom of the container is heated, causing TCE vapors (which are heavier than air) to rise into the middle and upper portions of the container. Near the top of the container are cooling coils that cause TCE vapors in the upper portions to condense, thereby inhibiting their escape from the container. Objects to be cleaned are placed in the TCE vapor zone that exists above the liquid TCE and below the cooling coils. Because the objects are initially at a lower temperature than the TCE vapors, TCE condenses on the objects, and dissolves oils, greases, and other soluble materials that may be present. The condensed TCE, along with dissolved materials, drips off the objects into the liquid layer below. When the objects reach the temperature of the vapors, condensation

[2] The most widely used vapor degreasing solvents other than TCE were PCE and 1,1,1-trichloroethane. The latter compound did not see significant use in vapor degreasing until roughly the mid-1960s due to difficulties with stabilization.

ceases. When the objects are removed from the degreaser, residual TCE quickly evaporates, leaving the work clean, dry and ready for further processing, such as painting, welding, inspection, or shipment.

Oils, greases, and other materials removed in the vapor degreasing process accumulated in the TCE reservoir at the bottom of the degreaser. As the concentrations of these impurities increased, the effectiveness of the solvent decreased. Therefore the TCE needed to be periodically replaced with fresh solvent, or distilled to remove impurities. The spent TCE and accumulated materials (or the still bottoms left after distillation) needed to be removed for recycling or disposal. Historical disposal practices of these wastes led to the introduction of TCE into soil and groundwater, as discussed further below.

Variations on the typical vapor degreasing process included the addition of steps where objects were immersed in liquid solvent, or subjected to a solvent spray. Some models included a distillation unit to recover used solvent. Because the vapor degreasing process involved exposure to metals at elevated temperatures, vapor degreasing grades of TCE included stabilizers to help prevent solvent degradation.

TCE, as well as other solvents, were used in ultrasonic cleaning machines that combined the solvent power of TCE with the agitation caused by high-frequency sound waves. These machines typically employed solvent in liquid rather than vapor form, and were best-suited to cleaning small parts with openings or recesses that were difficult to clean by other methods. The use of TCE in ultrasonic cleaners dates back to approximately 1953 (Business Week 1953).

1.2.3 Other Industrial, Commercial and Military Uses

In addition to its primary use as a metal cleaning agent, TCE was also used to a lesser degree as a refrigerant; a heat transfer medium; a cleaner for optical lenses and film; a solvent for fats, waxes, resins, oils, perfumes, greases, rubber, paints, and varnishes; and an ingredient in printing inks, paint strippers, lacquers, lubricants, pesticides, paints, adhesives, and rust preventors (Hardie 1964; Huff 1971; Doherty 2000).

During World War I, TCE was used extensively in Germany as a substitute for benzene and alcohol in the dissolution of fats. Four cases of poisoning at a German munitions plant were reported by Plessner in 1916. Exposure to TCE used as a substitute for carbon tetrachloride in the preparation of brake linings was cited as the cause of death of a New Jersey worker in 1923 (Hamilton 1925).

Carbon tetrachloride-based fire extinguishers, used for fighting liquid and electrical fires throughout the early-to-mid-twentieth century, could be utilized in conditions below carbon tetrachloride's −9 °F (−23 °C) freezing point by adding TCE. During World War II, the US Army Air Force recommended that TCE be added to carbon tetrachloride fire extinguishers to be used in cold conditions (HQ AAF 1944).

TCE was used by the chemical industry in the production of polyvinyl chloride, chloroacetic acid, hydrofluorocarbons, pharmaceuticals, insecticides, fungicides, fire retardants, fertilizer, and synthetic rubber (Leppart 1945; USEPA 1979; Doherty 2000). One of the major current uses for TCE is as a feedstock in the production of refrigerants (ICIS 2010).

1.2.4 Medical Uses

The use of TCE in the medical field included use as a general anesthetic and an analgesic in dental extractions, childbirth and other short surgical procedures (NIOSH 1975). Veterinary applications included use as an anesthetic for dogs, pigs, and cats, and as a disinfectant and detergent for surgical instruments and minor wounds (Huff 1971).

Reports of TCE's use in medicine date back to 1915, when it was used on a trial basis in Europe as an inhalant for the treatment of trigeminal neuralgia (Oppenheim 1915, cited in Parkhouse 1965). Reports of similar trials in the US date to 1928 (Oljenick 1928, cited in Parkhouse 1965). Due to the limited success of these trials and the finding that TCE had no specific effect on the trigeminal nerve, TCE was not widely used in the treatment of trigeminal neuralgia.

The use of TCE to anesthetize dogs was reported by Jackson in 1933 (Jackson 1934). In 1934, a pharmaceutical grade of TCE became available in England, and was recommended for cleaning wounds and burns (Hardie 1964). In 1935, Striker et al. reported on the use of TCE as an anesthetic and analgesic in over 300 cases (Striker 1935). In 1936, the Council on Pharmacy and Chemistry of the American Medical Association concluded that "the available evidence does not justify the acceptance of trichloroethylene for use as a general anesthetic" (JAMA 1936). However, the wartime need for a non-flammable anesthetic led to additional trials, and TCE became widely used for this purpose during the World War II years.

The use of TCE as an anesthetic was generally more common in Europe than in the US (Bundesen 1953). However, an estimated 35,000 l of TCE were used as an anesthetic and analgesic in the US in 1958 (Huff 1971). TCE's use as a general inhalational anesthetic decreased after the introduction of halothane in 1956 but it continued to be used throughout the 1960s (Aviado et al 1976). By 1975, it was estimated that no more than 60,000 patients per year were anesthetized using TCE (Seltzer 1975).

TCE was found to be an effective analgesic during proctoscopic examinations, for post-operative pain relief, for narcohypnosis, for angina pectoris, and for pregnant women in early stages of labor (Aviado et al 1976). TCE was used as an anesthetic during the birth of Queen Elizabeth's first child (Stafford 1952). Methods for self-administration of TCE by expectant mothers in labor were in use in the mid-1950s (Miles 1954). The widespread use of TCE in obstetrical analgesia was described by Parkhouse (1965), who described it as "the most convenient form of inhalation analgesia in terms of apparatus; it is certainly more effective that nitrous oxide/air, and involves less danger of maternal and fetal hypoxia."

1.2.5 Food Processing

TCE was used in the food processing industry primarily for extraction of fats, oils, and other substances from fish meal, meat meal, oil-containing seeds, soybeans, and coffee beans (Hardie 1964). Additional applications as of the mid-1950s included the extraction and purification of olive, maize, linseed, and other edible oils (Mellan 1957). It was also used as a fumigant for grains and other foodstuffs (OPM 1941), and in the preservation of eggs and fruit. The use of TCE for fat extraction dates back to at least 1916, when cattle poisonings were attributed to the use of TCE in defatting soybean oil meal animal feeds (Huff 1971). Extensive losses of cattle in Europe between 1923 and 1925 were attributed to the same source. Nevertheless, by 1927, "large and ever-increasing quantities" of TCE were being used by the food processing industry as an extraction solvent for natural fats and palm, coconut and soybean oils (Mertens 1991; Ind Chem 1927). In the US, the attribution of hemorrhagic diseases in cattle fed with TCE-treated soybean meal led to the voluntary withdrawal of the product in 1952 (Chem Wk 1953; Huff 1971).

Other reported uses of TCE included hop extraction and removal of oleoresins from spices (Mertens 1991; Seltzer 1975). TCE was one of a number of chemicals historically used to extract caffeine from coffee beans to produce decaffeinated coffee. As of 1975, the US Food and Drug Administration (FDA) limited TCE concentrations to 10 parts per million (ppm) in decaffeinated instant coffee, 25 ppm in decaffeinated ground coffee, and 30 ppm in spice oleoresins (Seltzer 1975). After NCI released its finding that TCE caused cancerous tumor growth in mice, the General Foods Corporation announced in July 1975 that it would substitute methylene chloride for TCE in the production of its decaffeinated coffee brands (C&EN 1975).

In 1977, the US FDA proposed a ban on the use of TCE in direct or indirect food production, cosmetics, and drug products (Mertens 1991; Conlon 1976). However, the ban was not enacted, and the 1975 residual concentration limits for TCE in decaffeinated coffee and spice oleoresins remain in effect (21 CFR 173.290 2012). The current allowable TCE residual in modified hop extract used in beer is set by the FDA at 150 ppm (21 CFR 172.560 2012).

1.2.6 Consumer Products

TCE was used in a wide variety of consumer products, including cleaning fluids, disinfectants, deodorizers, and adhesives. Huff (1971) provided a table of 26 commercially-available products that contained TCE, including spot remover, rug cleaner, air freshener, tree wound healer, chimney sweep cleaner, and false eyelash cleaner. Use as a wig cleaner, typewriter correction fluid, septic system cleaner, and mildew preventer has also been reported (Aviado et al 1976; US DHHS 1997; Kaplan 1983).

TCE was an ingredient in one formulation of Carbona, a widely-used household spot remover. Carbona, along with other cleaning and adhesive products that contained TCE, were commonly misused as narcotics during the 1950s, 1960s, and early 1970s by inhaling the vapors in an enclosed space. The resulting euphoric effects were often followed by nausea, vomiting, and in some cases, death (Huff 1971).

1.3 Environmental Impacts and Regulatory Development

The widespread use of TCE in degreasing coupled with the disposal practices prevalent during most of the twentieth century resulted in substantial releases of TCE to the environment. According to the Agency for Toxic Substances & Disease Registry (ATSDR), TCE is one of the most common contaminants found at Federal Superfund sites, having been detected at 852 of 1,430 sites as of 1997 (US DHHS 1997). The presence of TCE in soil, groundwater, and soil vapor at impacted locations has in some cases resulted in prolonged human exposure through both inhalation and ingestion routes. Due to TCE's relatively high persistence in the subsurface environment, it will likely continue to be present in these media (particularly groundwater) for decades to come.

Early instances of TCE contamination of groundwater were reported in the Reading, England area by Lyne and McLachlan (1949). Their short article described two cases of TCE contamination in wells near areas where TCE was released to the environment, and correctly noted that "it is evident that contamination by compounds of this nature is likely to be very persistent." A 1950 summary of Lyne and McLachlan's article in the American Chemical Society's *Chemical Abstracts* noted that "It often happens that wells near factories which use large quantities of C_2HCl_3 are rendered unfit for drinking by contamination by this liquid" (Hall 1950).

In addition to regulations mentioned in this chapter's Introduction (Rule 66 and the 1970 Clean Air Act), many other regulations affected the use of TCE. On October 21, 1976, the Resource Conservation and Recovery Act (RCRA) was enacted. EPA's press release announcing the act noted that "the contamination of groundwaters by substances leaching from disposal sites is a primary concern" (USEPA 1976). The original regulations promulgated under RCRA in 1980 included TCE in waste categories F001, F002 and U228. A fourth category (D040) was added by the Hazardous and Solid Waste Amendments of 1984.

TCE was one of the original 65 priority pollutants included in the 1977 Clean Water Act, which amended the 1972 Water Pollution Control Act to provide better control of discharges of toxic chemicals (Arbuckle et al 1991). Reportable quantities for spills of TCE and a variety of other chemicals were established under the 1980 Comprehensive Environmental Response, Compensation & Liability Act (CERCLA), which created the "Superfund" for the cleanup of the most serious hazardous waste sites, the majority of which were impacted to some degree by TCE.

To comply with the 1974 Safe Drinking Water Act's requirement to establish regulations for public water supplies, EPA proposed non-enforceable Maximum Contaminant Level Goals (MCLGs) for TCE and seven other chemicals on June 12, 1984 (USEPA 1984). Enforceable Maximum Contaminant Levels (MCLs) for the eight chemicals including TCE were proposed on November 13, 1985, and became effective January 9, 1989 (USEPA 1985 and USEPA 1987). The MCL for TCE was set at 5 parts per billion (ppb), and remains at that level to this day.

TCE contamination of groundwater has led to the closure of, or the need to provide treatment for, countless public and private water supply wells. Although a few notable examples of TCE contamination in the US are discussed herein, the scope of the problem is so broad that it is impractical to provide a comprehensive summary. While the number of Superfund sites impacted by TCE releases numbered less than 1,000 as of 1997, it is likely that the number of non-Superfund sites in the US (i.e., those regulated under state programs) far exceeds that number.

TCE was the primary contaminant in Wells G and H in Woburn, Massachusetts, the story of which was documented in the book (Harr 1995) and 1998 film "A Civil Action." A cluster of childhood leukemia cases, along with complaints of chemical tastes and odors from tap water, prompted residents to unsuccessfully lobby for shutdown of the Wells G and H in 1969. Ten years later, after 184 drums of polyurethane resin waste were found and removed from a nearby vacant lot, Massachusetts officials tested the wells and detected TCE in wells G and H at 267 and 183 ppb, respectively (Harr 1995). The wells were shut down in 1979, and the cleanup of the site is currently in its eighteenth year (USEPA 2013a).

In 1979, groundwater contamination by TCE, and, to a lesser degree, PCE and other chemicals, was discovered in wells in California's San Gabriel valley. The state's Department of Health Services (DHS) initiated a sampling program that led to the identification of 59 contaminated wells and, beginning in 1984, the inclusion of four locations on the Superfund National Priorities List (NPL). Cleanup at these Superfund sites is on-going and, based on the extent and difficulty of removing TCE from contaminated aquifers, is likely to continue for decades (USEPA 2011a and USEPA 2013b).

In the following year, DHS initiated a sampling program in the San Fernando Valley, a nearby area that is similar in terms of both geology and a history of heavy industrial usage. Large areas of the valley's aquifer were found to be contaminated, primarily with TCE and PCE. As a result, numerous water supply wells were taken out of service, and water was purchased from the Metropolitan Water District of Southern California. In 1986, four locations within the valley were declared Superfund sites (USEPA 2013b).

Beginning in 1980, testing of water supplies at the US Marine's Camp Lejeune in North Carolina for trihalomethanes (by-products of chlorination of water) indicated the presence of high levels of halogenated hydrocarbons (Barrett 2010).

By 1982, PCE, TCE, benzene and lesser concentrations of other chemicals had been identified, and the first of the impacted water supply wells was shut down in 1984. Cleanup activities are on-going (USEPA 2013c).

In recent years, increased attention has been focused on the impact of TCE-contaminated groundwater on indoor air. Migration of TCE vapors into buildings and subsurface structures has been noted at many sites impacted by TCE and other volatile organic compounds.

While the historic volume of TCE released to ambient (outdoor) air undoubtedly exceeded that released to soil and groundwater, TCE's half life in air is approximately 7 days, far shorter than typical half-lives in subsurface environments. According to the Toxics Release Inventory database (USEPA 2011a), TCE releases to air were 2.7 million pounds in 2011, down from a reported 9.8 million pounds in 2000 and over 40 million pounds in 1990. To a large extent these values reflect a decades-long trend of decreased TCE use by industry.

In September 2011, after nearly 25 years of evaluation and debate, EPA formally revised toxicity factors for TCE in response to data that indicated that previous factors underestimated risks to human health (USEPA 2011b). The revised toxicity factors may lead to a lowering of cleanup levels at TCE sites and a lowering of TCE's drinking water standard. The future effect of EPA's revisions may be even further reductions in the use of what once was a widely-used and readily available industrial, commercial, and household chemical.

References

21 CFR 172.560 (2012) Code of federal regulations: modified hop extract. Available at: www.accessdata.fda.gov. Accessed Apr 2013

Arbuckle JG, Vanderver TA, Randle RV (1991) Water pollution control. In: Environmental law handbook, 11th edn. Government Institutes, Rockville, pp 524–615

Aviado DM, Simaan JA, Zahkarai S, Ulsamer AG (1976) Methyl chloroform and trichloroethylene in the environment. CRC Press, Ohio, 49

Barbalace K (2013) Chemical database: trichloroethylene. http://environmental.chemistry.com/yogi/chemicals/cn/trichloroethylene.html. Accessed Apr 2013

Barrett B (2010) Warnings about Camp Lejeune's tainted water unheeded for years. Stars and Stripes website: http://www.stripes.com/news/warnings-about-camp-lejeune-s-tainted-water-unheeded-for-years-1.101004. Published 18 Apr 2010. Accessed Mar 2013

Bundesen HN (1953) Pain reliever now being used as minor surgery anesthetic. Bedford (PA) Gazette, 17 June 1953

Business Week (1953) Cleaning by sound. Business Week 1229, 21 Mar 1953

C&EN (1969) Trichloroethylene: not to Blame in Smog. Chem Eng News. 15 Sept 1969, 47:19

C&EN (1975) The use of trichloroethylene for coffee decaffeination is being stopped. Chem Eng News 28 July 1975, 53:14

CFR 173.290 (2012) Code of federal regulations: trichloroethylene. Available at: www.accessdata.fda.gov. Accessed Apr 2013

Chem Wk (1953) Tri-Per and Carbon Tet. Chem Wk. 2 May 1953, p 57

Chem Tr J Chem Eng (1912) New chlorinated carbon derivative. Chem Tr J Chem Eng. 21 Sept 1912, p 300

Conlon M (1976) FDA will ban potential cancer-causing chemical. Nashua (NH) Telegraph. 9 July 1976

Davidson WW (1938) Solvent degreasing. Trans Electrochem Soc 72:413–427

Doherty RE (2000) A history of the production and use of carbon tetrachloride, tetrachloroethylene, trichloroethylene and 1,1,1-trichloroethane in the United States: part 1 and part 2. J Environ Forensic 1:69–93

Doherty RE (2012) The manufacture, use, and supply of chlorinated solvents in the United States during World War II. Environ Forensic 13(1):7–26

Fischer E (1864) Ueber die Einwirkung von Wasserstoff auf EinfachChlorkohlenstoff. Jena Z Med Naturwiss 1:123

Gerhartz W (ed) (1986) Ullman's encyclopedia of industrial chemistry, vol A-6, 5th edn. Weinheim, New York, pp 299–302

Glauser J, Funda C (2012) CEH marketing research report: C2 chlorinated solvents. In: Chemical economics handbook. SRI International, Menlo Park, CA

Hall WT (1950) Water, sewage, and sanitation. Chemical abstracts. Am Chem Soc 44:776–777

Hamilton A (1925) Industrial poisons in the United States. MacMillan Co, New York

Hardie DWF (1964) Chlorocarbons and chlorohydrocarbons: trichloroethylene. In: Standen A (ed) Kirk-Othmer encyclopedia of chemical technology, vol 5, 2nd edn. Wiley, New York, pp 183–195

Harr J (1995) A civil action. Random House, New York

HQ AAF (1944) Aircraft Accessories-Fire Extinguishers; winterization of 1 quart type (carbon tetrachloride). TO 03-45B-2 Washington, DC, Apr 1944

Huff JE (1971) New evidence on the old problems of trichloroethylene. IMS Ind Med Surg 40(8):25–33

IARC (1997) Trichloroethene. In: IARC monographs on the evaluation of carcinogenic risks to humans-dry cleaning, some chlorinated solvents and other industrial chemicals, vol 63. WHO International Agency for Research on Cancer, Lyon, France

ICIS (2010) Inorganics news in brief, 25 October 2010. ICIS Chemical Business Magazine. Reed Bus Info Ltd. http://www.icis.com/Articles/2010/10/25/9403851/inorganics-news-in-brief.html. Accessed 24 Mar 2013

Ind Chem (1927) Trichloroethylene as a solvent for fats. Ind Chem. Sep 1927, pp 418–419

J City Post Tribune (1950) Hoarding of scarce materials banned. Jefferson City Post-Tribune, MO. 28 Dec 1950, 84(121)

Jackson DE (1934) A study of analgesia and anesthesia with special reference to such substances as trichloroethylene and vinesthene (divinyl ether), together with apparatus for their administration. Curr Res Anesth 13:198

JAMA (1936) The use of trichloroethylene for general anesthesia. JAMA 107:1302

Kaplan OB (1983) Some additives to septic tank systems may poison groundwater. J of Env'l Health. Mar/Apr 1983, p 259

Kirk R, Othmer D (eds) (1949) Encyclopedia of chemical technology, vol 3. Interscience Encyclopedia, Inc., New York, p 792

Leder A (1999) CEH abstract-C2 chlorinated solvents. Chem Ind News, CEH report, 13 Jan 1999

Lee G et al (2003) Synthetic organic chemicals. In: Kent JA (ed) Reigels handbook of industrial chemistry, 10th edn. Kluwer Academic Publishers, Oregon, p 821

Leppart J (1945) History of trichloroethylene-World War II. War Production Board. pp 1–6

Lyne FA, McLachlan T (1949) Contamination of water by trichloroethylene. Analyst 74:513

Mellan I (1957) Source book of industrial solvents, vol II, Halogenated hydrocarbons. Reinhold, New York

Mertens J (1991) Trichloroethylene. In: Kroschwitz J, Howe-Grant M (eds) Kirk-Othmer encyclopedia of chemical technology, vol 6, 4th edn. Wiley, New York, pp 40–49

Miles J (1954) When I had my baby, I gave myself the anesthetic. Am Wkly. 3 Oct 1954

Mirza T et al (2000) A study on the substitution of trichloroethylene as a spot remover in the textile industry. AIHAJ 61(3):431–438

National Cancer Institute (NCI) (1976) Carcinogenesis bioassay of trichloroethylene, NCI-CG-TR-2 Washington, DC
NIOSH (1975) Bulletin 2: trichloroethylene. Nat Inst Occup Safety Health. June 1975, pp 78–127
Office of Production Management (OPM) (1941) Division of priorities memorandum. 15 Oct 1941
Oljenick I (1928) Trichlorethylene treatment of trigeminal neuralgia. JAMA 91:1085
Oppenheim H (1915) Uber Trigeminuserkrankung infolge von Trichlorathylenvergiftung (discussion). Neurol Zbl 34:918
Parkhouse J (1965) Trichloroethylene. Br J ANA 37:681–687
PPG Ind Inc (1999) Trichloroethylene product literature. PPG Industries Inc. May 1999
Seltzer R (1975) Reactions grow to trichloroethylene alert. Chem Eng News 53:41
Stafford J (1952) Modern anesthetic due credit in historic operation. El Paso Herald-Post, TX. 19 Dec 1952
Striker C, Goldblatt S, Warm IS, Jackson DE (1935) Clinical experiences with the use of trichloroethylene in the production of over 300 analgesias and anesthesias. Curr Res Anesth 14:68
US Tariff Comm (1971) Census of dyes and other synthetic and organic chemicals. US Tariff Commission
US DHHS (1997) Toxicological profile for trichloroethylene (update) US Depart Health and Human Services. ATSDR 4:185–188
USEPA (1976) U.S. Environmental Protection Agency History Topics. New law to control hazardous wastes, end open dumping, promote conservation of resources. EPA Press Release 1976:http://www.epa.gov/history/topics/rcra/05.html. Accessed 2013
USEPA (1979) Status assessment of toxic chemicals-trichloroethylene. US Environmental Protection Agency (EPA) 600/2-79-210m 13
USEPA (1984) U.S. Environmental Protection Agency Federal Register. 12 June 1984, 49:24330
USEPA (1985) U.S. Environmental Protection Agency Federal Register. 13 Nov 1985, 50:46880
USEPA (1987) U.S. Environmental Protection Agency. Federal Register. 8 July 1987, 52:23690
USEPA (2011a) San Gabriel Valley Superfund Sites: progress report on San Gabriel Valley Ground Water Cleanup. U.S Environmental Protection Agency, Region 9. December 2011
USEPA (2011b) 2011 TRI National Analysis Report Now Available. U.S Environmental Protection Agency Toxic Release Inventory (TRI) Program website: http://www.epa.gov/tri/. Accessed Mar 2013
USEPA (2011c) Toxicological review of trichloroethylene. US Environmental Protection Agency EPA/635/R-09/011F. Sep 2011
USEPA (2013a) Waste site cleanup & reuse in New England. Wells G & H, Woburn. U.S. Environmental Protection Agency Region 1 Superfund web site: http://www.epa.gov/region1/superfund. Accessed Mar 2013
USEPA (2013b) San Gabriel Valley (all Areas) Superfund Site Fact Sheet: U.S. Environmental Protection Agency Pacific Southwest Region 9 Superfund website: http://yosemite.epa.gov/r9/sfund/r9sfdocw.nsf/84e3d3f748094337882572330079 4f02/0065ed704ae95ccc88257007005e941e!OpenDocument. Accessed Mar 2013
USEPA (2013c) Marine corps base Camp Lejeune: site summary profile. US Environmental Protection Agency Region 4 Superfund website: http://www.epa.gov/region4/superfund/sites/fedfacs/camplejnc.html. Accessed Mar 2013
USITC (2013) US International Trade Commission. www.usitc.gov. Accessed 22 Mar 2013

Chapter 2
Trichloroethylene and Autoimmunity in Human and Animal Models

Kathleen M. Gilbert

Abstract Based on likelihood of exposure and potential health impact trichloroethylene (TCE) is consistently ranked 16th out of 275 chemicals on the annual CERCLA (Comprehensive Environmental Response, Compensation, and Liability Act) list of hazardous substances. Although environmental contact with TCE in the water, air or soil, is generally thought to be risk-free, there is evidence that chronic exposure to TCE at levels too low to be overtly toxic can generate autoimmune diseases including lupus, scleroderma, and autoimmune liver disease. This chapter examines human exposure data. It also discusses the mechanistic information that has been provided by animal studies, and identifies some important gaps in our understanding. Since human exposure to TCE will continue for the foreseeable future, we need to understand and prevent the autoimmune-promoting effects of this toxicant.

Keywords Autoimmune disease • Immunotoxicity • $CD4^+$ T cells

2.1 Introduction to Autoimmune Disease

The immune system is supposed to be restricted to recognizing and attacking foreign antigens such as disease-causing micro-organisms. If the immune system instead attacks self-antigens chronic incurable disorders characterized as autoimmune diseases occur. There are over 80 different autoimmune diseases, and at least one for every organ system in the body. The NIH estimates up to 23.5 million Americans have at least one type of autoimmune disease. In comparison, cancer

K.M. Gilbert, PhD
Department of Microbiology and Immunology,
University of Arkansas for Medical Sciences, Arkansas Children's Hospital
Research Institute, 13 Children's Way, Little Rock, AR 72202, USA
e-mail: gilbertkathleenm@uams.edu

K.M. Gilbert, S.J. Blossom (eds.), *Trichloroethylene: Toxicity and Health Risks*, Molecular and Integrative Toxicology, DOI 10.1007/978-1-4471-6311-4_2,

affects up to 9 million and heart disease up to 22 million. The most prevalent of the more than 80 autoimmune diseases identified include Type 1 diabetes, rheumatoid arthritis, multiple sclerosis, systemic lupus erythematosus, Sjogren's Syndrome, and the several types of autoimmune thyroid disease. Most of these diseases are found much more often (3–10 fold) in women. Some autoimmune diseases are life-threatening; all are debilitating and require lifelong medical care.

There is much we do not know about autoimmune disease. We have been most successful at identifying the type of immune pathology (e.g. autoantibody vs T cell-mediated) associated with a particular autoimmune disease, and sometimes characterizing the specific autoantigens targeted. This information has been used to classify autoimmune diseases as type II, III or IV hypersensitivity reactions. However, this is only somewhat useful, since many autoimmune diseases involve more than one type of immune pathology. Even if we can document the type of immune pathology associated with a particular autoimmune disease, we don't know what initiates this pathology. Studies involving identical twins have provided some useful hints in this regard. Even though autoimmune diseases as a group affect between 5 and 8 % of the population in the US, the incidence for any one autoimmune disease is relatively rare. Thus, the fact that the concordance rate for developing a particular autoimmune disease in identical twins is much higher than the general population demonstrates the involvement of genetic susceptibility (He et al. 2001). On the other hand, the finding that the concordance rate is not 100 %, and is indeed usually much less than 50 % for any autoimmune disease, demonstrates that environmental factors also contribute to disease etiology. The environmental contribution to autoimmune disease is a relatively vague and wide-ranging concept that has come to include lifestyle (e.g. diet) and history of bacterial and/or viral infection. It also includes exposure to environmental chemicals which impact the immune system. One such chemical, trichloroethylene (TCE), will be examined here for its contribution to autoimmune disease. The current state of knowledge will be outlined as will the information gaps that need to be filled.

2.2 TCE and Autoimmunity/Hypersensitivity in Humans

As noted by a National Research Council report evidence on human health hazards from TCE exposure, either occupational or environmental, has strengthened in recent years (Committee on Human Health Risks of Trichloroethylne 2006). One of the predominant non-cancer outcomes associated with TCE exposure in humans is immunotoxicity, most notably the development of hypersensitivity responses. Although not all types of TCE-induced hypersensitivity has been classified as autoimmune, at least some of the hypersensitivity responses induced by TCE clearly mimic idiopathic autoimmune diseases.

The links between autoimmune disease and TCE were originally described in humans exposed to the chemical at work. Going back to the 1970s numerous case reports have correlated sometimes fatal systemic or localized sclerosis or diffuse fasciitis with industrial TCE exposure (Czirjak et al. 1994; Flindt-Hansen and Isager 1987; Karamfilov et al. 2003; Lockey et al. 1997; Pralong et al. 2009; Saihan et al. 1978; Waller et al. 1994). Systemic sclerosis, also known as scleroderma, is an autoimmune disease of normally unknown etiology. The autoimmune response targets connective tissue of the skin, internal organs and the walls of blood vessels. It is characterized by alterations of the microvasculature and by massive deposition of collagen and other matrix substances in the connective tissue. At least three case control studies of men or women with scleroderma identified TCE exposure in occupational or hobby settings as a likely risk factor (Diot et al. 2002; Garabrant et al. 2003; Nietert et al. 1998). Possible mechanisms by which TCE triggers scleroderma are not known.

Scleroderma is not the only autoimmune disease associated with TCE exposure. A cohort study of people living near a TCE-contaminated Superfund Site in New York demonstrated an increased prevalence of the autoimmune disease primary biliary cirrhosis (Ala et al. 2006). In another study TCE-exposed individuals from metal industries were shown to have increased urine levels of N-acetyl-beta-D-glucosaminidase, a marker of autoimmune lupus nephritis (Brogren et al. 1986). Case reports from around the world have also linked chronic occupational TCE exposure to sometimes fatal non-viral hepatitis that is worsened by rechallenge (Anagnostopoulos et al. 2004; Joron et al. 1955; McCunney 1988; Pantucharoensri et al. 2004; Schattner and Malnick 1990). Although the rechallenge exacerbation of this TCE-induced hepatitis suggests an immune component, this aspect of the disease was not tested.

There are several studies which have linked TCE exposure to the generation of autoantibodies, biomarkers of an autoimmune response if not actual autoimmune disease. Between 1964 and 1979 domestic water supplies in East Woburn, MA, were unknowingly contaminated with industrial solvents, with TCE as the main volatile organic found (267 ppb). Five years after the wells were closed individuals from East Woburn demonstrated increased numbers of total T cells (both $CD4^+$ and CD8) and increased incidence of anti-nuclear antibodies compared to controls (Byers et al. 1988). A cohort study of individuals exposed to TCE in contaminated well water in Arizona demonstrated significantly increased levels of anti-nuclear antibodies and increased ARA (American Rheumatism Association) scores for lupus (Kilburn and Washaw 1992). A recent serological proteome analysis showed that sera from patients with active TCE-induced hypersensitivity, unlike control sera, contained antibodies specific for several ontologically diverse self antigens including NM23 (nucleoside disphosphate kinase), and lactate dehydrogenase B (Liu et al. 2009). Interestingly, although TCE appeared to increase the levels of specific autoantibodies, it has also been shown to decrease serum levels of total IgG and IgM (Zhang et al. 2013). The mechanism by which TCE exposure activates specific antibodies or alters total immunoglobulin, and their functional significance, remains to be determined.

Even if overt autoimmune pathology was not revealed (in many cases not examined) other epidemiological studies have demonstrated TCE-induced immunotoxicity. Data collected from subjects who had worked at least 3 years in the in the printing industry showed that levels of TCE in the breathing zone and levels of a TCE metabolite in urine correlated with increased serum levels of T cell-derived cytokines IL-2 and IFN-γ and decreased levels of IL-4 (Iavicoli et al. 2005). TCE has also been shown to induce a hypersensitivity disorder that targets the skin and liver (Bond 1996; Xu et al. 2009a). The number of patients suffering from occupational TCE-related severe skin disorders has been increasing in areas where TCE is still widely used as a solvent, including the Philippines, Taiwan, Singapore, and the Guangdong Province, China. The clinical manifestations are different from irritating contact dermatitis caused by TCE defatting action. Instead, the subjects experience a relatively long period of exposure before disease onset, rash, fever, lymphadenopathy, liver dysfunction and recurrence after just minimal re-exposure (Nakajima et al. 2003). The TCE-induced dermatitis is considered to be a T cell-mediated type IV hypersensitivity disease. Although the pathology appears to be immune mediated, it is not clear whether the immune response is directed toward self. More information about this type of TCE-induced hypersensitivity will be provided in Chap. 3.

2.3 Xenobiotics and Autoimmunity in Animal Models

Defining toxicant exposure as a risk factor for a particular type of human disease, autoimmune or otherwise, is difficult. Many times people do not realize they have been in contact with a particular chemical such as TCE, and there are often few if any biomarkers of exposure. In addition, since people are never exposed to a single chemical how do you accurately assess the contribution of a single toxicant? These challenges make it difficult to define a direct cause and effect relationship between toxicant exposure and autoimmune disease. This has led to the popularity of animal models in which toxicant exposure can be controlled and monitored. Several animal models have been used to test the immunotoxicity of environmental chemicals. When testing chemicals such as TCE that are thought to inappropriately stimulate rather than suppress the immune system animal models with a genetic susceptibility to hypersensitivity are often selected. This is designed to mimic the similar ill-defined predisposition thought to be important for human idiopathic disease, and to increase the likelihood that toxicant-induced hypersensitivity can be detected.

There are several well-characterized mouse strains that are genetically predisposed to develop autoimmune disease. In some cases the diseases occur spontaneously, and in some cases they need to be triggered by administration of antigen or mitogen. Of the mouse models that develop disease spontaneously the most widely studied include NOD mice (type 1 diabetes), BXD1/TyJ (rheumatoid arthritis) and MRL/lpr, NZBWF1/J and BXSB/MpJ mice (lupus). Several of these models have been used to test the role of xenobiotics in autoimmune disease etiology. A recent excellent review describes the different animal models, and discusses environmental agents that have been shown to trigger or exacerbate autoimmune disease in these models (Germolec et al. 2012).

2.4 TCE-Induced Autoimmunity in Mice

2.4.1 Disease Characterization

In terms of TCE, its capacity to promote autoimmunity has been studied most extensively in the model consisting MRL+/+ mice. MRL+/+ mice are related to MRL/lpr mice which have a defect in Fas expression and spontaneously develop lupus within 3–4 months of age. Due to the rapidity of disease development in MRL/lpr mice, it can be difficult to test whether exposure to a toxicant exacerbates the response. In contrast to MRL/lpr mice, the genetically-similar but not identical MRL+/+ mice have normal Fas expression and spontaneously develop a relatively mild lupus-like disease late in life (50 % mortality at 17 months). MRL+/+ mice can also spontaneously develop other autoimmune disorders such as Sjogren's syndrome and T cell-infiltrating pancreatitis (Qu et al. 2002; Skarstein et al. 1997). The basis for the autoimmune predisposition in MRL+/+ mice is not known. Before they reach 1 year of age most female MRL+/+ mice do not exhibit autoimmune tissue pathology and indications of autoimmunity are minor. Thus, young adult female MRL+/+ mice, with their propensity for autoimmunity but absence of overt disease, make a good model to test whether TCE can boost autoimmunity.

In our initial study we expected TCE to accelerate the development of lupus in young adult female MRL+/+ mice. Instead, adding TCE at concentrations lower than sanctioned occupational exposure to drinking water at for 26 or 23 weeks generated a T cell-mediated liver disease commensurate with human idiopathic AIH (Griffin et al. 2000c). The TCE-induced AIH in the MRL+/+ mice was associated with several alterations in $CD4^+$ T cells, an immune subset that play a large role in driving autoimmune disease. One such alteration included decreased sensitivity to activation-induced apoptosis (Gilbert et al. 2006). Activation-induced apoptosis is supposed to keep $CD4^+$ T cells in check and thus help prevent autoimmune disease. This process occurs when autoreactive $CD4^+$ T cells repeatedly stimulated with self antigen co-express death receptors such as Fas as well as the ligand for the death receptor (e.g. FasL). Cross-linking of death receptors on the surface of susceptible T cells promotes the release of active caspase-8 thereby initiating apoptosis (Crispe 1994; Kischkel et al. 1995). Activation-induced cell death is widely believed to help the host protect itself against repeated stimulation and expansion of autoreactive $CD4^+$ T cells (Green et al. 2003; Marrack and Kappler 2004; Van Parijs et al. 1998).

Supporting the important protective effects of activation-induced apoptosis is the fact that defects in this process has been linked to the development of several idiopathic autoimmune diseases such as systemic lupus erythematosus, multiple sclerosis, and rheumatoid arthritis in both humans and mice (Bona et al. 2003; Kovacs et al. 1996; Semra et al. 2002; Sneller et al. 1997; Szodoray et al. 2003; Waiczies et al. 2002). On the other hand, therapies that facilitate Fas-mediated T-cell apoptosis can ameliorate autoimmune disease (Hong et al. 1998; Nishimura-Morita et al. 1997; Zhou et al. 1999). Events such as TCE exposure that inhibit this protective mechanism could thus promote autoimmunity by promoting the expansion of autoreactive CD4+ T cells. Perhaps because of its ability to decrease susceptibility to apoptosis TCE exposure in MRL+/+ mice also increased expansion of an activated/memory population ($CD62L^{lo}$ and/or $CD44^{hi}$) of $CD4^+$ T cells that produced more of the pro-inflammatory cytokine IFN-γ (Griffin et al. 2000c). Compared to naïve $CD4^+$ T cells activated/memory $CD4^+$ T cells have been shown to have a more robust effector function and cytokine production. Thus, work in our laboratory showed that chronic exposure of female MRL+/+ mice to TCE in drinking water induced T cell-mediated autoimmune hepatitis in association with several alterations in $CD4^+$ T cells that align with increased autoreactivity.

Other laboratories have also studied the effects of TCE on autoimmune disease in the MRL+/+ mouse model. A series of important studies conducted by researchers at the University of Texas at Galveston showed that long term exposure to TCE (0.5 mg/ml) in drinking water increased production of lupus-associated autoantibodies as well as promoting the generation of autoimmune hepatitis (Cai et al. 2008; Khan et al. 1995). The autoantibodies induced by TCE encompassed nuclear proteins as well as lipid peroxidation products (Khan et al. 2001; Wang et al. 2007). More about these TCE-induced antibodies will be described in Chap. 4.

2.4.2 Need for Metabolism

The toxicity of many chemicals requires their metabolism. TCE can be metabolized by a glutathione-dependent pathway in the kidney. However, in both mice and humans the majority of TCE absorbed into the circulation is metabolized by an oxidative pathway in the liver (Lipscomb et al. 1996). In this pathway cytochrome P450s (CYPs) rapidly converts TCE to trichloroacetaldehyde (TCAA; also known as chloral), which in solution is in equilibrium with trichloroacetaldehyde hydrate (TCAH ;also known as chloral hydrate) (Fig. 2.1). Once formed, TCAA and TCAH are converted to trichloroacetic acid (TCA), or trichloroethanol (TCOH) which is excreted as the alcohol glucuronide [see review(Lash et al. 2000)]. This later pathway is regulated by alcohol dehydrogenase that works to convert TCOH back to aldehyde. Thus, the level of TCAH depends on the activity of several metabolizing enzymes, all of which display considerable genetic variation in both humans and mice. For example, MRL+/+ mice have much higher levels of alcohol dehydrogenase than C3H/HeJ mice (Teichert-Kuliszewska et al. 1988), and may therefore be expected to have an increased steady-state level of TCAH if exposed to the chemical.

It appears that many of the $CD4^+$ T cell modulating effects of TCE are in fact induced by its metabolite TCAH. It was shown that immune dysfunction induced by TCE in MRL+/+ mice could be blocked by suppressing the activity of CYP2E1 (Griffin et al. 2000a). Similarly, MRL+/+ mice exposed to TCAH instead of TCE in their drinking water developed the same alterations in $CD4^+$ T cells as mice exposed to the parent compound (Blossom et al. 2007b). In humans, TCE-induced

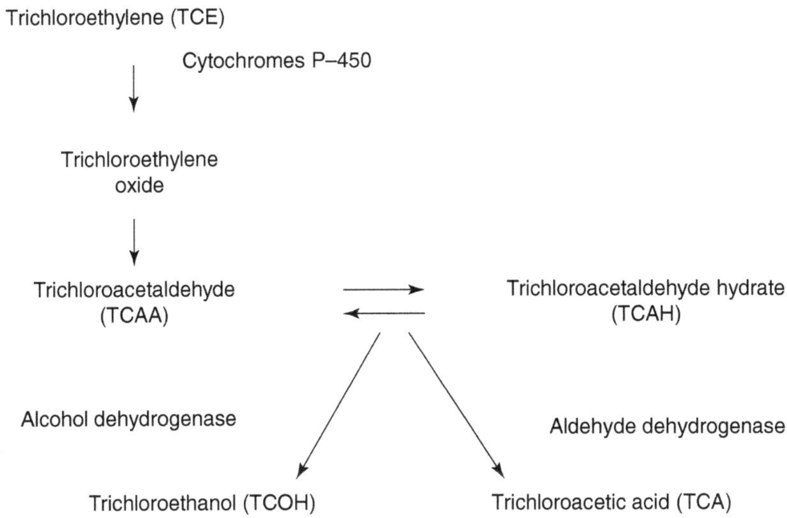

Fig. 2.1 Metabolism of trichloroethylene

hypersensitivity dermatitis has been linked to single nucleotide polymorphisms of TCE metabolizing enzymes CYP2E1 and CYP1A1 (Xu et al. 2009b). CYP2E1 was also shown to be the main enzyme involved in TCE-induced hepatoxicity in mice (Ramdhan et al. 2008). Taken together, it seems likely that similar to many of its other toxic effects TCE-induced immunotoxicity requires its metabolism.

2.4.3 Mechanisms of Immune Alteration

How TCE, seemingly via TCAH, alters $CD4^+$ T cell function is not clear. However, the structure of TCAH may provide a clue. As an aldehyde TCAH has the capacity to form a chemical reaction known as Schiff base, a transient covalent bond between nucleophiles on proteins (e.g. amino group on lysine) and electrophilic carbonyl carbons of aldehydes. As it turns out, Schiff base formation is the foundation for some of the stimulatory interactions that normally occur between specific molecules on the $CD4^+$ T cell surface and associated ligands on the surface of accessory cells such as dendritic cells or endothelial cells (Chen et al. 1997). These interactions between $CD4^+$ T cell and accessory cells are crucial for many aspects of $CD4^+$ T cell activation and effector function. The role of Schiff base formation in these interactions means that certain small Schiff-base-forming compounds may be able to bypass the need for ligand-bearing accessory cells and co-stimulate $CD4^+$ T cells directly. One such compound, tucaresol, is being clinically tested as a drug capable of stimulating T cells to combat neoplasia and opportunistic infection (Charo et al. 2004; Rhodes et al. 1995). The ultimate effect of Schiff base formation on $CD4^+$ T

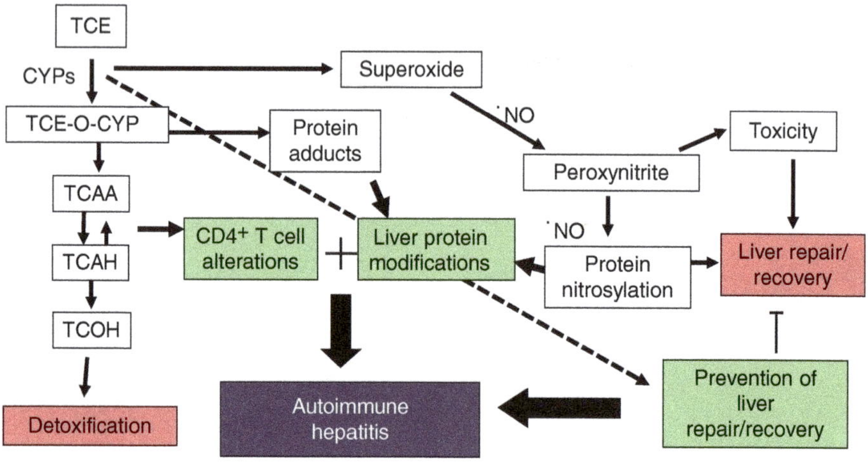

Fig. 2.2 Possible mechanism of TCE-induced autoimmune hepatitis

cells may depend on the existing baseline immune response; in immunosuppressed individuals this event may be beneficial, while in individuals with a predisposition for hypersensitivity, it may be enough to trigger autoimmune disease.

The stimulatory Schiff base-forming compounds identified thus far are aldehydes, similar to TCAH. Schiff base formation by TCAH should be a major reaction because of the electron withdrawing of the three chloro groups on the adjacent carbon. *In vitro* experiments demonstrated that TCAH could form a functionally-active Schiff base with molecules on the surface of $CD4^+$ T cells (Gilbert et al. 2004). This interaction triggered signaling events in the $CD4^+$ T cells similar to those initiated by interaction with ligand-bearing accessory cells. A better understanding of the signaling events triggered in $CD4^+$ T cells by this chemical interaction is required. This includes identifying the molecules on the $CD4^+$ T cell surface that are altered by the TCAH-induced Schiff base formation. In addition, the possibility that these signaling events encompass epigenetic alterations by TCE in $CD4^+$ T cells is being studied, and will be discussed in more detail in Chap. 10.

2.4.4 Liver Events

Although TCE appeared to induce $CD4^+$ T cells alterations commensurate with autoimmunity, it was not clear why the pathology targeted the liver instead of some other organ in the MRL+/+ mice. Simultaneous events in the liver such as nitrosative/oxidative stress and/or adduct formation which are induced by TCE may represent a second requirement for disease pathology that involves protein modification (Fig. 2.2). It has been proposed that formation of chemically-modified self-proteins capable of triggering an immune response represents a mechanism by which chemicals could initiate autoimmunity.

Very early in its metabolism in the liver (prior to TCAH formation) TCE produces a highly reactive intermediate (TCE-O-CYP) that can form adducts with nearby proteins. TCE has been shown to form adducts with a number of liver proteins, most predominantly CYP2E1 (Halmes et al. 1997). Some of these adducted proteins are immunogenic; antibodies specific for TCE-protein adducts have been found in TCE-treated MRL+/+ mice (Griffin et al. 2000b). We have demonstrated a time-dependent increase in the repertoire of liver microsomal proteins recognized by antibodies in the sera of TCE-treated MRL+/+ mice as compared to age-matched untreated MRL+/+ mice. Interestingly, the antibodies in the sera recognized liver microsomal protein from control mice. This indicated that even if chemically-altered liver protein was required to initiate the autoimmune response, the resulting antibody reaction recognized non-modified liver protein. The liver protein epitopes targeted by the TCE-induced autoantibodies, and the specificity of the $CD4^+$ T cells that promote the autoantibody production remains to be determined.

TCE-induced nitrosative/oxidative stress may also increase the immunogenicity of liver proteins. Nitrosative/oxidative stress occurs when the generation of reactive oxygen species (ROS) and/or reactive nitrogen species (RNS) is not balanced by the appropriate detoxification by various antioxidant compounds (e.g. glutathione or vitamin E) and enzymes (e.g. glutathione peroxidase). One important consequence of nitrosative/oxidative stress is the generation of the superoxide anion which can interact with another free radical nitric oxide to form the extremely reactive peroxynitrite. Peroxynitrite can trigger a variety of cellular responses ranging from lipid peroxidation, protein tyrosine nitration, DNA damage and cell death. Tissue damage associated with increased levels of inducible nitric oxide synthase (iNOS) (an enzyme which produces nitric oxide), and/or the accumulation of nitrotyrosine residues has been found in a variety of autoimmune diseases in humans, including autoimmune hepatitis (Pemberton et al. 2004; Sanz-Cameno et al. 2002). Similarly, both iNOS and nitrotyrosine accumulation in the liver have been found in TCE-treated mice (Wang et al. 2007). In addition, investigators have shown that sera from TCE-treated MRL+/+ mice contain antibodies specific for lipid peroxidation-derived aldehydes such as malondialdehyde (MDA) and 4-hydroxynonenal (HNE) (Khan et al. 2001). More about these antibodies, and the $CD4^+$ T cells that support their production, will be discussed in Chap. 4.

2.5 Susceptibility Factors for Toxicant-Induced Autoimmunity

It has not been possible for researchers to define a single causative agent for idiopathic autoimmune disease. Even those linked to contact with a particular toxicant only occur in a fraction of people with the seemingly same level of exposure. Consequently, etiology is largely suspected of having several contributing factors

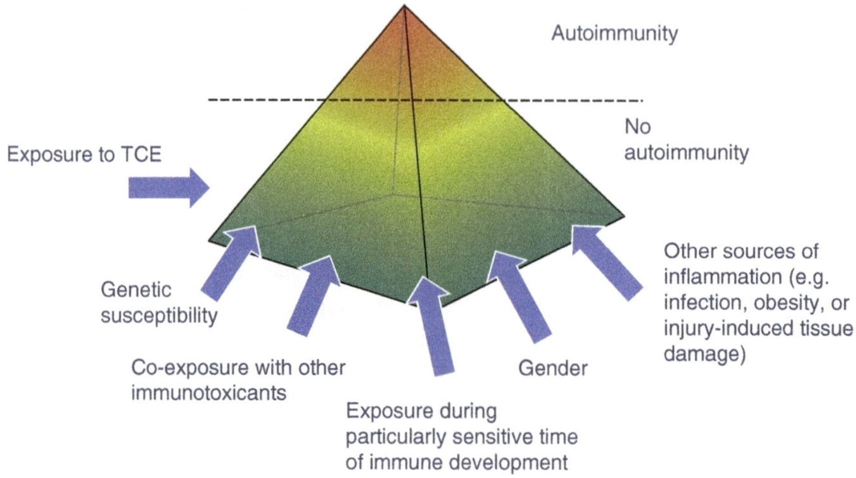

Fig. 2.3 Possible contributors to TCE-induced autoimmune disease

which may be additive or synergistic in nature. Defining these factors, and estimating their relative contribution to autoimmunity is proving to be a huge challenge. Some of the susceptibility factors that appear to contribute to TCE-induced autoimmune disease (Fig. 2.3) are described below.

2.5.1 Genetics

As mentioned above the high concordance rate for identical twins developing the same autoimmune disease demonstrates the importance of genetic susceptibility in human idiopathic autoimmune disease (Jarvinen and Aho 1994; Tomer and Davies 1997). If one assumes that some idiopathic autoimmunity is actually triggered by undetected chemical exposure this implies that genetics also plays a role in toxicant-induced autoimmune disease and other forms of hypersensitivity. Genetic predisposition has also been found in certain hypersensitivity disorders such as asthma, atopic eczema, drug hypersensitivity and food allergies that represent a response to exogenous irritants (Dreskin 2006; Pirmohamed 2006; Mohrenschlager et al. 2006; Meurer et al. 2006). The mechanism for this increased sensitivity is not known.

Aside from MRL+/+ mice, other strains of mice [i.e. (NZBxNZW)F1, and female BXSB] with ill-defined genetic patterns that make them "autoimmune-prone", have been used to test the disease-promoting capacity of xenobiotics. The insecticide chlordecone was shown to accelerate the development of lupus in (NZBxNZW)F1 mice, but had no effect in non-autoimmune-prone BALB/c mice (Sobel et al. 2006). Similarly, exposure to low doses of mercuric chloride has been shown to promote autoimmunity in female BXSB mice, but not in MHC-compatible non-autoimmune-prone C57BL/6 mice (Pollard et al. 2001). Thus, a propensity for autoimmunity appears to be a requirement for at least some types of xenobiotic-induced autoimmunity in mice.

The need for genetic susceptibility in TCE-induced autoimmunity in mice is still being debated. Keil et al. reported that chronic exposure to TCE in drinking water slightly increased renal pathology in non-autoimmune-prone B6C3F1 mice but not autoimmune-prone NZBWF1 mice (Keil et al. 2009). Similarly, TCE was shown to increase serum levels of lupus-related autoantibodies (i.e. anti-ds and anti-ss DNA) at only early time points in NZBWF1 mice, but at multiple time points in B6C3F1 mice. However, the levels of autoantibodies and renal pathology at every time point were higher in the untreated NZBWF1 mice than in the TCE-treated B6C3F1 mice. Since the levels of lupus-associated autoantibodies increase spontaneously in most lupus-prone strains of mice as they age, a high baseline response may mask a TCE-induced effect on this type of autoantibodies. Regardless of the effect of TCE in the NZBWF1 mice, the immunotoxicity of the pollutant could be observed in non-autoimmune-prone mice.

Although we have not examined the effects of TCE itself in non-autoimmune-prone mice we have examined whether its metabolite TCAH required a genetic predisposition to be effective. Chronic exposure to TCAH induced autoimmunity and $CD4^+$ T cell alterations in female MRL+/+ mice, but had much more modest effects when added to the drinking water of non-autoimmune-prone but genetically-related female C3H/HeJ mice (Blossom et al. 2007a). This result would suggest that at least in terms of some parameters, an autoimmune-predisposition increased susceptibility to TCE immunotoxicity.

In order to better define a possible autoimmune-predisposition to TCE we performed a transciptomic analysis comparing unstimulated splenic $CD4^+$ T cells from untreated female age-matched MRL+/+ and C3H/HeJ mice. Table 2.1 provides a list of some of the most robustly altered functionally significant genes flagged in the transcriptomic analysis, and confirmed by qRT-PCR. Also included in Table 2.1 are qRT-PCR results for genes examined as specificity controls for the flagged genes, or because they that had been identified through other assays. The expression levels of some genes were not surprising; a mutation in *Tlr1* is a hallmark of C3H/HeJ mice. Not as predictable, $CD4^+$ T cells from female MRL+/+ mice expressed higher constitutive levels of *Stra6*, which acts as a high-affinity cell-surface receptor for the complex comprised of retinol (main metabolite of Vitamin A) and retinol-binding protein. Stra6 removes retinol from the complex and transports it across the cell membrane where it has been shown to promote $CD4^+$ T cell differentiation and recruitment to inflammatory sites (Pino-Lagos et al. 2011). $CD4^+$ T cells from the MRL+/+ mice also expressed higher baseline levels of *Spp1*, a gene that encodes for pro-inflammatory cytokine osteopontin (OPN). Interestingly, quantitative trait loci (QTL) analysis conducted by others revealed that the locus for susceptibility to lupus nephritis in MRL mice corresponded to the OPN gene, and that allelic polymorphism of OPN caused the functional differences in antibody production between MRL and C3H strains (Miyazaki et al. 2005).

Also flagged by the transcriptomic analysis was the differential expression of a member of the tumor necrosis factor receptor superfamily (Tnfrsf). The $CD4^+$ T

Table 2.1 Differential gene expression in T cells from MRL+/+ and C3H/HeJ mice

Gene ID	Gene name	Gene description	Transcriptomics fold change (p-value)	qRT-PCR fold change ± SD
Splenic CD4⁺ T cells				
NM_030682	*Tlr1*	Toll-like receptor 1	11.221 (0.0001)	
NM_009291	*Stra6*	Stimulated by retinoic acid gene 6	9.0145 (0.0002)	1.89 ± 0.30
NM_009263	*Spp1*	Secreted phosphoprotein 1; osteopontin	7.6511 (0.0001)	4.18 ± 1.34
NM_011838	*Lynx1*	Ly6/neurotoxin 1	5.2374 (0.0001)	13.24 ± 6.52
NM_013599	*Mmp9*	Matrix metallopeptidase 9	4.6446 (0.0007)	5.34 ± 0.51
NM_007399	*ADAM10*	A disintegrin and metalloproteinase domain-containing protein 10		1.14 ± 0.26
NM_178589	*Tnfrsf21*	Tumor necrosis factor receptor superfamily, member 21	0.3198 (0.0001)	0.32 ± 0.10
NM_013869	*Tnfrsf19*	Tumor necrosis factor receptor superfamily, member 19	16.406 (0.0001)	10.88 ± 5.70
NM_178931	*Tnfrsf14*	Tumor necrosis factor receptor superfamily, member 14		1.79 ± 0.88
	Lap	Intracisternal A particle		0.04 ± 0.02
Y12713	*Muerv*	Murine endogenous retrovirus		1.1 ± 0.18
NM_010066	*Dnmt1*	DNA methyltransferase 1		1.05 ± 0.26
NM_007872	*Dnmt3a*	DNA methyltransferase 3 alpha		0.83 ± 0.35
Thymocytes				
NM_013869	*Tnfrsf19r*	Tumor necrosis factor receptor superfamily, member 19		3.24 ± 0.52
NM_054039	*Foxp3*	Forkhead box P3		1.85 ± 0.44
NM_009646	*Aire*	Autoimmune regulator		1.51 ± 0.26

All shaded results were statistically different from results obtained from control CD4+ T cells or thymocytes

cells from MRL+/+ mice expressed comparatively lower levels of *Tnfrsf21*, a gene that encodes for a protein known as death receptor 6 (DR6). Interestingly, a decrease in DR6 in CD4⁺ T cells has been shown to enhance proliferation and production of IL-2 (Liu et al. 2001), effects which may enhance expansion and effector function of autoreactive CD4⁺ T cells. *Tnfrsf14* was not differentially expressed in CD4⁺ T cells from MRL+/+ and C3H/HeJ mice. On the other hand, *Tnfrsf19* was highly increased (>10 fold) in CD4⁺ T cells from MRL+/+ mice. The protein encoded by *Tnfrsf19* is highly expressed during embryonic development, but in adults is primarily expressed in pulmonary and ductal epithelium, but can be detected in lymphocytes. The functional significance of its increased expression in the CD4⁺ T cells from MRL+/+ mice is not known. However, the cell-inappropriate expression of *Tnfrsf19* suggests some kind of epigenetic mechanism.

Several epigenetic mechanisms have been shown to regulate $CD4^+$ T cell activity and autoimmune pathology in MRL/lpr mice (Pan et al. 2010; Sawalha and Jeffries 2007; Yang et al. 2013). Although distinct from the MRL+/+ mice used to test the effects of TCE, MRL/lpr are closely related. Consequently, as a preliminary look at epigenetics in our model we compared expression of the retrotransposon *Iap* (intracisternal A particle) in the $CD4^+$ T cells from MRL+/+ mice and C3H/HeJ mice. Since *Iap* expression is largely dependent on DNA methylation, its expression is often used as an indirect measurement of this epigenetic process. As shown in Table 2.1 expression of *Iap* was dramatically suppressed in $CD4^+$ T cells from the MRL+/+ mice. This finding suggests that epigenetics plays a role in at least some of the differential gene expression in $CD4^+$ T cells from the MRL+/+ and C3H/HeJ mice. The effects of DNA methylation on susceptibility to toxicant-induced autoimmunity is currently being investigated in and will be described in more detail in Chap. 10.

Taken together, the results demonstrated that $CD4^+$ T cells from MRL+/+ mice differentially express several genes which may make them more likely to be autoreactive. Defining the relative contribution of these alterations to the autoimmune-prone phenotype of the MRL+/+ mice, not to mention determining how they impact the response to TCE exposure, constitutes an important challenge.

2.5.2 Gender

Aside from an ill-defined genetic predisposition autoimmune disease in humans is also regulated by sex. At least 75 % of people with autoimmune disease are women, with the male/female ratios varying among disease. For example, type 1 diabetes is found in both sexes at about the same ratio, while most thyroid autoimmune diseases (e.g. Graves' disease) occur ten times more often in women. In most mouse models of spontaneous lupus nephritis females are more susceptible. In the MRL+/+ mouse model both sexes develop lupus, but the disease is more robust in females which die at about 73 weeks of age compared to males which die at about 93 weeks of age. There is not much known about the role of sex differences in toxicant-induced autoimmune disease in humans. This is largely due to the fact that in epidemiology studies the particular exposure being studied is often gender biased, either toward men (e.g. occupational exposure to toxicants such as silica) or women (e.g. exposure to cosmetics). Animal studies have also been conducted primarily using females except in those few cases in which autoimmunity primarily occurs in males (e.g. BXSB mice). A recent review highlights what is known about gender differences in autoimmunity induced by chemical exposure (Pollard 2012).

In terms of TCE, very little is known about sex-specific immunotoxicity. With regard to adult exposure one meta-analysis of case–control studies concluded that although scleroderma affected women predominantly, among subjects with occupational exposure to solvents (which included TCE) men were at higher risk for developing the disease (Kettaneh et al. 2007). Developmental exposure to TCE at 14,000 ppb reportedly decreased all thymic T cell subsets in male but not female MRL+/+ mice (Peden-Adams et al. 2008). Developmental exposure to lower levels of TCE has been shown to induce subtle differences in double-negative lineage thymocytes in male and female MRL+/+ mice (Blossom and Doss 2007). How the sex-specific thymocyte alterations induced by developmental TCE exposure impact the peripheral immune phenotype during the lifespan of the mice is not clear. Based on the paucity of information, it is currently impossible to predict whether TCE does in fact induce sex-specific alterations in immune function. This represents a gap in the knowledge base that needs to be filled.

2.5.3 Age of Exposure

Most epidemiological studies of immunotoxicity have focused on adult occupational contact with the particular chemical since it is easier to document and often involves relatively higher exposure levels. However, the developing immune system is especially sensitive to environmental perturbation. A recent review compared early vs adult exposure to several immunosuppressive toxins including lead and tributyltin in animal models (Luebke et al. 2006). In all cases sensitivity was greater if exposure occurred during development. In fact, immune suppression in developmentally exposed offspring often occurred at doses that were ineffective in adults. Developmental sensitivity to toxicants has also been found in humans. For example, prenatal exposure to polychlorinated biphenyls decreased the immune response to standard immunizations (Heilmann et al. 2006). Prenatal exposure to polybrominated diphenyl ethers produced a persistent decrease in lymphocyte numbers (Leijs et al. 2009). Aside from immune suppression, there is increasing evidence that adult onset autoimmunity can be triggered by pre- and early post-natal toxicant exposure to environmental factors such as cigarette smoke or organochlorines (Colebatch and Edwards 2011; Langer et al. 2008).

Developmental exposure to TCE is not uncommon; one study showed that 100 % of breast milk samples from 4 US urban areas had detectable levels of TCE (Pellizzari et al. 1982). TCE exposure is also a possible concern for bottle-fed infants because they ingest more water on a bodyweight basis than adults. Gestational and early-life TCE exposure has primarily been examined for its neurotoxicity rather than immunotoxicity (Gist and Burg 1995). However, children continuously exposed for 3–19 years beginning *in utero* to a water supply contaminated with solvents [with TCE being the predominant toxicant (267 ppb)] had altered ratios of T cell subsets and increased levels of autoantibodies (Byers et al. 1988). Autoimmune disease was not assessed. Blossom et al. have shown that continuous exposure to TCE in mice (gestation, lactation and early life) generated $CD4^+$ T cell alterations and early signs of tissue inflammation (Blossom and Doss 2007). More information about the effects of developmental exposure to TCE is available in the Chap. 7. The published experiments concerning developmental effects of TCE on immunotoxicity in MRL+/+ mice to TCE were not extended past 6–8 weeks of age, and although they detected early signs of liver inflammation, they did not assess actual autoimmune disease. The experiments to make such an assessment are currently underway in our laboratory.

2.5.4 Toxicant Co-exposure

In addition to genetics, gender and age of the immunotoxic response to TCE may also be influenced by chemical co-exposure. Anyone exposed to TCE, either as an adult or during development, is also exposed to other chemicals with defined or potential immunotoxicity. Co-exposure to another immunotoxicant with additive or synergistic effects may promote autoimmunity at concentrations that would be harmless for either chemical alone. One such toxicant is mercury.

Human exposure to mercury (#3 CERCLA Priority List of Hazardous substances) is common due to its existence as a natural element and its anthropogenic release from industrial use. Blood mercury analyses in the 1999–2000 National Health and Nutrition Examination Survey for 16–49 year old women showed that approximately 8 % of women in the survey had blood mercury concentrations greater than 5.8 μg/L (which is a blood mercury level equivalent to the current RfD). TCE and mercury are often found together; 44 % of the active Superfund sites on the National Priorities List contaminated with TCE also list mercury as a contaminant. In addition, since both are listed in the top 20 chemicals of the almost 300 chemicals on the CERCLA list based in part on the likelihood of human contact, co-exposure is probable. Mercury has been implicated as a co-factor in systemic human autoimmunity, where studies showed that mercury exposure increased levels of anti-nucleolar antibodies (Cooper et al. 2004; Gardner et al. 2010). The ability of mercury to promote autoimmunity has been especially well-documented in mice where it promotes autoantibodies specific for fibrillarin and other nuclear antigens such as chromatin, and induces immune complex-mediated lupus nephritis (Hultman et al. 1996; Pollard et al. 2001).

A recent study examined the combined effects of mercury and TCE exposure on the induction of autoimmune disease in adult female MRL+/+ mice. Mice received either 0, 0.1 or 2.0 mg/ml TCE in their drinking water for 8 weeks. Some mice were injected sc twice per week for 8 weeks with 40 μg $HgCl_2$. Exposure to TCE alone at these concentrations for only 8 weeks was not expected to induce autoimmune hepatitis. And indeed the livers of mice exposed to $HgCl_2$ or TCE alone exhibited no significant pathology. However, based on cumulative scores of mononuclear cell infiltration, fibrosis, and hepatocellular enlargement, liver pathology in mice exposed to $HgCl_2$ and either 0.1 or 2.0 mg/ml TCE was significantly increased from that of control mice, indicating the early stages of autoimmune hepatitis. In addition, TCE and heavy metals have been shown by others to have an additive effect on antioxidant endpoints (Tabrez and Ahmad 2011), an effect that can promote immunotoxicity.

Documenting the ability of TCE to augment the activity of other known or suspected immunotoxicants is needed to accurately evaluate the role of TCE in promoting seemingly idiopathic autoimmune disease. This however, represents a large challenge. There seems to be little consensus in how to study mixtures regardless of the outcome measured. Should we use a labor-intensive and expensive full-factorial design in a mouse model starting with TCE-containing binary mixtures and expanding to include other immunotoxicants? Or, should we study the murine effects of complex TCE-containing mixtures selected because they include those found most commonly in ground water or Superfund sites? Alternatively, should we identify the chemical mixtures to which a particular patient subset (e.g. children recently diagnosed juvenile autoimmune hepatitis) are most commonly exposed, and test those mixtures for immunotoxicity in mice? The fact that there are pros and cons associated with all of these approaches should not preclude selecting one to further this important area of toxicity assessment.

2.6 Challenges

Both epidemiological studies and work with animal models provide evidence that TCE exposure contributes to autoimmunity. However, as mentioned in the text there are several aspects of TCE-induced immunotoxicity that remain to be explored. Studying the ability of a toxicant such as TCE to promote autoimmune disease is complicated by the difficulty of determining exposure in humans, and by the contribution of risk factors such as ill-defined genetic susceptibility, differential sensitivity based on sex or age, and possible augmentation or antagonism by co-exposure to other environmental triggers (e.g. additional chemicals, diet, infections, changes in the microbiome or obesity). Similarly, determining which of the TCE-induced alterations in $CD4^+$ T cells (e.g. susceptibility to apoptosis, expansion of memory/activated populations, and skewing of cytokine production) are actually required for pathology represents another challenge. Nevertheless, in view of the increased prevalence of certain autoimmune diseases in humans, and the widespread nature of TCE exposure, determining the contribution of the later to the former should remain a priority. Once the mechanism(s) of action have been more clearly defined it should be possible to circumvent at least some of these processes, and thereby decrease the likelihood of TCE-induced autoimmunity. For example, dietary interventions that combat DNA methylation could be used until TCE remediation is more advanced. The changes induced in $CD4^+$ T cells by often undetected TCE exposure may make the host more sensitive to a variety of hypersensitivity/autoimmune responses. Instead of focusing on a few specific syndromes such as scleroderma we should consider the possibility that TCE and other immunotoxic chemicals contribute to a wide array of idiopathic chronic inflammatory diseases.

References

Ala A, Stanca CM, Bu-Ghanim M, Ahmado I, Branch AD, Schiano TD, Odin JA, Bach N (2006) Increased prevalence of primary biliary cirrhosis near Superfund toxic waste sites. Hepatology 43(3):525–531

Anagnostopoulos G, Sakorafas GH, Grigoriadis K, Margantinis G, Kostopoulos P, Tsiakos S, Arvanitidis D (2004) Hepatitis caused by occupational chronic exposure to trichloroethylene. Acta Gastroenterol Belg 67(4):355–357

Blossom SJ, Doss JC (2007) Trichloroethylene alters central and peripheral immune function in autoimmune-prone MRL(+/+) mice following continuous developmental and early life exposure. J Immunotoxicol 4(2):129–141

Blossom SJ, Doss JC, Gilbert KM (2007a) Ability of environmental toxicant trichloroethylene to promote immune pathology is strain-specific. J Immunotoxicol 3:179–187

Blossom SJ, Doss JC, Gilbert KM (2007b) Chronic exposure to a trichloroethylene metabolite in autoimmune-prone MRL+/+ mice promotes immune modulation and alopecia. Toxicol Sci 95(2):401–411

Bona G, Defranco S, Chiocchetti A, Indelicato M, Biava A, DiFranco D, Dianzani I, Ramenghi U, Corrias A, Weber G, De Sanctis V, Iughetti L, Radetti G, Dianzani U (2003) Defective function of Fas in T cells from paediatric patients with autoimmune thyroid diseases. Clin Exp Immunol 133(3):430–437

Bond GR (1996) Hepatitis, rash and eosinophilia following trichloroethylene exposure: a case report and speculation on mechanistic similarity to halothane induced hepatitis. J Toxicol Clin Toxicol 34(4):461–466

Brogren CH, Christensen JM, Rasmussen K (1986) Occupational exposure to chlorinated organic solvents and its effect on the renal excretion of N-acetyl-beta-D-glucosaminidase. Arch Toxicol Suppl 9:460–464

Byers VS, Levin AS, Ozonoff DM, Baldwin RW (1988) Association between clinical symptoms and lymphocyte abnormalities in a population with chronic domestic exposure to industrial solvent-contaminated domestic water supply and a high incidence of leukemia. Cancer Immunol Immunother 27:77–82

Cai P, Konig R, Boor PJ, Kondraganti S, Kaphalia BS, Khan MF, Ansari GA (2008) Chronic exposure to trichloroethene causes early onset of SLE-like disease in female MRL +/+ mice. Toxicol Appl Pharmacol 228(1):68–75

Charo J, Lindencrona JA, Carlson LM, Hinkula J, Kiessling R (2004) Protective efficacy of a DNA influenza virus vaccine is markedly increased by the coadministration of a Schiff base-forming drug. J Virol 78(20):11321–11326

Chen H, Hall S, Heffernan B, Thompson NT, Rogers MVF, Rhodes J (1997) Convergence of Schiff base costimulatory signaling and TCR signaling at the level of mitogen-activated protein kinase ERK2. J Immunol 159:2274–2281

Colebatch AN, Edwards CJ (2011) The influence of early life factors on the risk of developing rheumatoid arthritis. Clin Exp Immunol 163(1):11–16

Committee on Human Health Risks of Trichloroethylne NRC (2006) Assessing the human health risks of trichloroethylene: key scientific issues. National Research Council, Washington, DC: The National Academies Press

Cooper GS, Parks CG, Treadwell EL, St Clair EW, Gilkeson GS, Dooley MA (2004) Occupational risk factors for the development of systemic lupus erythematosus. J Rheumatol 31(10):1928–1933

Crispe IN (1994) Fatal interactions: Fas-induced apoptosis of mature T cells. Immunity 1:347–349

Czirjak L, Pocs E, Szegedi G (1994) Localized scleroderma after exposure to organic solvents. Dermatology 189(4):399–401

Diot E, Lesire V, Guilmot JL, Metzger MD, Pilore R, Rogier S, Stadler M, Diot P, Lemarie E, Lasfargues G (2002) Systemic sclerosis and occupational risk factors: a case–control study. Occup Environ Med 59(8):545–549

Dreskin SC (2006) Genetics of food allergy. Curr Allergy Asthma Rep 6(1):58–64
Flindt-Hansen H, Isager H (1987) Scleroderma after occupational exposure to trichloroethylene and trichlorethane. Toxicol Lett 95:173–181
Garabrant DH, Lacey JV Jr, Laing TJ, Gillespie BW, Mayes MD, Cooper BC, Schottenfeld D (2003) Scleroderma and solvent exposure among women. Am J Epidemiol 157(6):493–500
Gardner RM, Nyland JF, Silva IA, Ventura AM, de Souza JM, Silbergeld EK (2010) Mercury exposure, serum antinuclear/antinucleolar antibodies, and serum cytokine levels in mining populations in Amazonian Brazil: a cross-sectional study. Environ Res 110(4):345–354
Germolec D, Kono DH, Pfau JC, Pollard KM (2012) Animal models used to examine the role of the environment in the development of autoimmune disease: findings from an NIEHS Expert Panel Workshop. J Autoimmun 39(4):285–293
Gilbert KM, Whitlow AB, Pumford NR (2004) Environmental contaminant and disinfection by-product trichloroacetaldehyde stimulates T cells in vitro. Int Immunopharmacol 4(1):25–36
Gilbert KM, Pumford NR, Blossom SJ (2006) Environmental contaminant trichloroethylene promotes autoimmune disease and inhibits T-cell apoptosis in MRL+/+ mice. J Immunotoxicol 3:263–267
Gist GL, Burg JR (1995) Trichloroethylene – a review of the literature from a health effects perspective. Toxicol Ind Health 11(3):253–307
Green DR, Droin N, Pinkoski M (2003) Activation-induced cell death in T cells. Immunol Rev 193:70–81
Griffin JD, Gilbert KM, Pumford NR (2000a) Inhibition of CYP2E1 reverses CD4+ T cell alterations in trichloroethylene-treated MRL+/+ mice. Toxicol Sci 54:384–389
Griffin JM, Blossom SJ, Jackson SK, Gilbert KM, Pumford NR (2000b) Trichloroethylene accelerates an autoimmune response in association with Th1 T cell activation in MRL+/+ mice. Immunopharmacology 46:123–137
Griffin JM, Gilbert KM, Lamps LW, Pumford NR (2000c) CD4+ T cell activation and induction of autoimmune hepatitis following trichloroethylene treatment in MRL+/+ mice. Toxicol Sci 57:345–352
Halmes NC, Samokyszyn VM, Pumford NR (1997) Covalent binding and inhibition of cytochrome P4502E1 by trichloroethylene. Xenobiotica 27(1):101–110
He XS, Ansari AA, Gershwin ME (2001) Xenobiotic considerations for the development of autoimmune liver diseases: bad genes and bad luck. Rev Environ Health 16:191–202
Heilmann C, Grandjean P, Weihe P, Nielsen F, Budtz-Jorgensen E (2006) Reduced antibody responses to vaccinations in children exposed to polychlorinated biphenyls. PLoS Med 3(8):e311
Hong NM, Masuko-Hongo K, Sasakawa H, Kato T, Shirai T, Okumura K, Nishioka K, Kobata T (1998) Amelioration of lymphoid hyperplasia and hypergammaglobulinemia in lupus-prone mice (gld) by Fas-ligand gene transfer. J Autoimmun 11(4):301–307
Hultman P, Turley SJ, Enestrom S, Lindh U, Pollard KM (1996) Murine genotype influences the specificity, magnitude and persistence of murine mercury-induced autoimmunity. J Autoimmun 9(2):139–149
Iavicoli I, Marinaccio A, Carelli G (2005) Effects of occupational trichloroethylene exposure on cytokine levels in workers. J Occup Environ Med 47(5):453–457
Jarvinen P, Aho K (1994) Twin studies in rheumatic diseases. Semin Arthritis Rheum 24(1):19–28
Joron GE, Cameron DG, Halpenny GW (1955) Massive necrosis of the liver due to trichlorethylene. Can Med Assoc J 73(11):890–891
Karamfilov T, Buslau M, Durr C, Weyers W (2003) Pansclerotic porphyria cutanea tarda after chronic exposure to organic solvents. Hautarzt 54(5):448–452
Keil DE, Peden-Adams MM, Wallace S, Ruiz P, Gilkeson GS (2009) Assessment of trichloroethylene (TCE) exposure in murine strains genetically-prone and non-prone to develop autoimmune disease. J Environ Sci Health A Tox Hazard Subst Environ Eng 44(5):443–453
Kettaneh A, Al MO, Tiev KP, Chayet C, Toledano C, Fabre B, Fardet L, Cabane J (2007) Occupational exposure to solvents and gender-related risk of systemic sclerosis: a metaanalysis of case–control studies. J Rheumatol 34(1):97–103

Khan MF, Kaphalia BS, Prabhakar BS, Kanz MF, Ansari GAS (1995) Trichloroethylene-induced autoimmune response in female MRL +/+ mice. Toxicol Appl Pharmacol 134:155–160
Khan MF, Wu X, Ansari GA (2001) Anti-malondialdehyde antibodies in MRL+/+ mice treated with trichloroethene and dichloroacetyl chloride: possible role of lipid peroxidation in autoimmunity. Toxicol Appl Pharmacol 170(2):88–92
Kilburn KH, Washaw RW (1992) Prevalence of symptoms of systemic lupus erythematosus (SLE) and of fluorescent antinuclear antibodies associated with chronic exposure to trichloroethylene and other chemicals in well water. Environ Res 57:1–9
Kischkel FC, Hellbardt S, Behrmann I, Germer M, Pawlita M, Krammer PH, Peter ME (1995) Cytotoxicity-dependent APO-1 (Fas/CD95)-associated proteins form a death-inducing signaling complex (DISC) with the receptor. EMBO J 14(22):5579–5588
Kovacs B, Vassilopoulos D, Vogelgesang SA, Tsokos GC (1996) Defective CD3-mediated cell death in activated T cells from patients with systemic lupus erythematosus: role of decreased intracellular TNF-alpha. Clin Immunol Immunopathol 81(3):293–302
Langer P, Kocan A, Tajtakova M, Koska J, Radikova Z, Ksinantova L, Imrich R, Huckova M, Drobna B, Gasperikova D, Sebokova E, Klimes I (2008) Increased thyroid volume, prevalence of thyroid antibodies and impaired fasting glucose in young adults from organochlorine cocktail polluted area: outcome of transgenerational transmission? Chemosphere 73(7): 1145–1150
Lash LH, Fisher JW, Lipscomb JC, Parker JC (2000) Metabolism of trichloroethylene. Environ Health Perspect 108:177–200
Leijs MM, Koppe JG, Olie K, van Aalderen WM, de Voogt P, ten Tusscher GW (2009) Effects of dioxins, PCBs, and PBDEs on immunology and hematology in adolescents. Environ Sci Technol 43(20):7946–7951
Lipscomb JC, Mahle DA, Brashear WT, Garrett CM (1996) A species comparison of chloral hydrate metabolism in blood and liver. Biochem Biophys Res Commun 227(2):340–350
Liu J, Na S, Glasebrook A, Fox N, Solenberg PJ, Zhang Q, Song HY, Yang DD (2001) Enhanced CD4+ T cell proliferation and Th2 cytokine production in DR6-deficient mice. Immunity 15(1):23–34
Liu J, Xing X, Huang H, Jiang Y, He H, Xu X, Yuan J, Zhou L, Yang L, Zhuang Z (2009) Identification of antigenic proteins associated with trichloroethylene-induced autoimmune disease by serological proteome analysis. Toxicol Appl Pharmacol 240(3):393–400
Lockey JE, Kelly CR, Cannon GW, Colby TV, Aldrich V, Livingston GK (1997) Progressive systemic sclerosis associated with exposure to trichloroethylene. J Occup Med 29:493–496
Luebke RW, Holsapple MP, Ladics GS, Luster MI, Selgrade M, Smialowicz RJ, Woolhiser MR, Germolec DR (2006) Immunotoxicogenomics: the potential of genomics technology in the immunotoxicity risk assessment process. Toxicol Sci 94(1):22–27
Marrack P, Kappler J (2004) Control of T cell viability. Annu Rev Immunol 22:765–787
McCunney RJ (1988) Diverse manifestations of trichloroethylene. Br J Ind Med 45(2):122–126
Meurer JR, Lustig JV, Jacob HJ (2006) Genetic aspects of the etiology and treatment of asthma. Pediatr Clin North Am 53(4):715–725
Miyazaki T, Ono M, Qu WM, Zhang MC, Mori S, Nakatsuru S, Nakamura Y, Sawasaki T, Endo Y, Nose M (2005) Implication of allelic polymorphism of osteopontin in the development of lupus nephritis in MRL/lpr mice. Eur J Immunol 35(5):1510–1520
Mohrenschlager M, Darsow U, Schnopp C, Ring J (2006) Atopic eczema: what's new? J Eur Acad Dermatol Venereol 20(5):503–511, 513
Nakajima T, Yamanoshita O, Kamijima M, Kishi R, Ichihara G (2003) Generalized skin reactions in relation to trichloroethylene exposure: a review from the viewpoint of drug-metabolizing enzymes. J Occup Health 45(1):8–14
Nietert PJ, Sutherland SE, Silver RM, Pandey JP, Knapp RG, Hoel DG, Dosemeci M (1998) Is occupational organic solvent exposure a risk factor for scleroderma? Arthritis Rheum 41:1111–1119
Nishimura-Morita Y, Nose M, Inoue T, Yonehara S (1997) Amelioration of systemic autoimmune disease by the stimulation of apoptosis-promoting receptor Fas with anti-Fas mAb. Int Immunol 9(12):1793–1799

Pan W, Zhu S, Yuan M, Cui H, Wang L, Luo X, Li J, Zhou H, Tang Y, Shen N (2010) MicroRNA-21 and microRNA-148a contribute to DNA hypomethylation in lupus CD4+ T cells by directly and indirectly targeting DNA methyltransferase 1. J Immunol 184(12):6773–6781

Pantucharoensri S, Boontee P, Likhitsan P, Padungtod C, Prasartsansoui S (2004) Generalized eruption accompanied by hepatitis in two Thai metal cleaners exposed to trichloroethylene. Ind Health 42(3):385–388

Peden-Adams MM, Eudaly JG, Lee AM, Miller J, Keil DE, Gilkeson GS (2008) Lifetime exposure to trichloroethylene (TCE) does not accelerate autoimmune disease in MRL +/+ mice. J Environ Sci Health A Tox Hazard Subst Environ Eng 43(12):1402–1409

Pellizzari ED, Hartwell TD, Harris BS III, Waddell RD, Whitaker DA, Erickson MD (1982) Purgeable organic compounds in mother's milk. Bull Environ Contam Toxicol 28(3):322–328

Pemberton PW, Aboutwerat A, Smith A, Burrows PC, McMahon RF, Warnes TW (2004) Oxidant stress in type I autoimmune hepatitis: the link between necroinflammation and fibrogenesis? Biochim Biophys Acta 1689(3):182–189

Pino-Lagos K, Guo Y, Brown C, Alexander MP, Elgueta R, Bennett KA, De Vries V, Nowak E, Blomhoff R, Sockanathan S, Chandraratna RA, Dmitrovsky E, Noelle RJ (2011) A retinoic acid-dependent checkpoint in the development of CD4+ T cell-mediated immunity. J Exp Med 208(9):1767–1775

Pirmohamed M (2006) Genetic factors in the predisposition to drug-induced hypersensitivity reactions. AAPS J 8(1):E20–E26

Pollard KM (2012) Gender differences in autoimmunity associated with exposure to environmental factors. J Autoimmun 38(2–3):J177–J186

Pollard KM, Pearson DL, Hultman P, Deane TN, Lindh U, Kono DH (2001) Xenobiotic acceleration of idiopathic systemic autoimmunity in lupus-prone bxsb mice. Environ Health Perspect 109:27–33

Pralong P, Cavailhes A, Balme B, Cottin V, Skowron F (2009) Diffuse systemic sclerosis after occupational exposure to trichloroethylene and perchloroethylene. Ann Dermatol Venereol 136(10):713–717

Qu WM, Miyazaki T, Terada M, Okada K, Mori S, Kanno H, Nose M (2002) A novel autoimmune pancreatitis model in MRL mice treated with polyinosinic: polycytidylic acid. Clin Exp Immunol 129(1):27–34

Ramdhan DH, Kamijima M, Yamada N, Ito Y, Yanagiba Y, Nakamura D, Okamura A, Ichihara G, Aoyama T, Gonzalez FJ, Nakajima T (2008) Molecular mechanism of trichloroethylene-induced hepatotoxicity mediated by CYP2E1. Toxicol Appl Pharmacol 231(3):300–307

Rhodes J, Chen H, Hall SR, Beesley JE, Jenkins DC, Collins P, Zheng B (1995) Therapeutic potentiation of the immune system by costimulatory Schiff-base-forming drugs. Nature 377:71–75

Saihan EM, Burton JL, Heaton KW (1978) A new syndrome with pigmentation, scleroderma, gynaecomastia, Raynaud's phenomenon and peripheral neuropathy. Br J Dermatol 99:437–440

Sanz-Cameno P, Medina J, Garcia-Buey L, Garcia-Sanchez A, Borque MJ, Martin-Vilchez S, Gamallo C, Jones EA, Moreno-Otero R (2002) Enhanced intrahepatic inducible nitric oxide synthase expression and nitrotyrosine accumulation in primary biliary cirrhosis and autoimmune hepatitis. J Hepatol 37(6):723–729

Sawalha AH, Jeffries M (2007) Defective DNA methylation and CD70 overexpression in CD4+ T cells in MRL/lpr lupus-prone mice. Eur J Immunol 37(5):1407–1413

Schattner A, Malnick SD (1990) Anicteric hepatitis and uveitis in a worker exposed to trichloroethylene. Postgrad Med J 66(779):730–731

Semra YK, Seidi OA, Sharief MK (2002) Disease activity in multiple sclerosis correlates with T lymphocyte expression of the inhibitor of apoptosis proteins. J Neuroimmunol 122:159–166

Skarstein K, Johannessen AC, Holmdahl R, Jonsson R (1997) Effects of sialadenitis after cellular transfer in autoimmune MRL/lpr mice. Clin Immunol Immunopathol 84(2):177–184

Sneller MC, Wang J, Dale JK, Strober W, Middelton LA, Choi Y, Fleisher TA, Lim MS, Jaffe ES, Puck JM, Lenardo MJ, Straus SE (1997) Clinical, immunologic, and genetic features of an autoimmune lymphoproliferative syndrome associated with abnormal lymphocyte apoptosis. Blood 89(4):1341–1348

Sobel ES, Wang F, Butfiloski E, Croker B, Roberts SM (2006) Comparison of chlordecone effects on autoimmunity in (NZBxNZW) F(1) and BALB/c mice. Toxicology 218(2–3):81–89

Szodoray P, Jellestad S, Nakken B, Brun JG, Jonsson R (2003) Programmed cell death in rheumatoid arthritis peripheral blood T-cell subpopulations determined by laser scanning cytometry. Lab Invest 83(12):1839–1848

Tabrez S, Ahmad M (2011) Some enzymatic/nonenzymatic antioxidants as potential stress biomarkers of trichloroethylene, heavy metal mixture, and ethyl alcohol in rat tissues. Environ Toxicol 26(2):207–216

Teichert-Kuliszewska K, Israel Y, Cinader B (1988) Alcohol dehydrogenase is not a major determinant of alcohol preference in mice. Alcohol 5(1):45–47

Tomer Y, Davies TF (1997) The genetic susceptibility to Graves' disease. Baillieres Clin Endocrinol Metab 11(3):431–450

Van Parijs L, Peterson DA, Abbas AK (1998) The Fas/Fas lignad pathway and Bcl-2 regulate T cell responses to model self and foreign antigens. Immunity 8:265–274

Waiczies S, Weber A, Lunemann JD, Aktas O, Zschenderlein R, Zipp F (2002) Elevated Bcl-X(L) levels correlate with T cell survival in multiple sclerosis. J Neuroimmunol 126:213–220

Waller PA, Clauw D, Cupps T, Metcalf JS, Silver RN, Leroy EC (1994) Fasciitis (not scleroderma) following prolonged exposure to an organic solvent (trichloroethylene). J Rheumatol 21:1567–1570

Wang G, Cai P, Ansari GA, Khan MF (2007) Oxidative and nitrosative stress in trichloroethene-mediated autoimmune response. Toxicology 229(3):186–193

Xu X, Yang R, Wu N, Zhong P, Ke Y, Zhou L, Yuan J, Li G, Huang H, Wu B (2009a) Severe hypersensitivity dermatitis and liver dysfunction induced by occupational exposure to trichloroethylene. Ind Health 47(2):107–112

Xu XY, Chen GH, Wu N, Yu L, Huang F, Yang LQ (2009b) Relationship between gene polymorphism of CYP2E1, CYP1A1, IL-4 and medicamentosa-like dermatitis induced by trichloroethylene. Zhonghua Lao Dong Wei Sheng Zhi Ye Bing Za Zhi 27(2):71–73

Yang ML, Gee AJ, Gee RJ, Zurita-Lopez CI, Khare S, Clarke SG, Mamula MJ (2013) Lupus autoimmunity altered by cellular methylation metabolism. Autoimmunity 46(1):21–31

Zhang L, Bassig BA, Mora JL, Vermeulen R, Ge Y, Curry JD, Hu W, Shen M, Qiu C, Ji Z, Reiss B, McHale CM, Liu S, Guo W, Purdue MP, Yue F, Li L, Smith MT, Huang H, Tang X, Rothman N, Lan Q (2013) Alterations in serum immunoglobulin levels in workers occupationally exposed to trichloroethylene. Carcinogenesis 34:799–802

Zhou T, Song L, Yang P, Wang Z, Lui D, Jope RS (1999) Bisindolylmaleimide VIII facilitates Fas-mediated apoptosis and inhibits T cell-mediated autoimmune diseases. Nat Med 5(1):42–48

Chapter 3
Hypersensitivity Dermatitis and Hepatitis

Michihiro Kamijima, Hailan Wang, Osamu Yamanoshita, Yuki Ito, and Tamie Nakajima

Abstract Occupational exposure to trichloroethylene (TCE) rarely induces severe generalized hypersensitivity dermatitis accompanying grave hepatitis, referred to as occupational TCE hypersensitivity syndrome (HS), in susceptible workers. TCE HS resembles delayed-type severe cutaneous adverse reactions to drugs and is totally different from solvent-induced irritating contact dermatitis. Importantly, human herpesvirus 6, which remains latent within the body after primary infection during infancy, is reactivated in most patients, and the reactivation affects the clinical course of this disease. Lines of evidence have established the current notion that TCE has sensitization potency. Though human leucocyte antigen (HLA)-B*13:01 has been identified as a marker of individual susceptibility, appropriate occupational hygiene practices to reduce the exposure and the biological monitoring of urinary TCE metabolites can be crucial to preventing this disease. Since the causal relationship between TCE exposure and this life-threatening occupational disease can be overlooked, the disease needs special attention from occupational health professionals and clinicians.

Keywords Trichloroethylene • Hypersensitivity • Dermatitis • Hepatitis • Human herpesvius 6 • Drug-induced hypersensitivity syndrome (DIHS) • HLA-B*13:01 • Urine • Trichloroacetic acid

M. Kamijima (✉) • Y. Ito
Department of Occupational and Environmental Health,
Nagoya City University Graduate School of Medical Sciences,
1 Kawasumi, Mizuho-cho, 4678601 Mizuho-ku, Nagoya, Japan

H. Wang
Guangdong Province Hospital for Occupational Disease Prevention and Treatment,
Guangzhou, P.R. of China

O. Yamanoshita • T. Nakajima
Chubu University College of Life and Health Sciences, Kasugai, Japan

K.M. Gilbert, S.J. Blossom (eds.), *Trichloroethylene: Toxicity and Health Risks*,
Molecular and Integrative Toxicology, DOI 10.1007/978-1-4471-6311-4_3,

3.1 Occupational Trichloroethylene Hypersensitivity Syndrome: Hypersensitivity Dermatitis Due to Trichloroethylene Exposure

Trichloroethylene (TCE) (CAS number: 79-01-6) is still an important industrial solvent today, used for degreasing during production of metal parts. Although the production volume of TCE has been decreasing, especially in developed countries, health problems related to TCE exposure remain an important occupational health issue at least in some Asian countries since there are a large number of exposed workers. The annual production of TCE in China and Japan was 160,900 and 47,745 t in 2010, respectively.

In workshops where organic solvent is used, skin problems, mostly irritant contact dermatitis, are in general the frequently encountered occupational health issues. The dermatitis is primarily attributable to irritation due to the local defatting action of the solvent, in which skin surface lipids, the lipid material in the stratum corneum, and the fatty fraction of the cell membranes are dissolved (Wahlberg and Adams 1999). However, workers engaging in a job exposed to TCE could also suffer from idiosyncratic generalized dermatitis accompanying grave hepatitis, which is totally different from the irritant contact dermatitis (Huang et al. 2002; Kamijima et al. 2007). The mortality rate of this disease is surprisingly high, i.e., 9–13 %. Liver failure, infections, and the resulting sepsis are the principal causes of fatalities (Phoon et al. 1984; Pantucharoensri et al. 2004; Kamijima et al. 2007).

This generalized dermatitis resembles severe drug hypersensitivities, and is designated as occupational TCE hypersensitivity syndrome (HS). It is mediated by a delayed-type hypersensitivity mechanism, but is not classified as allergic contact dermatitis, which only involves the areas of skin directly contacting TCE liquid. It could involve whole body surface, even the mucous membrane in the oral cavity and genitalia. In addition to the characteristics of the rash, about a 1-month duration from the commencement of exposure to the disease onset, fever, abnormally increased leukocyte number, lymphadenopathy, liver dysfunction and the resulting fatalities, and the recurrence just after the minimal re-exposure overlap characteristics of a disease entity referred to as severe cutaneous adverse reactions to drugs, namely drug-induced hypersensitivity syndrome (DIHS) or drug reaction with eosinophilia and systemic symptoms (DRESS) (Huang et al. 2006; Watanabe et al. 2010; Watanabe 2011; Kamijima et al. 2013). Since TCE still plays a significant role as a degreasing solvent today and this life-threatening occupational disease can be misdiagnosed as generalized drug eruptions, the disease needs particular attention of occupational health professionals and clinicians.

3.2 Epidemiology of Trichloroethylene Hypersensitivity Syndrome

The incidence rate of TCE HS is by far lower than TCE-induced irritant contact dermatitis. Though the exact incidence rate of the disease remains unclear today, the disease prevalence estimated from previous reports ranges between 0.25 and 12.5 % (Kamijima et al. 2007), which suggests that the prevalence depends on exposure dose and the existence of one or more certain individual susceptibility factors. The first description of TCE HS can be found in an American textbook on occupational skin disorders that was published in 1947 as follows: "It (TCE) is also a sensitizer and can cause a more or less generalized acute eczematoid type of dermatitis which begins as an erythema, becomes papular, then vesicular, and is followed by oozing, crusting, and desquamation" (Schwartz et al. 1947). However, the authors did not mention the possibly accompanying hepatitis, fever, hematological abnormalities, lymphadenopathy and mucosal lesions, which are important features of this disease and will be described in more detail later in this chapter. After this publication, less than ten cases of TCE HS were reported in each decade between 1960 and 1990 from a limited number of industrialized countries, i.e., USA, Japan, Singapore and Spain (Bauer and Rabens 1974; Conde-Salazar et al. 1983; Phoon et al. 1984; Goh and Ng 1988; Nakayama et al. 1988; Hisanaga et al. 2002). In contrast, the reported number increased dramatically after the mid-1990s mainly in industrializing Asia, particularly in China where more than 300 cases have been reported (Huang et al. 2002; Dai et al. 2004; Kamijima et al. 2007). In China, patients were found in some provinces, but the number of reported cases was by far the largest in a southern part of China, Guangdong Province. This growing number of disease occurrences was considered to be partly attributable to the rapid economic development in that area and the resulting increase in the use of TCE and exposed populations (Huang et al. 2002), especially after the conclusion of the Montreal Protocol to phase out the use and production of chlorofluorocarbons and 1,1,1-trichloroethane. Patients were also reported in Korea, Singapore, Thailand, Philippines, USA, and Japan until this day (Bond 1996; Chittasobhaktra et al. 1997; Tan et al. 1997; Estrella-Gust et al. 1999; Goon et al. 2001; Pantucharoensri et al. 2004; Kamijima et al. 2007; Ikeoka et al. 2009; Watanabe et al. 2010; Jung et al. 2012). Those patients were mostly engaged in degreasing, especially cleaning metal-made products, machines, plastic toys, or electronics parts. Degreasing work using TCE thus seems to carry a higher risk of suffering from this disease.

It should be noted that an occupational history linked to TCE exposure can be overlooked in a patient exhibiting generalized rash if the clinician does not ask the patient questions focusing on solvent exposure (Watanabe et al. 2010). There may be more latent patients even in developed countries today. However, the number of workers engaging in solvent-exposed work in factories is generally small in these countries, and poor working environments in terms of TCE exposure are encountered in small-scale enterprises rather than in large-scale ones. Given the very low incidence rate of TCE HS, workplace-based epidemiological studies to clarify the precise rate are practically difficult to carry out.

3.3 Characteristics and Pathophysiology of Trichloroethylene Hypersensitivity Syndrome

3.3.1 Clinical Features Common to Severe Cutaneous Adverse Reactions to Drugs

As mentioned in the introductory part of this review, TCE HS shows characteristics common to delayed-type severe drug hypersensitivities. The characteristics of the disease are summarized in Table 3.1. Duration of exposure is one of the most important features of this disease. The duration is 4 weeks on average, and almost all within 3 months. If a heavily-exposed worker does not suffer from the disease during this period, he/she is not susceptible. The initial symptom is fever (>38 °C) or rash, or both. The fever is often considered a sign of any infectious disease, and an antibiotic and/or an antipyretic can be prescribed, following which a generalized rash appears and possibly be misdiagnosed as drug eruption. Jaundice is often seen from the early stage of the clinical course (Kamijima et al. 2007).

The dermatitis starts as a diffuse erythematous maculopapular rash on the extremities, face, neck or trunk, and spreads to the entire body surface within one to several days. Chinese researchers in the field of occupational and clinical medicine, focusing on the rash's similarity to generalized drug eruption based on their cumulative experience of treating many patients, classified the rash phenotypes observed during the clinical course into the following categories: exfoliative dermatitis (ED), erythema multiforme (EM), and Stevens-Johnson syndrome (SJS)/toxic epidermal necrolysis (TEN) (Ministry of Health of the People's Republic of China 2007) (Fig. 3.1). Though the rash was the same as that of severe cutaneous adverse reactions to drugs, 3/4 of the patients did not receive medication before the onset of the rash. The remaining 1/4 took medicines for cold because of the initial feverish symptoms (Huang et al. 2006). Of the rash phenotypes most prevalent one is ED. In severe cases, the rash develops into generalized edematous erythema, with facial swelling accompanied by exudates and incrustation, sometimes involving the oral mucous membrane. The rash darkens with increasing desquamation; the scales may be thick on the palm, and can be exfoliated like torn gloves (Ministry of Health of the People's Republic of China 2007).

Importantly, human herpesvirus 6 (HHV6) is reactivated in patients suffering from TCE HS. HHV6 is a causative of exanthema subitum during infancy, and remains latent within the body after the primary infection in most persons in all healthy populations (Yoshikawa et al. 1989; Asano et al. 1989; Yamanishi et al. 1988).

Table 3.1 Clinical characteristics of trichloroethylene hypersensitivity syndrome

Clinical features	Note
Incidence	Less than 1–13 % of the occupationally exposed population (Kamijima et al. 2007). What determines the incidence remains unclear. Both exposure dose (average and peak exposure concentration) and susceptible gene polymorphism (see below) may be involved
Interval from commencement of exposure to disease onset	Four weeks on average, mostly 2–6 weeks, and almost all within 3 months (Huang et al. 2002, 2006; Kamijima et al. 2007)
Initial symptoms	Fever and/or rash (Huang et al. 2002, 2006; Kamijima et al. 2007)
Fever	Frequent (>38 °C) (Huang et al. 2002, 2006; Kamijima et al. 2007)
Rash	Different phenotypes were reported: exfoliative dermatitis (ED) type, erythema multiforme (EM) type, Stevens-Johnson syndrome (SJS)/toxic epidermal necrolysis (TEN) type. ED type is most frequently observed (Huang et al. 2002, 2006; Kamijima et al. 2007). ED and TEN are associated with higher-level human herpesvirus 6 (HHV6) reactivation and stronger proinflammatory cytokine responses (Kamijima et al. 2013)
Mucosal lesion	Possible (Huang et al. 2002, 2006; Kamijima et al. 2007)
Lymphadenopathy	Frequent (cervical nodes, axillary nodes, inguinal nodes and others) (Huang et al. 2002, 2006; Kamijima et al. 2007)
Eosinophilia ($>1.5\times10^9$/L)	Possible. In cases whose eosinophilia is not evident, leucocytosis ($>11\times10^9$/L) is frequently observed (Kamijima et al. 2007; Huang et al. 2006)
Hepatitis	Most patients suffer and it can be a cause of fatality. Jaundice is frequently observed (Huang et al. 2002, 2006; Kamijima et al. 2007)
Possible other organ involvement	Heart, lung, spleen, adrenal gland, larynx and brain (encephalitis) (Kamijima et al. 2007; Huang et al. 2006)
Reactivation of latent viruses/co-infection	HHV6 and cytomegalovirus following HHV6 reactivation (Huang et al. 2006; Watanabe et al. 2010; Watanabe 2011). Other betaherpesvirinae of which the reactivation has been reported in DIHS/DRESS patients can be reactivated (Watanabe et al. 2010; Watanabe 2011; Yanagiba 2007). Increase in IgM antibody titer against measles virus was also reported (Huang et al. 2006)
Percent fatal	9–13 % (Kamijima et al. 2007). Extensive use of corticosteroid can decrease the fatality(Kamijima et al. 2013)
Factor(s) determining individual susceptibility	*HLA-B*13:01* and *HLA-B*44* are the reported major determinant (Li et al. 2007). See Table 3.2 for more details
Positive skin patch test results	Trichloroethanol (0.005 %) (Watanabe et al. 2010; Nakayama et al. 1988), trichloroacetic acid (5 %) (Watanabe et al. 2010), chloral hydrate (5 %) (Watanabe et al. 2010)

Milder cases may lack the above typical features

The reactivation frequency in TCE HS patients has been reported to be at least about 90 % (Kamijima et al. 2013). This finding indicated that the pathophysiology of occupational TCE HS was exactly the same as that of DIHS or DRESS (Huang et al. 2006; Watanabe et al. 2010; Watanabe 2011; Kamijima et al. 2013). DIHS is characterized by (1) maculopapular rash developing >3 weeks after starting with a limited number of drugs, (2) prolonged clinical symptoms 2 weeks after discontinuation of

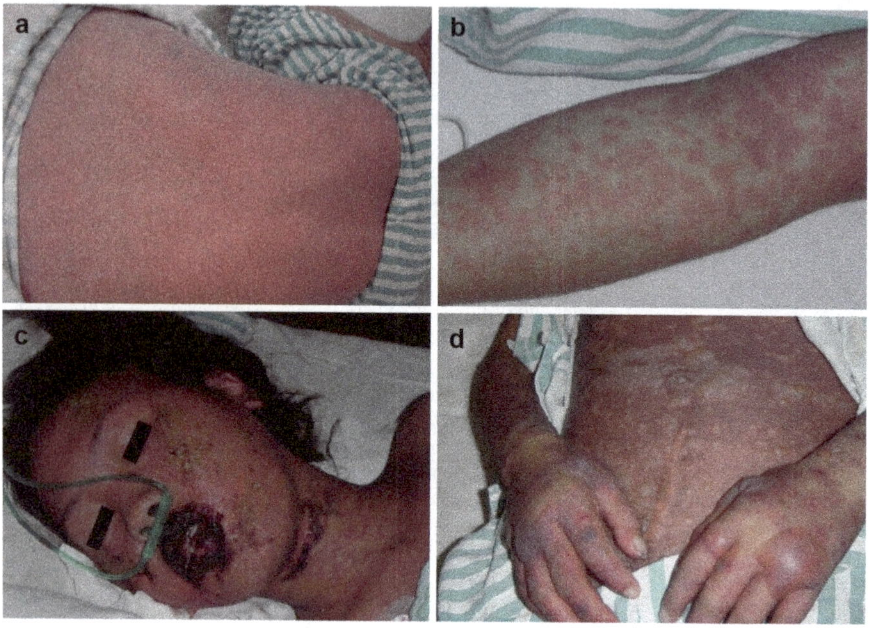

Fig. 3.1 Representative cutaneous manifestations of trichloroethylene-induced generalized hypersensitivity dermatitis. (**a**) Exfoliative dermatitis type, (**b**) Erythema multiforme type, (**c**) Stevens-Johnson syndrome type, (**d**) Toxic epidermal necrolysis type. Rash phenotypes were classified according to the 'Diagnostic criteria of occupational medicamentose-like dermatitis due to trichloroethylene' (GBZ 185–2006) developed by the Ministry of Health of the People's Republic of China (2007) (This figure originally appeared in a previous article (Huang et al. 2006) and was reproduced with permission)

the causative drug, (3) fever (>38°C), (4) liver abnormalities, (5) leucocyte abnormalities being exhibited as either leucocytosis, atypical lymphocytosis or eosinophilia, (6) lymphademopathy, and (7) HHV6 reactivation. A patient is diagnosed as typical DIHS when these seven criteria are met and as atypical DIHS when five (1–5) criteria are met (Shiohara et al. 2007). Thus, most of the cases reported in the past meet the diagnostic criteria of DIHS (Watanabe 2011), although our previous study was the first to detect HHV6 reactivation (Huang et al. 2006). The notion has been fully established that TCE is a causative agent of DIHS (Huang et al. 2006; Watanabe 2011; Kamijima et al. 2013).

3.3.2 *Human Herpesvirus 6 Reactivation and Inflammation-Related Cytokines*

Interestingly, HHV6 viral load in the blood was associated with the rash phenotype; its viremia was more frequently observed and the maximum HHV6 DNA copy numbers were higher in patients with ED than in those with EM type. Patients with TEN type rash also showed a higher-level of reactivation (Kamijima et al. 2013).

The reactivation of HHV6 in the clinical course of DIHS/DRESS is one of the major concerns for achieving a better treatment outcomes since the prognosis of the disease depends on whether or not the reactivation can be controlled (Hashimoto et al. 2003; Shiohara et al. 2006). In patients with TCE HS, levels of blood cytokines, i.e., tumor necrosis factor (TNF) -α, interferon (IFN)-γ, interleukin (IL)-5, IL-6, and IL-10, are significantly and remarkably higher than in healthy TCE-exposed workers (Kamijima et al. 2013; Okamura et al. 2007; Ito et al. 2007). Jia et al. also reported that serum levels of TNF-α, IL-1β, IL-6 and IL-8 in TCE HS patients were significantly higher than in TCE-exposed workers and non-exposed controls (Jia et al. 2012). Their team found that a TCE metabolite trichloroethanol, but not trichloroacetic acid (TCA), increased the release of IL-1α and IL-6 in a keratinocyte cell line in a dose-dependent manner (Jia et al. 2012). In contrast, healthy workers occupationally exposed to TCE reportedly showed lower IL-4 and higher IFN-γ and IL-2 levels in their serum (Iavicoli et al. 2005). IL-10 concentrations in exposed workers were also lower than in unexposed controls whereas there were no significant differences for TNF-α and IL-6 (Bassig et al. 2013). In inflammatory responses of TCE HS, a contribution by the gene polymorphism of TNF-α (G-308A) to disease predisposition was reported (Dai et al. 2004). A TNF-α concentration upon hospitalization higher than 3 standard deviations above the mean value of healthy exposed workers was significantly associated with a subsequent or simultaneous increase in HHV6 DNA in the clinical course (odds ratio 22.4) (Kamijima et al. 2013). Elevation of TNF-α and IL-6 levels reportedly preceded HHV6 reactivation in DIHS (Yoshikawa et al. 2006). It was also shown that HHV6 upregulates the production of TNF-α in peripheral blood mononuclear cells (Flamand et al. 1991). Thus, it may be likely that TCE exposure affects the immunological condition of exposed workers, and that once the TCE HS has occurred in a susceptible worker, the increased TNF-α might induce HHV6 reactivation, or vice versa, resulting in the manifestation of a specific rash phenotype. In addition to the reactivation of HHV6, that of cytomegalovirus (Watanabe et al. 2010) and HHV7 (Yanagiba 2007) can be detected as well. However, co-infection/reactivation of viruses other than betaherpesvirinae is not usually observed. In patients with DIHS, flaring of symptoms such as fever and hepatitis was closely related to HHV6 reactivation during the clinical course after cessation of the causative medication (Tohyama et al. 2007). The same phenomenon is observed in the clinical courses of patients suffering from TCE HS, even after TCA has become undetectable in their urine.

3.3.3 *Liver Dysfunction*

Another important characteristic of TCE HS is liver dysfunction that is observed in most of the patients. Systematic investigation of possible risk factors for hepatitis, e.g., hepatitis A, B, and C viruses, alcohol consumption, drug abuse, use of sanitary chemicals, drinking unsanitary water, and past and family history of allergic, immunological or hepatocystic diseases, ruled out the involvement of these factors in the disease. C-reactive protein and the erythrocyte sedimentation rate usually show negative results (Huang et al. 2006), as does the antinuclear antibody detection test (Kamijima et al. 2013), which is different from those of active autoimmune diseases such as systemic lupus erythematosus.

Some of the chlorinated hydrocarbons are known to have hepatotoxicity although their inherent toxicity varies depending on the chemicals. Exposure to high concentrations of TCE in the air could induce liver dysfunction (World Health Organization (WHO) 1985), but the effects on the liver are generally not massive in those who were anesthetized with TCE or who were occupationally exposed to TCE (Agency for Toxic Substances and Disease Registry 1997), which could have been due to activation of peroxisome proliferator-activated receptor α (Ramdhan et al. 2008; Ramdhan et al. 2010). In contrast, the hepatitis observed in TCE HS is induced by exposure at much lower concentrations and can be a life-threatening fulminant one; aspartate aminotransferase (AST) and alanine aminotransferase (ALT) levels can increase to several thousand U/L. Unfortunately, biopsy results which could address the immune-mediated component of the hepatotoxicity have not been reported in the literature.

The two types of TCE-induced hepatitis mentioned above have different pathophysiologies. An experimental rodent study clearly showed that TCE can induce two different types of hepatitis in guinea pigs, i.e., acute-type toxic hepatitis and immune-mediated hepatitis (Tang et al. 2008). The former type was observed in a dose-dependent manner; in an experiment where guinea pigs received intradermal TCE injections of 0, 167, 500, 1,500 or 4,500 mg/kg, the increase in AST was significant at and above TCE dosages of 1,500 mg/kg, and the increase in ALT and evident fatty degeneration, hepatic sinusoid dilation and inflammatory cell infiltration were observed at 4,500 mg/kg (Tang et al. 2008). A mechanistic study using knockout mice lacking cytochlome P450 (CYP) 2E1 suggested that CYP 2E1-mediated bioactive TCE metabolite(s) might induce nuclear factor kappa B (NFκB) (p52), leading to hepatic inflammation due to the high TCE exposure (Ramdhan et al. 2008). Some intermediate metabolite of TCE to chloral hydrate is considered to play a major role in the hepatotoxicity (Nakajima et al. 1988). In contrast, histopathological findings of the animals sensitized with TCE in guinea pig maximization test (GPMT) were different from the above findings; they were characterized by diffuse ballooning changes without lymphocyte infiltration and necrotic hepatocytes. In addition, 90.9 % (30/33) of sensitized guinea pigs revealed either ALT or AST levels higher than the upper limits of reference values, while 88.2 % (15/17) of guinea pigs without sensitization showed both AST and ALT levels below the limits (Tang et al. 2008). Thus, it appears that low-dose exposure to TCE can induce hepatic damage only by means of immune-mediated mechanisms, which is under investigation in ongoing studies. This experimental evidence suggests that the mechanisms of hepatotoxicity due to TCE are not unique, and that the hepatitis observed in TCE HS, which can be life-threatening, should be treated from the viewpoint of an immune-mediated mechanism. Use of an extensive dose of corticosteroid is reportedly effective (Table 3.1). For example, a series of patients have been treated with methylprednisolone at an initial daily dosage of 80–250 mg with progressive tapering, depending on the clinical severity and response to the treatment (Kamijima et al. 2013).

3.4 Trichloroethylene as a Causative Agent of Hypersensitivity Syndrome

Although a considerable number of reported cases suggested a causal relationship between TCE exposure and the disease, the question of whether the causative agent was TCE itself or the impurities/stabilizers/contaminant(s) of the solvent was a controversial issue, because commercial solvent products used in workplaces are generally not 100 % pure. Impurities depend on the manufacturing route, the type and quality of feed stock used, the type of distillation equipment, and the level of compliance with technical specifications (World Health Organization (WHO) 1985). Stabilizers, some of which have skin sensitization potency (Wahlberg and Adams 1999), are added to prevent the solvent from breaking down into hydrochloric acid, which can corrode the parts being cleaned and the cleaning equipment itself (Mohr 2001). However, today there are the following lines of evidence establishing the causal relationship between TCE itself and TCE HS (Kamijima et al. 2008). First, TCE metabolites, especially TCA, were detected in all the patients' urine when urine was sampled within the period of several times its biological half-life of 57.6 h (Ikeda and Imamura 1973). Second, a comprehensive survey of the solvent constituents including impurities, stabilizers, and metals, as well as the airborne chemicals, showed that no chemical except for TCE was commonly detected in the patients' workplaces. Third, it was confirmed that TCE had a strong sensitization potential in GPMT. This model for type IV hypersensitivity (Kimber et al. 2002) revealed sensitization rates of 66–71 %, showing erythema and skin edema and immune-mediated liver injury at doses below those inducing acute toxic liver injury (Tang et al. 2002; Tang et al. 2008). Skin patch test conducted in a limited number of patients showed positive results at least for TCE metabolites at low concentrations (0.005 % trichloroethanol) (Table 3.1) (Watanabe et al. 2010; Nakayama et al. 1988).

One important viewpoint in the field of occupational health is that some patients themselves did not use TCE but worked close to degreasing tubs (Goon et al. 2001; Kamijima et al. 2013). In a case in China, a patient after recovery suffered from rash again only from a short-time visit to the person's workplace to pick up the belongings. This indicates that skin contact with liquid TCE is not essential for the onset of this skin disorder, which is completely different from solvent-induced irritant dermatitis due to its defatting action (Kamijima et al. 2007). This notion is supported by the fact that oral administration of chloral hydrate, a metabolite of TCE used as a sedative, could also induce generalized skin eruption equivalent to TCE HS (Lindner et al. 1990).

3.5 Susceptible Population

An important feature of TCE HS is the remarkable difference in individual susceptibility to the disease. When an exposed worker does not suffer from it within 3 months after commencement of an extensive exposure to TCE, it can be said that he/she is tolerant to TCE HS. Thus, there should be genetic risk factors leading to

Table 3.2 Genetic factors reported as candidate biomarkers of individual susceptibility

Gene polymorphisms	OR (95 % Cl)	Note	References
*HLA-B*13:01*	27.5 (13.5–55.7)	*B*13:01 or B*44* : OR 36.8 (95%CI 17.8–76.1)	Li et al. (2007)
*HLA-B*44*	20.1 (2.6–157.5)		
NAT1 (SS)	1.10 (0.52–2.34)	*NAT1* (SS) and *NAT2* (FS+SS): OR 2.71 (95%CI 1.29–5.70)	Dai et al. (2009)
NAT2 (FS+SS)	2.01 (1.14–3.54)		
*ALDH2 * 1/ * 2* and *ALDH2 * 2/ * 2*	0.5 (0.29–0.85)		Li et al. (2006)
TNF A I/II (TNF-α *-308* site)	0.398 (0.164–0.967)	*TNF A I*: wild-type allele	Dai et al. (2004)

Abbreviations. *ALDH* aldehyde dehydrogenase, *95 % CI* 95 % confidence interval, *FS + SS* intermediate or slow acetylators, *HLA* human leucocyte antigen, *NAT* N-Acetyltransferases, *OR* odds ratio

ways to search for a biomarker of individual susceptibility, focusing on gene polymorphisms (Dai et al. 2004; Nakajima et al. 2003; Li et al. 2007). One approach was to focus on polymorphisms of drug-metabolizing enzymes (Nakajima et al. 2003). The polymorphisms investigated so far were CYP1A1, CYP2E1, glutathione S-transferase (GST), alcohol dehydrogenase (ADH), aldehyde dehydrogenase (ALDH), and two genes (*NAT1* and *NAT2*) that encode *N*-acetyltransferases (NATs) (Table 3.2). Another focus was human leukocyte antigen (HLA). Associations of HLA-DM and HLA-B with TCE HS have been reported, and the investigation of the latter revealed remarkable success. Li et al. reported HLA-B*13:01 and HLA-B*44 as biomarkers of individual susceptibility with an odds ratio of 36.8 in a patient group having either of the HLA loci (Table 3.2) (Li et al. 2007). This is a finding analogous to carbamazepine-induced SJS/TEN and allopurinol-induced severe cutaneous adverse reactions in which strong associations with HLA-B*15:02 (Chung et al. 2004) and HLA-B*58:01 (Hung et al. 2005), respectively, were reported. Today, candidate peptides bound to the HLA-B*13:01 molecule have been identified (Zhang et al. 2013).

3.6 Preventive Measures of Trichloroethylene Hypersensitivity Syndrome from the Viewpoint of Dose–Response Relationship

Generally speaking, ambient air monitoring is a good measure for assessing external exposure to a solvent. In 2007, the Time-Weighted-Average (TWA) Threshold Limit Value (TLV) of TCE, set by the American Conference of Governmental Industrial Hygienists (ACGIH), was reduced from 50 ppm (270 mg/m^3) to 10 ppm (54 mg/m^3) to protect against effects on the central nervous system (CNS), as well as other potential effects including renal toxicity and cancer. A TLV-short-term exposure limit (STEL) of 25 ppm (135 mg/m^3) was also recommended because the CNS effects of TCE appeared to be related to peak exposures (American Conference of Governmental Industrial Hygienists 2007). However, prevention of TCE HS is

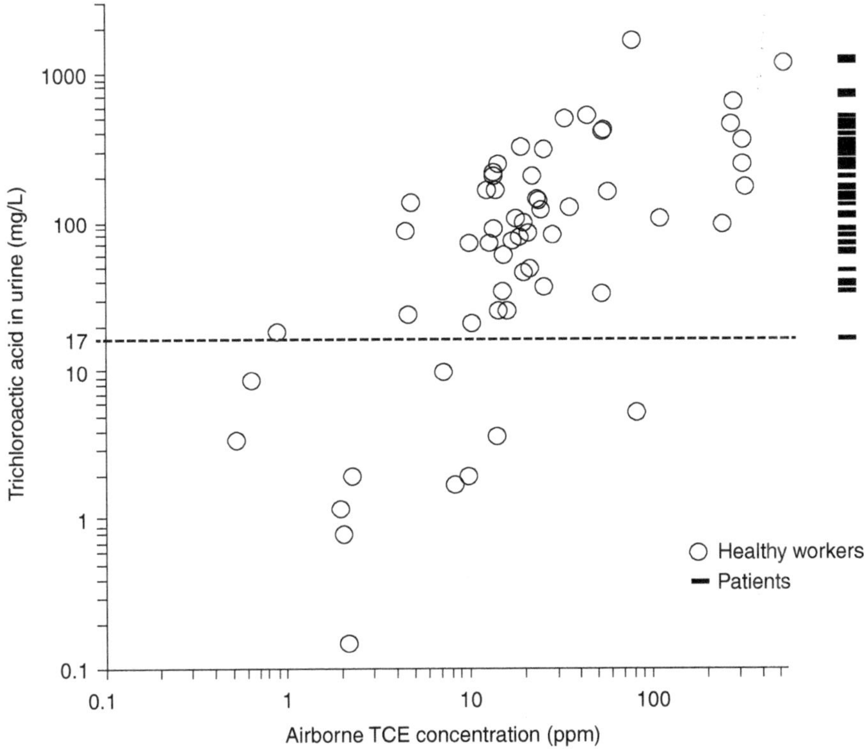

Fig. 3.2 End-of-shift trichloroacetic acid (TCA) concentrations (mg/L, *vertical axis*) in urine of patients (n = 42, *bars*) and healthy workers (n = 59, *circles*) occupationally (directly or indirectly) exposed to TCE. For healthy workers, 8-h time-weighted average personal exposure concentrations (ppm, *horizontal axis*) were also shown, but concentrations were not available for patients. End-of-shift TCA concentrations of the patients were estimated from the concentrations on the day of hospitalization (Kamijima et al. 2008; Kamijima et al. 2013)

not considered in the current documentation. Thus, this section aims to primarily address TCE exposure dose that could elicit hypersensitivity skin reactions.

In the previous case reports, airborne personal exposure concentrations of each patient were usually not available. The only way to directly assess a patient's personal exposure was to measure TCE metabolites in urine. However, measurements were not conducted in most of the case reports published in the last century probably because the importance of biological monitoring at the initial clinical stage of the disease was overlooked. Another problem was that urine available from patients were usually not end-of-shift, so as to enable a comparison of the metabolite concentration with its biological exposure index (BEI). Then we investigated airborne personal exposure concentrations of TCE and TCA concentrations in end-of-shift urine of the healthy workers who worked in the same workplaces as the patients (Fig. 3.2) (Kamijima et al. 2008). On the other hand, we estimated the end-of-shift concentration of hospitalized patients' urine based on information of the time period between the end of shift and the urine collection, on the assumption that the biological half-life of urinary TCA did not vary between patients. The estimated average urinary TCA concentration at the end of their shift was 206 mg/L (95 % confidence interval 78–542 mg/L) (Kamijima et al. 2008). Another estimation from a different patient group showed similar results: 238 mg/L (arithmetic mean) or 153 mg/L (geometric mean) (Kamijima et al. 2013). Thus, it was revealed that many of the patients suffering from TCE HS were extensively exposed to TCE. However, Fig. 3.2 clearly shows that the concentration ranges of the patients and healthy workers widely overlap. The mechanism by which susceptible individuals suffer from the disease should be clarified further.

The next discussion is about the threshold TCA concentration of the patients, which provides useful information for preventing TCE HS. As shown in Fig. 3.2, the lowest end-of-shift concentrations estimated using the reported biological half-life of 57.6 h (Ikeda and Imamura 1973) was 17 mg/L (Kamijima et al. 2013). It is recommended that urinary TCA concentrations be kept below the ACGIH BEI value of 15 mg/L to better prevent TCE HS in susceptible workers.

3.7 Future Directions: Issues Remaining to Be Solved

As discussed above, substantial knowledge about TCE HS has accumulated during this century. The following are examples of the remaining questions. The first concerns the threshold exposure dose that could elicit the disease. More comprehensive studies may be necessary to answer the question of whether the current BEI value is effective to protecting susceptible individuals, especially according to the HLA-B*13:01 status. The second question is whether or not TCE exposure is the only trigger of the sensitization process in susceptible individuals. This question remains to be answered since clusters of disorder occurrence in a workshop during a short period, even within a half month in some cases, have been reported from several countries so far (Hisanaga et al. 2002; Lin et al. 2003; Kamijima et al. 2007). Unrevealed work-related factor(s) other than TCE exposure and genetic factors might have played a supplementary role in causing or developing the disease.

3.8 Conclusions

Occupational diseases are theoretically preventable. It is a great pity that dozens of predisposed workers lost their lives due to TCE HS. The fatalities resulted from lack of knowledge that this disease is immune-mediated. The sensitized victims continued working in the same workplace even after onset of the disease, i.e., they continued being exposed to TCE, or returned to the same work after they recovered from the illness. Such a tragedy should not be repeated in any part of the world.

At present, if replacement of TCE by a safer substitute is not a realistic measure, the best preventive strategy for this disease is to control the exposure under the understanding that TCE is an allergen. A sensitizer notation is necessary for the occupational hygiene standards of TCE. Available evidence suggests that biological monitoring of urinary TCA, which is superior to trichloroethanol because the biological half-life of TCA is longer, is preferable to environmental measurement. It is recommended that the exposure be controlled so as to keep the urinary TCA concentration below the ACGIH BEI value of 15 mg/L in the end-of-shift urine in order to reduce the risk of TCE HS. Education to enforce better occupational hygiene practices is also necessary to reduce the exposure.

For early detection of the disease, occupational health professionals should pay careful attention to initial symptoms of the disease, especially fever, rash and jaundice. Exposed workers should be informed of the initial symptoms of the disease as well. Clinicians are requested to ask about solvent exposure history from patients who suffer from generalized rash like drug hypersensitivities.

References

Agency for Toxic Substances and Disease Registry (1997) Toxicological profile for trichloroethylene. U.S. Department of Health and Human Services, Atlanta

American Conference of Governmental Industrial Hygienists (2007) Trichloroethylene. In: TLVs and BEIs 2007, Cincinnati: ACGIH.

Asano Y, Yoshikawa T, Suga S, Yazaki T, Hata T, Nagai T, Kajita Y, Ozaki T, Yoshida S (1989) Viremia and neutralizing antibody response in infants with exanthem subitum. J Pediatr 114:535–539

Bassig BA, Zhang L, Tang X, Vermeulen R, Shen M, Smith MT, Qiu C, Ge Y, Ji Z, Reiss B, Hosgood HD 3rd, Liu S, Bagni R, Guo W, Purdue M, Hu W, Yue F, Li L, Huang H, Rothman N, Lan Q (2013) Occupational exposure to trichloroethylene and serum concentrations of IL-6, IL-10, and TNF-alpha. Environ Mol Mutagen 54:450–454

Bauer M, Rabens SF (1974) Cutaneous manifestations of trichloroethylene toxicity. Arch Dermatol 110:886–890

Bond GR (1996) Hepatitis, rash and eosinophilia following trichloroethylene exposure: a case report and speculation on mechanistic similarity to halothane induced hepatitis. J Toxicol Clin Toxicol 34:461–466

Chittasobhaktra T, Wannanukul W, Wattanakrai P, Pramoolsinsap C, Sohonslitdsuk A, Nitiyanant P (1997) Fever, skin rash, jaundice and lymphadenopathy after trichloroethylene exposure: a case report. J Med Assoc Thai 80 Suppl 1:S144–S148

Chung WH, Hung SI, Hong HS, Hsih MS, Yang LC, Ho HC, Wu JY, Chen YT (2004) Medical genetics: a marker for Stevens-Johnson syndrome. Nature 428:486

Conde-Salazar L, Guimaraens D, Romero LV, Sanchez YE (1983) Subcorneal pustular eruption and erythema from occupational exposure to trichloroethylene. Contact Dermatitis 9:235–237

Dai Y, Leng S, Li L, Niu Y, Huang H, Cheng J, Zheng Y (2004) Genetic polymorphisms of cytokine genes and risk for trichloroethylene-induced severe generalized dermatitis: a case-control study. Biomarkers 9:470–478

Dai Y, Leng S, Li L, Niu Y, Huang H, Liu Q, Duan H, Cheng J, Zheng Y (2009) Effects of genetic polymorphisms of N-acetyltransferase on trichloroethylene-induced hypersensitivity dermatitis among exposed workers. Ind Health 47:479–486

Estrella-Gust D, Cucueco M, Granadillos N, Dumayag C (1999) Outbreak of Stevens-Johnson syndrome (SJS) in an electronics company in the Philippines. In: Proceeding of 16th Asian conference on occupational health. Cebu

Flamand L, Gosselin J, D'Addario M, Hiscott J, Ablashi DV, Gallo RC, Menezes J (1991) Human herpesvirus 6 induces interleukin-1 beta and tumor necrosis factor alpha, but not interleukin-6, in peripheral blood mononuclear cell cultures. J Virol 65:5105–5110

Goh CL, Ng SK (1988) A cutaneous manifestation of trichloroethylene toxicity. Contact Dermatitis 18:59–61

Goon AT, Lee LT, Tay YK, Yosipovitch G, Ng SK, Giam YC (2001) A case of trichloroethylene hypersensitivity syndrome. Arch Dermatol 137:274–276

Hashimoto K, Yasukawa M, Tohyama M (2003) Human herpesvirus 6 and drug allergy. Curr Opin Allergy Clin Immunol 3:255–260

Hisanaga N, Jonai H, Yu X, Ogawa Y, Mori I, Kamijima M, Ichihara G, Shibata E, Takeuchi Y (2002) Stevens-Johnson syndrome accompanied by acute hepatitis in workers exposed to trichloroethylene or tetrachloroethylene. Sangyo Eiseigaku Zasshi 44:33–49 (In Japanese with English abstract)

Huang HL, Li LY, Chen BJ, Huang JX, Kuang SR (2002) New problems caused by occupational trichloroethylene exposure. Int J Immunopathol Pharmacol 15:30–32

Huang H, Kamijima M, Wang H, Li S, Yoshikawa T, Lai G, Huang Z, Liu H, Chen J, Takeuchi Y, Nakajima T, Li L (2006) Human herpesvirus 6 reactivation in trichloroethylene-exposed workers suffering from generalized skin disorders accompanied by hepatic dysfunction. J Occup Health 48:417–423

Hung SI, Chung WH, Liou LB, Chu CC, Lin M, Huang HP, Lin YL, Lan JL, Yang LC, Hong HS, Chen MJ, Lai PC, Wu MS, Chu CY, Wang KH, Chen CH, Fann CS, Wu JY, Chen YT (2005) HLA-B*5801 allele as a genetic marker for severe cutaneous adverse reactions caused by allopurinol. Proc Natl Acad Sci U S A 102:4134–4139

Iavicoli I, Marinaccio A, Carelli G (2005) Effects of occupational trichloroethylene exposure on cytokine levels in workers. J Occup Environ Med 47:453–457

Ikeda M, Imamura T (1973) Biological half-life of trichloroethylene and tetrachloroethylene in human subjects. Int Arch Arbeitsmed 31:209–224

Ikeoka T, Saito T, Hosaka Y, Takahara N, Suzuki R, Fukushima T, Sato T, Oshima H, Kuramochi S, Chong T (2009) Drug-induced hypersensitivity syndrome caused by trichloroethylene exposure. J Jpn Soc Int Med 98:1120–1123 (In Japanese)

Ito Y, Kamijima M, Yanagiba Y, Okamura A, Yamanoshita O, Nakajima T (2007) Generalized skin disorders and accompanying hepatitis in trichloroethylene-exposed workers: 2nd report – relationship between HHV-6 reaction and changes of serum or plasma cytokines. Sangyo Eiseigaku Zasshi 49(Suppl):426 (In Japanese)

Jia Q, Zang D, Yi J, Dong H, Niu Y, Zhai Q, Teng Y, Bin P, Zhou W, Huang X, Li H, Zheng Y, Dai Y (2012) Cytokine expression in trichloroethylene-induced hypersensitivity dermatitis: an in vivo and in vitro study. Toxicol Lett 215:31–39

Jung HG, Kim HH, Song BG, Kim EJ (2012) Trichloroethylene hypersensitivity syndrome: a disease of fatal outcome. Yonsei Med J 53:231–235

Kamijima M, Hisanaga N, Wang H, Nakajima T (2007) Occupational trichloroethylene exposure as a cause of idiosyncratic generalized skin disorders and accompanying hepatitis similar to drug hypersensitivities. Int Arch Occup Environ Health 80:357–370

Kamijima M, Wang H, Huang H, Li L, Shibata E, Lin B, Sakai K, Liu H, Tsuchiyama F, Chen J, Okamura A, Huang X, Hisanaga N, Huang Z, Ito Y, Takeuchi Y, Nakajima T (2008) Trichloroethylene causes generalized hypersensitivity skin disorders complicated by hepatitis. J Occup Health 50:328–338

Kamijima M, Wang H, Yamanoshita O, Ito Y, Xia L, Yanagiba Y, Okamura A, Huang Z, Qiu X, Song X, Cai T, Liu L, Ge Y, Deng Y, Naito H, Yoshikawa T, Tohyama M, Li L, Huang H, Nakajima T (2013) Occupational trichloroethylene hypersensitivity syndrome: human herpesvirus 6 reactivation and rash phenotypes. J Dermatol Sci 72:218–224

Kimber I, Basketter DA, Gerberick GF, Dearman RJ (2002) Allergic contact dermatitis. Int Immunopharmacol 2:201–211

Li HS, Dai YF, Huang HL, Sun YF (2006) Polymorphisms of aldehyde and alcohol dehydrogenase genes associated with susceptibility to trichloroethylene-induced medicamentosa-like dermatitis. Wei Sheng Yan Jiu 35:149–151 (in Chinese)

Li H, Dai Y, Huang H, Li L, Leng S, Cheng J, Niu Y, Duan H, Liu Q, Zhang X, Huang X, Xie J, Feng Z, Wang J, He J, Zheng Y (2007) HLA-B*1301 as a biomarker for genetic susceptibility to hypersensitivity dermatitis induced by trichloroethylene among workers in China. Environ Health Perspect 115:1553–1556

Lin B, Chen S, Wang D (2003) An investigation on three factories with consecutive occurrences of trichloroethylene-induced medicamentosa-like dermatitis. Chin Occup Med 30:65–66 (in Chinese)

Lindner K, Prater E, Schubert H, Siegmund S (1990) Drug exanthema due to chloral hydrate. Dermatol Monatsschr 176:483–485 (in German with English abstract)

Ministry of Health of the People's Republic of China. Diagnostic criteria of occupational medicamentose-like dermatitis due to trichloroethylene. National Occupational Health Standard GBZ 185–2006 of P.R. of China: 2007. (In Chinese)

Mohr TK (2001) Solvent stabilizers: white paper. Santa Clara Valley Water District, San Jose, p 55

Nakajima T, Okino T, Okuyama S, Kaneko T, Yonekura I, Sato A (1988) Ethanol-induced enhancement of trichloroethylene metabolism and hepatotoxicity: difference from the effect of phenobarbital. Toxicol Appl Pharmacol 94:227–237

Nakajima T, Yamanoshita O, Kamijima M, Kishi R, Ichihara G (2003) Generalized skin reactions in relation to trichloroethylene exposure: a review from the viewpoint of drug-metabolizing enzymes. J Occup Health 45:8–14

Nakayama H, Kobayashi M, Takahashi M, Ageishi Y, Takano T (1988) Generalized eruption with severe liver dysfunction associated with occupational exposure to trichloroethylene. Contact Dermatitis 19:48–51

Okamura A, Kamijima M, Ito Y, Yanagiba Y, Yamanoshita O, Wang H, Huang H, Li L, Yoshikawa T, Nakajima T (2007) Generalized skin disorders and accompanying hepatitis intrichloroethylene-exposed workers: 1st report – course of blood cytokine concentrations after occurrence of the disorders. Sangyo Eiseigaku Zasshi 49(Suppl):425 (In Japanese)

Pantucharoensri S, Boontee P, Likhitsan P, Padungtod C, Prasartsansoui S (2004) Generalized eruption accompanied by hepatitis in two Thai metal cleaners exposed to trichloroethylene. Ind Health 42:385–388

Phoon WH, Chan MO, Rajan VS, Tan KJ, Thirumoorthy T, Goh CL (1984) Stevens-Johnson syndrome associated with occupational exposure to trichloroethylene. Contact Dermatitis 10:270–276

Ramdhan DH, Kamijima M, Yamada N, Ito Y, Yanagiba Y, Nakamura D, Okamura A, Ichihara G, Aoyama T, Gonzalez FJ, Nakajima T (2008) Molecular mechanism of trichloroethylene-induced hepatotoxicity mediated by CYP2E1. Toxicol Appl Pharmacol 231:300–307

Ramdhan DH, Kamijima M, Wang D, Ito Y, Naito H, Yanagiba Y, Hayashi Y, Tanaka N, Aoyama T, Gonzalez FJ, Nakajima T (2010) Differential response to trichloroethylene-induced hepatosteatosis in wild-type and PPARalpha-humanized mice. Environ Health Perspect 118:1557–1563

Schwartz L, Tulipan L, Birmingham DJ (1947) Analysis of skin hazards in various occupations. In: Occupational diseases of the skin. Lea and Febiger, Philadelphia, p 771

Shiohara T, Inaoka M, Kano Y (2006) Drug-induced hypersensitivity syndrome (DIHS): a reaction induced by a complex interplay among herpesviruses and antiviral and antidrug immune responses. Allergol Int 55:1–8

Shiohara T, Iijima M, Ikezawa Z, Hashimoto K (2007) The diagnosis of a DRESS syndrome has been sufficiently established on the basis of typical clinical features and viral reactivations. Br J Dermatol 156:1083–1084

Tan HH, Tsu-Li Chan M, Goh CL (1997) Occupational skin disease in workers from the electronics industry in Singapore. Am J Contact Dermat 8:210–214

Tang XJ, Li LY, Huang JX, Deng YY (2002) Guinea pig maximization test for trichloroethylene and its metabolites. Biomed Environ Sci 15:113–118

Tang X, Que B, Song X, Li S, Yang X, Wang H, Huang H, Kamijima M, Nakajima T, Lin Y, Li L (2008) Characterization of liver injury associated with hypersensitive skin reactions induced by trichloroethylene in the guinea pig maximization test. J Occup Health 50:114–121

Tohyama M, Hashimoto K, Yasukawa M, Kimura H, Horikawa T, Nakajima K, Urano Y, Matsumoto K, Iijima M, Shear NH (2007) Association of human herpesvirus 6 reactivation with the flaring and severity of drug-induced hypersensitivity syndrome. Br J Dermatol 157:934–940

Wahlberg JE, Adams RM (1999) Solvents. In: Adams RM (ed) Occupational skin disease. W.B. Saunders, Philadelphia, pp 484–500

Watanabe H (2011) Hypersensitivity syndrome due to trichloroethylene exposure: a severe generalized skin reaction resembling drug-induced hypersensitivity syndrome. J Dermatol 38:229–235

Watanabe H, Tohyama M, Kamijima M, Nakajima T, Yoshida T, Hashimoto K, Iijima M (2010) Occupational trichloroethylene hypersensitivity syndrome with human herpesvirus-6 and cytomegalovirus reactivation. Dermatology 221:17–22

World Health Organization (WHO) (1985) Environmental health criteria 50. Trichloroethylene. World Health Organization, Geneva

Yamanishi K, Okuno T, Shiraki K, Takahashi M, Kondo T, Asano Y, Kurata T (1988) Identification of human herpesvirus-6 as a causal agent for exanthem subitum. Lancet 1:1065–1067

Yanagiba Y: Generalized skin disorders and accompanying hepatitis in trichloroethylene-exposed workers. 3rd report. the relation between HHV6 genotype and HHV7. Sangyo Eiseigaku Zasshi. 49(Suppl): 427, 2007 (In Japanese)

Yoshikawa T, Suga S, Asano Y, Yazaki T, Kodama H, Ozaki T (1989) Distribution of antibodies to a causative agent of exanthem subitum (human herpesvirus-6) in healthy individuals. Pediatrics 84:675–677

Yoshikawa T, Fujita A, Yagami A, Suzuki K, Matsunaga K, Ihira M, Asano Y (2006) Human herpesvirus 6 reactivation and inflammatory cytokine production in patients with drug-induced hypersensitivity syndrome. J Clin Virol 37 Suppl 1:S92–S96

Zhang J, Yang H, Li H, Liu F, Jia Q, Duan H, Niu Y, Bin P, Zheng Y, Dai Y (2013) Peptide-binding motifs and characteristics for HLA -B*13:01 molecule. Tissue Antigens 81:442–448

Chapter 4
Trichloroethylene-Induced Oxidative Stress and Autoimmunity

M. Firoze Khan and Gangduo Wang

Abstract Trichloroethylene (trichloroethene, TCE) is a widely used organic solvent and a common environmental and occupational contaminant. Apart from diseases like cancer and heart defects, TCE exposure has also been implicated in the development of various autoimmune diseases (ADs), such as systemic lupus erythematosus (SLE), systemic sclerosis and fasciitis, both from occupational and environmental exposures. Experimental studies using MRL+/+ mice as an animal model also support an association between TCE exposure and autoimmunity. Increasing evidence suggests that free radical-mediated reactions could play a potential role in the pathogenesis of ADs, and TCE exposure is known to cause oxidative stress both in vivo and in vitro. Recent studies have contributed to the understanding of the role of oxidatively modified proteins, especially lipid peroxidation-derived aldehyde (LPDA)-modified proteins in TCE-induced autoimmune response. These studies support that oxidative modification of endogenous proteins leads to structural alterations, resulting in the formation of neoantigens which elicit autoimmune responses by stimulating T and/or B lymphocytes, particularly Th1 and Th17 lymphocytes. More detailed studies to understand the distinct pathways by which oxidative stress contributes to autoimmunity, especially mapping of gene expression, analyzing proteome, blocking/inhibiting specific signal transduction pathways will also unravel critical mechanisms in TCE-mediated autoimmunity.

Keywords Trichloroethylene • Autoimmunity • Oxidative stress • MDA/HNE-protein adducts • Th1 cells • Th17 cells

M.F. Khan (✉) • G. Wang
Department of Pathology, University of Texas Medical Branch, Galveston, TX 77555, USA
e-mail: mfkhan@utmb.edu

K.M. Gilbert, S.J. Blossom (eds.), *Trichloroethylene: Toxicity and Health Risks*, Molecular and Integrative Toxicology, DOI 10.1007/978-1-4471-6311-4_4,

4.1 Introduction

Trichloroethylene (trichloroethene, TCE) is a widely used organic solvent and a common environmental and occupational contaminant (Diot et al. 2002; Hardin et al. 2005; Bakke et al. 2007; Moran et al. 2007; ATSDR 2010; Purdue et al. 2011). About 3.5 million people are occupationally exposed to TCE in the United States mainly through its use in degreasing operation, but also through dry cleaning, textile scouring, and in handling adhesives, drugs, paints, leather and other products (Wu and Schaum 2000; Bakke et al. 2007; ATSDR 2010).

Environmental exposure to TCE occurs through air, contaminated ground water and drinking water. Most TCE used in the United States is released to the atmosphere from vapor degreasing operations, and its release to air also occurs at sewage treatment and disposal facilities, water treatment facilities, and landfills (Wu and Schaum 2000; Bakke et al. 2007; ATSDR 2010). TCE has been detected in the air throughout the United States, and the 1998 air levels across all 115 monitors ranged between 0.01–3.9 μg/m^3 with a mean of 0.88 μg/m^3 (Wu and Schaum 2000; Bakke et al. 2007; ATSDR 2010). TCE was detected in 28 % of 9,295 surface water reporting stations nationwide, and it is the most frequently reported organic contaminant in groundwater with up to 34 % of the drinking water supplies in USA contaminated with TCE (IARC 1995; Wu and Schaum 2000; ATSDR 2010).

TCE has also been identified in 72 food items in the US Food and Drug Administration's Total Diet Study, including fruits, beverages and many foods prepared with oils and fats (Wu and Schaum 2000; ATSDR 2010). Because of its widespread commercial use and improper disposal, TCE has become a major occupational and environmental toxicant, and is one of the most abundant organic contaminants (NTP 1990; Bourg et al. 1992; Ashley et al. 1994; Hardin et al. 2005; Moran et al. 2007; ATSDR 2010). Therefore, there is clearly a need to extensively study the potential adverse health effect of TCE.

4.2 TCE Exposure and Autoimmune Response: Human Studies

TCE exposure has been associated with a variety of human diseases. Apart from diseases like cancer and heart defects (Boyer et al. 2000; Rhomberg 2000; Caldwell and Keshava 2006; Drake et al. 2006; Purdue et al. 2011), TCE has also been implicated in the development of various autoimmune diseases (ADs), such as systemic lupus erythematosus (SLE), systemic sclerosis and fasciitis, both from occupational (Phoon et al. 1984; Flindt-Hansen and Isager 1987; Lockey et al. 1987; Yáñez Díaz et al. 1992; Waller et al. 1994; Nietert et al. 1998; Cooper et al. 2009) and environmental exposures (Haustein and Ziegler 1985; Byers et al. 1988; Kilburn and Warshaw 1992; Hayashi et al. 2000 ; Albert et al. 2005 ; Cooper et al. 2009). The involvement of TCE exposure in ADs was first reported as early as in 1957 (Reinl 1957), and in recent years an increasing number of reports have further implicated TCE in the development of various ADs. Kilburn and Warshaw (1992) examined the prevalence of connective tissue disease symptoms and ANA, by comparing 362 residents of Tucson to 158 residents of another area of Southwest Arizona. The prevalence of some self-reported symptoms (malar rash, arthritis/arthalgias, Raynaud syndrome, skin lesions, and seizure or convulsion) and ANA levels were higher in Tucson residents (Kilburn and Warshaw 1992). Reports have shown that occupational TCE exposure is also associated with scleroderma (Flindt-Hansen and Isager 1987; Lockey et al. 1987; Yáñez Díaz et al. 1992; Nietert et al. 1998; Diot et al. 2002; Pralong et al. 2009) and fasciitis (Waller et al. 1994). Some case-control studies provided data specifically about TCE exposure, based on industrial hygienist review of job history data. Three of these studies are of scleroderma (Nietert et al. 1998; Diot et al. 2002; Garabrant et al. 2003), one is of undifferentiated connective tissue disease (Lacey et al. 1999), and one is of small vessel vasculitis involving anti-neutrophil cytoplasmic autoantibodies (Beaudreuil et al. 2005). Occupational TCE-induced Stevens-Johnson syndrome and other skin disorders have also drawn attention (Phoon et al. 1984; Huang et al. 2006; Kamijima et al. 2008; Jia et al. 2012).

4.3 TCE Exposure and Autoimmune Response: In Vivo Studies

Khan and his colleagues were first to propose and use MRL+/+ mice as an animal model to provide direct evidence of an association between TCE exposure and autoimmunity (Khan et al. 1995). This association was further substantiated by their subsequent studies and reports from other laboratories using MRL+/+ mice (Gilbert et al. 1999; Griffin et al. 2000a; Khan et al. 2001; Wang et al. 2007a, b, 2008, 2009, 2012a; Cai et al. 2008). MRL+/+ mice, therefore, have been the most often used animal models in experimental studies of TCE exposure.

Several studies in MRL+/+ mice have reported autoimmunity-related effects following exposure to TCE via drinking water (Blossom et al. 2004, 2007; Cai et al. 2008; Gilbert et al. 1999; Griffin et al. 2000a, b, c; Wang et al. 2007a) or ip injection (Cai et al. 2006; Khan et al. 1995; Wang et al. 2007b, 2008, 2009, 2012a). The initial drinking water studies used relatively high TCE concentrations of 2.5 and 5 mg/mL, with serologic measurements of ANA and IgG levels and assays for the activation of $CD4^+$ T cells from spleen (Gilbert et al. 1999; Griffin et al. 2000a). Subsequent studies focused on examining TCE effects at lower exposure levels (0.1, 0.5, and 2.5 mg/mL) (Griffin et al. 2000b; Wang et al. 2007a, 2012a; Cai et al. 2008), also showed an accelerated autoimmune response. The effects observed by Griffin et al. (2000a) with respect to formation of TCE-protein adducts and $CD4^+$ T cell activation was blocked by inhibiting CYP2E1 metabolic pathway (Griffin et al. 2000b), suggesting the role of TCE activation and its metabolites. In another chronic exposure study (0.5 mg/mL TCE in drinking water), Cai et al. (2008) found evidence of systemic inflammation as determined by serum cytokines measured after 36–48 weeks of exposure.

Some chronic oral exposure studies in the MRL+/+ mice, with exposure periods of 32–48 weeks, reported the presence of distinct clinical effects in exposed mice. One of these effects was characterized as an autoimmune hepatitis (Griffin et al. 2000b; Cai et al. 2008). Griffin et al. (2000b) found an inflammatory focal areas in the 0.5- and 2.5-mg/mL TCE-treated mice, with a dose-related effect on severity hepatic infiltrate in the portal tracts and lobular scores seen at 32 weeks. Cai et al. (2008) found similar liver lymphocytic infiltrates at 36 and 48 weeks in a study using 0.5 mg/mL TCE exposure through drinking water, and also infiltrates in the pancreas, lungs, and kidneys at 48 weeks. Wang et al. (2007a) observed increased autoantibodies in another study using 0.5 mg/mL TCE via drinking water for 48 weeks. In a 40-week study using trichloroacetaldehyde hydrate, Blossom et al. (2007) reported diffuse alopecia and skin inflammation and ulceration. Studies with other animal models also demonstrated the potential of TCE in inducing autoimmune response. For example, a chronic (26-week) drinking water exposure study in NZB × NZW mice reported increased level of proteinuria and prevalence of renal pathology with TCE exposure of 10,000 ppb via drinking water (Gilkeson et al. 2004). Production of anti-dsDNA and other antibodies was increased following 1,400 ppb TCE exposure for 19 weeks.

Several studies also evaluated the involvement of one or more metabolites of TCE in the induction of an autoimmune response observed in MRL+/+ mice. These include studies of dichloroacetyl chloride (Khan et al. 1995, 2001; Cai et al. 2006), trichloroacetaldehyde hydrate (Blossom et al. 2004, 2007; Blossom and Gilbert 2006; Gilbert et al. 2006), and trichloroacetic acid (Blossom et al. 2004). Effects were similar to those found with TCE in terms of accelerated autoantibody expression, T cell activation, and secretion of inflammatory cytokines.

4.4 Role of Oxidative Stress in Autoimmune Diseases

ADs such as SLE, rheumatoid arthritis and scleroderma are chronic and life-threatening disorders that affect ~3 % of United States population, and contribute disproportionately to morbidity and mortality among young to middle-aged women (Jacobson et al. 1997; Walsh and Rau 2000). Despite high prevalence of these diseases, molecular mechanisms underlying systemic autoimmune response remain largely unknown. In recent years, increasing evidence suggests that free radical-mediated reactions could play a potential role in the pathogenesis of ADs (Khan et al. 2001; Hadjigogos 2003; Frostegard et al. 2005; Kurien and Scofield 2008; Wang et al. 2008, 2010a; Vasanthi et al. 2009; Iuchi et al. 2010). Indeed increased oxidative stress is reported in various ADs (Grune et al. 1997; Frostegard et al. 2005; Tam et al. 2005; Morgan et al. 2009; Vasanthi et al. 2009; Shah et al. 2010; Wang et al. 2010a; Al-Shobaili and Rasheed 2012; Al-Shobaili et al. 2013).

Reactive oxygen species (ROS) including superoxide anion ($O_2^{\cdot-}$), hydrogen peroxide (H_2O_2), hydroxyl radical ($^{\cdot}OH$) and other reactive molecules containing oxygen formed during aerobic metabolism in cells as well as due to phagocyte and neutrophil activation during inflammation, have potential to initiate cellular damage to lipids, proteins and DNA (Biemond et al. 1984; Halliwell and Gutteridge 1984;

Finkel 2011). A variety of ROS-mediated modifications of proteins have been reported in ADs and aging (Stadtman and Berlett 1998; Oates et al. 1999; Beal 2002; Morgan et al. 2005; Sheikh et al. 2007; Wang et al. 2010a; Al-Shobaili and Rasheed 2012; Al-Shobaili et al. 2013). Sheikh et al. (2007) observed increased protein carbonyls and recognition of ROS-modified human serum albumin by circulating SLE autoantibodies in SLE patients. Recently, Al-Shobaili et al. (2013) reported higher level of anti-oxidized-catalase (CAT)-antibodies in SLE patients with varying levels of disease activity according to SLE Disease Activity Index (SLEDAI). These antibodies showed strong relation with the SLEDAI, disease induction and progression (Al-Shobaili and Rasheed 2012; Al-Shobaili et al. 2013), suggesting oxidized protein may be a useful biomarker in evaluating the progression of SLE and in elucidating the mechanisms of disease pathogenesis.

Reactive nitrogen species (RNS) are nitrogen-containing molecules, i.e., nitric oxide (·NO), peroxynitrite ($ONOO^-$) and nitroxyl anion (HNO^-) (Hill et al. 2010). Like ROS, RNS could also play a significant role in the pathogenesis of SLE and other ADs, and have drawn considerable attention in recent years. ·NO, generated by the enzyme inducible nitric oxide synthase (iNOS), is one of the most important and widely studied RNS. The potential of ·NO in disease pathogenesis lies largely to the extent of its production and generation of $O_2^{\cdot-}$, leading to formation of peroxynitrite ($ONOO^-$). $ONOO^-$ is a potent nitrating and oxidizing agent which can react with tyrosine residues to form nitrotyrosine (NT; Weinberg et al. 1994; Xia and Zweier 1997; Khan et al. 2003). In addition, $ONOO^-$-mediated modifications of endogenous proteins and DNA may enhance their immunogenicity, leading to a break in immune tolerance (Khan et al. 2003; Ohmori and Kanayama 2005; Kurien et al. 2006). Accumulating evidence in murine lupus shows increasing iNOS activity with the development and progression of ADs, and studies using competitive inhibitors suggest that iNOS could play a pathogenic role in murine ADs (Weinberg et al. 1994; Xia and Zweier 1997; Karpuzoglu and Ahmed 2006; Wang et al. 2009). Also elevated presence of nitrated proteins, particular NT, a stable end product of increased RNS production, has been found in many diseases including ADs (Oates et al. 1999; Morgan et al. 2005; Khan et al. 2006; Ohmori and Kanayama 2005). Growing observational data in humans also suggest that overexpression of iNOS and increased production of $ONOO^-$ may contribute to glomerular and vascular pathology and in the pathogenesis of many other ADs (Wanchu et al. 1998; Nagy et al. 2007a; Morgan et al. 2009). There is appreciable evidence that NT and other markers of protein oxidation are enhanced in diabetes and many other ADs, and may contribute to the pathogenesis of these diseases (Stadtman and Berlett 1998; Oates et al. 1999; Martín-Gallán et al. 2003; Morgan et al. 2005; Ohmori and Kanayama 2005; Khan et al. 2006; Khan and Ali 2006; Renke et al. 2007; Wang et al. 2010a).

Reactive lipid species (RLS) are usually derived from unsaturated lipids, including lipid peroxidation-derived aldehydes (LPDAs) and reactive prostaglandins of A- and J-series (Higdon et al. 2012). LPDAs such as malondialdehyde (MDA) and 4-hydroxynonenal (HNE) are highly reactive and can bind covalently to proteins resulting in their structural modifications and may elicit an autoimmune response and contribute to disease pathogenesis (Khan et al. 1997, 1999; Januszewski et al. 2005; Reed et al. 2009; Wang et al. 2010a; Ben Mansour et al. 2010). Indeed higher levels of MDA-/HNE-modified proteins have been observed in AD patients (Grune et al. 1997; Kurien

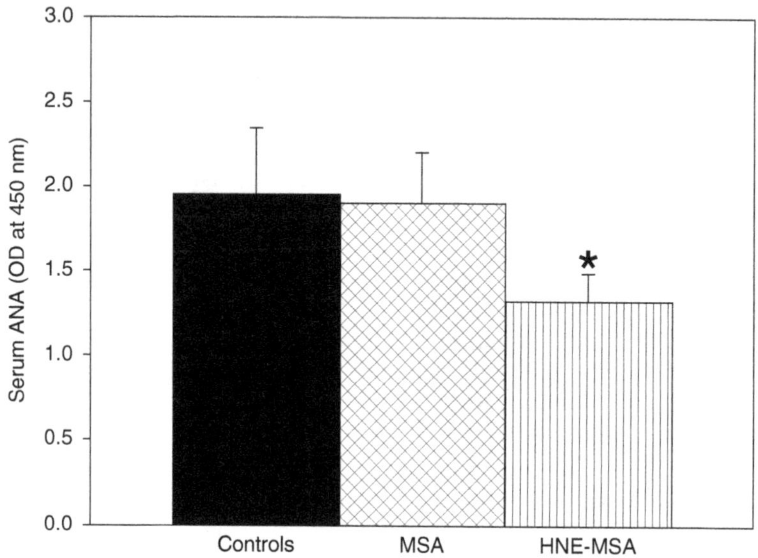

Fig. 4.1 Inhibitory effect of HNE-mouse serum albumin (*MSA*) on ANA binding to nuclear antigens. Serum from 18-week MRL/lpr mice was pre-incubated with HNE-MSA followed by ANA determination using ELISA kit. The values are means ± SD. $*p<0.05$ vs. controls (serum only) (Adapted from Informa Healthcare)

and Scofield 2003; Frostegard et al. 2005; D'souza et al. 2008; Ben Mansour et al. 2010; Wang et al. 2010a), suggesting a potential role for these oxidatively modified proteins in ADs. Wang et al. (2010a) analyzed the sera from 72 SLE patients with varying levels of disease activity (according to the SLEDAI) and 36 age- and gender-matched healthy controls for oxidative stress markers and showed significantly higher levels of both MDA/HNE protein adducts and anti-MDA/anti-HNE protein adduct antibodies in SLE patients compared with healthy controls. Interestingly, not only was there an increased number of subjects positive for anti-MDA or anti-HNE antibodies, but also the levels of both of these antibodies were significantly higher among SLE patients whose SLEDAI scores were ≥6 as compared with SLE patients with lower SLEDAI scores (<6). In addition, a significant correlation was observed between the levels of anti-MDA or anti-HNE antibodies and the SLEDAI score, suggesting a possible causal relationship between these antibodies and SLE. The stronger response observed in serum samples from patients with higher SLEDAI scores suggests that markers of oxidative stress may be useful in evaluating the progression of SLE and in elucidating the mechanisms of disease pathogenesis. Recently, Wang et al. (2012b) observed age-related increases in the formation of MDA-/HNE-protein adducts, their corresponding antibodies and MDA-/HNE-specific immune complexes in MRL/lpr mice, the widely used model for SLE. Interestingly, HNE-MSA adducts mimic nuclear antigens and cause significant inhibition in ANA binding to nuclear antigens (Fig. 4.1; Wang et al. 2012b), suggesting that LPDA-modified proteins could be important sources of autoantibodies and CICs in these mice, and thus contribute to autoimmune disease pathogenesis.

4.5 TCE Exposure and Oxidative Stress

Free radical-mediated reactions have drawn increasing attention as the potential mechanism in the pathogenesis of ADs and other diseases (Khan et al. 2001; Hadjigogos 2003; Karpuzoglu and Ahmed 2006; Kurien et al. 2006; Cuzzocrea 2006; Nagy et al. 2007b). TCE has been shown to generate free radicals and induce oxidative stress both in vivo and in vitro (Ogino et al. 1991; Channel et al. 1998; Khan et al. 2001; Zhu et al. 2005; Wang et al. 2007a, b, 2008, 2012a). A list of studies leading to TCE-induced oxidative stress are summarized in Table 4.1. Several studies have reported an association between TCE exposure and increased oxidative stress, especially lipid peroxidation (Ogino et al. 1991; Channel et al. 1998; Khan et al. 2001; Zhu et al. 2005; Wang et al. 2007a, b, 2008, 2012a). Most of earlier studies (Cojocel et al. 1989; Ogino et al. 1991; Channel et al. 1998; Toraason et al. 1999) used high doses of TCE (125–2,000 mg/kg) to demonstrate TCE-induced lipid peroxidation which contributes to toxic response in the livers or kidneys. Wang et al. (2007b, 2008) reported that TCE exposure for 6 or 12 weeks led to significantly increased formation of MDA-/HNE-protein adducts in the livers of TCE-treated female MRL +/+ mice at both 6 and 12 weeks, but with greater response at 12 weeks. Further characterization of these adducts in liver microsomes showed increased formation of MDA-protein adducts with molecular masses of 86, 65, 56, 44, and 32 kD, and of HNE-protein adducts with molecular masses of 87, 79, 46, and 17 kD in TCE-treated mice (Wang et al. 2007b). In addition, significant induction of anti-MDA- and anti-HNE-protein adduct-specific antibodies was observed in the sera of TCE-treated mice, and showed a pattern similar to MDA- or HNE-protein adducts. TCE-induced formation of MDA-/HNE-protein adducts and their respective antibodies were also observed in mice exposed to a relatively lower dose of TCE (Wang et al. 2007a, 2012a).

The potential of TCE in inducing nitrosative stress has also drawn attention recently (Wang et al. 2007a, 2009; Blossom et al. 2012). TCE exposure resulted in increased formation of NT and induction of iNOS in the serum of female MRL +/+ mice. TCE treatment also led to greater NT formation, and iNOS protein and mRNA expression in the livers and kidneys (Wang et al. 2009). TCE-induced formation of NT was also observed at relatively lower dosages of TCE (Wang et al. 2007a; Blossom et al. 2012), which could potentially contribute to TCE-induced autoimmune response.

The potential of TCE exposure leading to carbonylation of proteins has also been examined. TCE exposure (10 mmol/kg, i.p., every fourth day) in female MRL+/+ mice resulted in increased (~3 fold) serum protein carbonyls (a marker of protein oxidation) at both 6 and 12 weeks. Increased protein carbonyls were also observed in the livers and kidneys (2.1 and 1.3 fold, respectively) at 6 weeks, and to a greater extent at 12 weeks (3.5 and 2.1 fold, respectively) following TCE treatment (Wang et al. 2009). Increased protein carbonyls were also observed in ovaries, oocytes, sperms and kidneys if rats or mice exposed to TCE via drinking water (DuTeaux et al. 2004; Wu and Berger 2007; Fan et al. 2012). Fan et al. (2012) analyzed and

Table 4.1 Experimental studies of TCE exposure and oxidative stress

References	Studies	TCE exposure	Effect
Huang et al. (2012)	Humans	Occupational exposure	Oxidative stress
Blossom et al. (2012)	In vivo, female MRL+/+ mice	0.0 or 0.1 mg/ml in drinking water with 1 % of EL-620 for 6 weeks	Increased nitrotyrosine
Wang et al. (2012a)	In vivo, female MRL+/+ mice	0.5, 1.0 or 2.0 mg/ml in drinking water with 1 % of EL-620 for 12, 24, 36 weeks	Increased MDA-/HNE-protein adducts and antibodies; increased autoantibodies
Ali and Sultana (2012)	In vivo, male Swiss albino mice	200 µl of TCE (80 %, v/v, dissolved in acetone)	Depletion in GSH and SOD activity, induction of iNOS expression
Tabrez and Ahmad (2011)	In vivo, Swiss albino rats	1,000 mg/kg, i.p., in corn oil	Increased lipid peroxidation (MDA levels) and GST activity
Gharib (2009)	In vivo, male albino rats	1,000 mg/kg, oral, for 2 weeks	Increased MDA and NO, and decreased GSH
Khan et al. (2009)	In vivo, male Wistar rats	1,000 mg/kg/day, i.p., in corn oil for 25 days	Increased lipid peroxidation and declined SOD activity
Wang et al. (2009)	In vivo, female MRL+/+ mice	10 mmol/kg, i.p., every fourth day for 6 or 12 weeks	Increased MDA-/HNE-protein adducts & their antibodies; increased autoantibodies
Blossom et al. (2008)	In vivo, female MRL+/+ mice	0.1 mg/ml in drinking water with 1 % of EL-620 for 26 weeks	Increased ROS and decreased GSH
Wang et al. (2008)	In vivo, female MRL+/+ mice	10 mmol/kg, i.p., every fourth day for 4 weeks	Increased MDA-/HNE-protein adducts, their antibodies and autoantibodies
Shen et al. (2008)	Skin exposure, BALB/c hairless mice	50 µl of TCE dissolved in olive oil, skin exposure for 4 h, twice daily for 2 weeks	Increased MDA levels and inhibition of SOD activities
Hu et al. (2008)	In vitro, human HepG2 cells	0.5–4 mM	Increased 8-OhdG and increased lipid peroxidation (TBARS)
Shen et al. (2007)	In vitro, epidermal keratinocytes	0.125, 0.25, 0.50, 1.0 and 2.0 mM	Dose-dependent increases of iNOS activities and mRNA expression
Wang et al. (2007b)	In vivo, female MRL+/+ mice	10 mmol/kg, i.p., every fourth day for 6 or 12 weeks	Increased nitrityrosine, protein carbonyls with increased autoantibodies

Wang et al. (2007a)	In vivo, female MRL+/+ mice	0.5 mg/ml in drinking water with 1 % of EL-620for 48 weeks	Increased anti-MDA-/HNE-protein adduct antibodies, nitrotyrosine along with increased autoantibodies
Wu and Berger (2007)	In vivo, albino rats	0.45 % TCE (v/v) in 3 % Tween in drinking water for 4–5 days	Increased protein carbonyls
DuTeaux et al. (2004)	In vivo, Sprague-Dawley rats	0, 0.2 % or 0.4 % (v/v) in drinking water for 14 days	Increased protein carbonyls
Zhu et al. (2005)	In vitro, epidermal keratinocytes	0.01–31.6 mM for 1, 2, 3, 4 h	Time- and concentration-dependent increases of MDA and decreases in GSH
Chen et al. (2002)	In vitro, cell lines: H460, H1299	0.9–6.5 μl/cm^2 for 24 h	Increased TBARS and MDA; decreased GSH
Khan et al. (2001)	In vivo, female MRL+/+ mice	10 mmol/kg, i.p., every fourth day for 6 weeks	Increased anti-MDA specific antibodies
Toraason et al. (1999)	In vivo, Fisher rats	0, 100, 500, 1,000 mg/kg, i.p., in 1:4 (v/v) of Alkamuls/water	Increased TBARS and 8OHdG formation
Channel et al. (1998)	In vivo, B6C3F1 mice	0, 400, 800, and 1,200 mg/kg in corn oil, orally, once daily for 8 weeks	Increased TBARS and 8OHdG formation
Ogino et al. (1991)	In vivo, male Wistar rats	2,000 mg/kg, i.p., in olive oil	Increased MDA
Cojocel et al. (1989)	In vivo, Male NMRI mice	125–150 mg/kg, i.p., in sesame oil	Increased MDA

characterized the carbonylated proteins by using two-dimensional (2D) gel electrophoresis, Western blot along with matrix-assisted laser-desorption ionization time-of-flight mass spectrometry (MALDI TOF/TOF MS/MS) in the kidneys following TCE exposure in female MRL+/+ mice (2 mg/ml via drinking water) for 36 weeks. TCE treatment led to significantly increased protein carbonyls in the kidney protein extracts. Interestingly, among 18 identified carbonylated proteins, 10 were found only in the kidneys of TCE-treated mice, whereas other eight were present in the kidneys of both control and TCE-treated mice. The identified carbonylated proteins represent skeletal proteins, chaperones, stress proteins, enzymes, plasma protein, and proteins involved in signaling pathways. Huang et al. (2012) examined the serum proteome in the TCE-induced hypersensitivity dermatitis patients via 2D gel coupled with MALDI-TOF-TOF/MAS and also found that inflammatory responses and oxidative stress might contribute to TCE-induced hypersensitivity dermatitis.

In vitro studies (Chen et al. 2002; Zhu et al. 2005; Shen et al. 2007; Hu et al. 2008) using a variety of human cell lines such as human epidermal keratinocytes, human lung cancer H460 and p54-null H1299 cells have shown that TCE can induce oxidative stress, particularly lipid peroxidation in a time- and concentration-dependent pattern, and GSH, an intracellular antioxidant, provided protection against TCE-induced oxidative damage.

4.6 TCE-Induced Oxidative Stress and Induction of Autoimmunity

The role of oxidative stress in the TCE-induced autoimmune response was first proposed by Khan and his colleagues (Khan et al. 2001). That led to a series of studies examining the contributions of oxidative stress, especially the LPDAs, in TCE-induced autoimmune response by his research group (Wang et al. 2007a, b, 2008, 2009, 2012a; Fan et al. 2012). Their observations led them to hypothesize (Fig. 4.2) that TCE-induced oxidative stress leads to a variety of RONS-mediated structural modifications of the endogenous proteins, such as increased formation of LPDA-protein adducts (e.g. MDA-/HNA-protein adducts), carbonylation and nitration of proteins, which could potentially lead to generation of neoantigens. After antigen processing, these neoantigens could elicit autoimmune response by stimulating T and B lymphocytes, especially Th1 and Th17 cells (Wang et al. 2008, 2012a).

Wang et al. (2007b) detected increased formation of MDA- and HNE-protein adducts in the livers of MRL+/+ mice treated with TCE (10 mmol/kg, i.p., every fourth day) for 6 and 12 weeks. Significant induction of anti-MDA- and anti-HNE-protein adduct specific antibodies was also observed in the sera of TCE-treated mice, which showed a pattern similar to MDA-/HNE-protein adducts. More importantly, the increases in anti-MDA- and anti-HNE-protein adduct antibodies were associated with significant elevation in serum anti-nuclear (ANA)-, anti-ssDNA- and anti-dsDNA-antibodies at 6 weeks and, to a greater extent at 12 weeks. Those studies, even though at a relatively high dose, served as the basis for further evaluation of TCE-induced oxidative

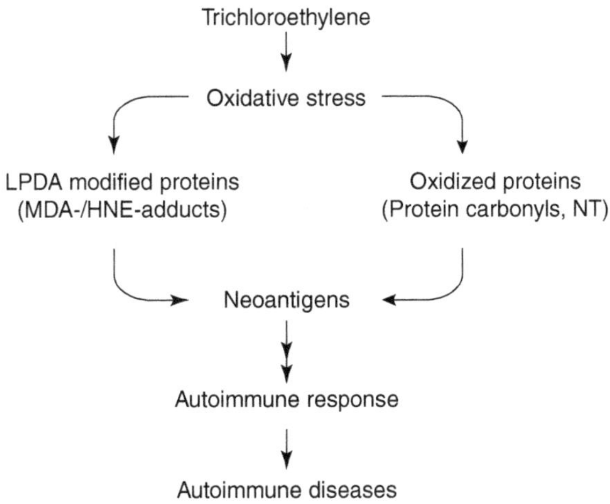

Fig. 4.2 Projected sequence of events leading to ADs following TCE exposure

modification of proteins at occupationally relevant doses. Wang et al. (2007a, 2012a) observed that occupationally-relevant doses of TCE (0.5, 1.0 or 2.0 mg/ml in drinking water for 12, 24, 36, 48 weeks) led to dose- and time-related increases in MDA-/HNE-protein adducts and their corresponding antibodies in the sera. Furthermore, strong relationship between the increases in MDA-/HNE-protein adducts and significant elevation in serum ANA and anti-ssDNA-antibodies, suggested an association between TCE-induced oxidative stress and autoimmune response. Interestingly, stimulation of cultured splenic lymphocytes from both control and TCE-treated female MRL +/+ mice (10 mmol/kg, i.p., every fourth day for 4 weeks) with MDA-adducted mouse serum albumin (MDA-MSA) or HNE-MSA for 72 h showed significant proliferation of $CD4^+$ T cells in TCE-treated mice as analyzed by flow cytometry. Also, splenic lymphocytes from TCE-treated mice released more IFN-γ and IL-2 into cultures when stimulated with MDA-MSA or HNE-MSA, suggesting a Th1 cell activation (Wang et al. 2008). Similarly, after female MRL +/+ mice were orally exposed to TCE (0.5, 1.0 or 2.0 mg/ml in drinking water) for 12, 24, 36 weeks, the splenocytes from mice treated with TCE for 24 weeks secreted significantly higher levels of IL-17 and IL-21 than did splenocytes from controls after stimulation with MDA-MSA or HNE-MSA adducts. The increased release of these cytokines was dose-dependent and more pronounced in mice treated with TCE for 36 weeks (Wang et al. 2012a; Fig. 4.3). These studies provide evidence that MDA- and or HNE-modified proteins contribute to TCE-mediated autoimmunity, which may be via activation of Th1, Th17 cells (Wang et al. 2008, 2012a). Recent studies in groups of female MRL +/+ mice treated with TCE, NAC or TCE plus NAC for 6 weeks (TCE, 10 mmol/kg, i.p., every fourth day; NAC, 250 mg/kg/day through drinking water), showed that NAC supplementation not only attenuated the TCE-induced formation of anti-MDA/-HNE-protein adduct antibodies and increased carbonylation of serum proteins, but also increases in serum levels of ANA, anti-Sm- and anti-dsDNA-antibodies, evidenced by their reduced levels in the sera of TCE plus NAC treated mice, further supporting a role of oxidatively modified proteins in TCE-induced autoimmune response (Wang et al. 2010b).

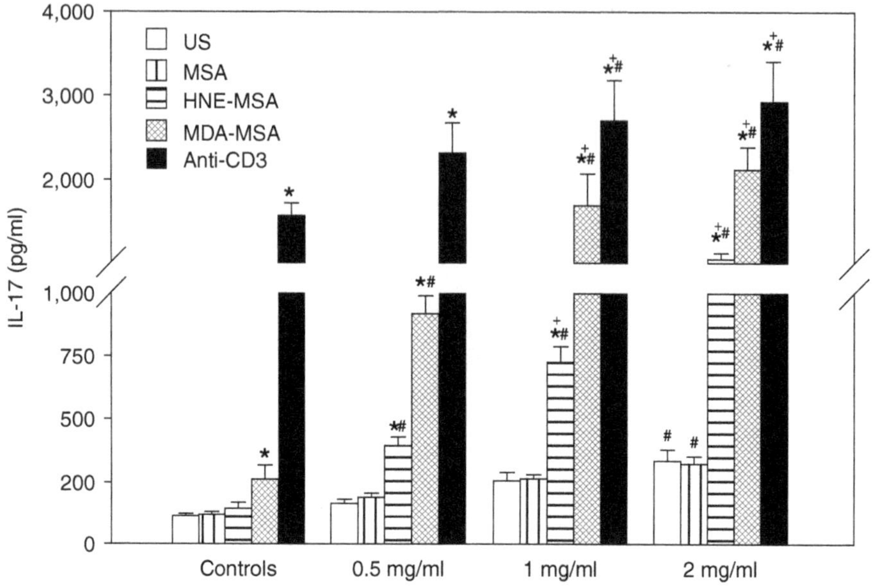

Fig. 4.3 IL-17 release in the culture supernatants of splenocytes from control and TCE-treated mice (0.5, 1.0, 2.0 mg/ml of TCE in drinking water for 36 weeks). Splenocytes were stimulated with MSA alone, HNE-MSA, MDA-MSA or anti-CD3 antibody for 72 h. *US* un-stimulated cells. $^{*}p<0.05$ vs. US; $^{\#}p<0.05$ vs. stimulated control group; $^{+}p<0.05$ vs. stimulated lower dose groups (0.5 and 1 mg/ml) (Adapted from Elsevier)

Recent studies have also explored the contribution of protein oxidation (carbonylation and nitration) in the induction of TCE-induced autoimmune response (Wang et al. 2007a, 2009, 2010b, 2013). TCE exposure (10 mmol/kg, i.p., every fourth day) in female MRL +/+ mice for 6 or 12 weeks (10 mmol/kg, i.p., every fourth day), led to time-dependent increases in carbonylation and nitration of proteins with enhanced iNOS activity. More importantly, the increases in TCE-induced protein oxidation (carbonylation and nitration) were associated with significant increases in Th1-specific cytokine (IL-2, IFN-γ) release into splenocyte cultures (Wang et al. 2009). These data along with the evidence that TCE induces autoimmune response (Khan et al. 1995; Wang et al. 2007a, 2008), suggest an association between oxidative modification of proteins and autoimmunity. The modification of proteins, such as nitration or carbonylation, may alter immunogenicity of self-antigens (converting them to neoantigens), and may lead to an autoimmune response by stimulating T cells (especially activation of Th1 cells; Wang et al. 2009). Lower dose of TCE exposure (0.5 mg/ml via drinking water) also led to significant increases of serum NT along with elevation of ANA and anti-dsDNA antibodies (Wang et al. 2007a). Interestingly, TCE treatment in iNOS-null female MRL+/+ mice even though still led to increases in serum ANA and anti-dsDNA, but the increases in these autoantibodies induced by TCE were significantly less pronounced compared to that in MRL+/+ mice (Wang et al. 2013). These results suggest an association between protein oxidation and induction/exacerbation of autoimmune response, and present a potential mechanism by which oxidatively modified proteins could contribute to TCE-induced autoimmune response (Wang et al. 2007a, 2009, 2010b, 2013).

4.7 Conclusions and Future Direction

Recent studies have contributed to the understanding of the role of oxidatively modified proteins, especially LPDAs modified proteins, in TCE-induced autoimmune response. These studies support that oxidative modification of proteins (e.g., MDA-/HNE-protein adducts, nitration and carbonylation of proteins) cause structural alterations to endogenous proteins, resulting in the formation of neoantigens which elicit autoimmune responses by stimulating T and/or B lymphocytes, particularly Th1 and Th17 lymphocytes (Khan et al. 2001; Wang et al. 2007a, b, 2008, 2009, 2012a). These studies not only demonstrated that TCE exposure leads to increased formation of MDA-/HNE-protein adducts, nitration and carbonylation of proteins, and formation of anti-MDA-/HNE-protein adduct antibodies, but more importantly, observed a significant association between formation of these modified proteins, corresponding antibodies and increased autoantibodies. Furthermore, MDA-/HNE-MSA stimulated greater release of IFN-γ, IL-2, IL-17 and IL-21, suggesting the contribution of oxidatively modified proteins in TCE-mediated autoimmune responses. Further detailed studies to unravel the distinct pathways by which oxidative stress contributes to autoimmunity, especially mapping of gene expression, analyzing proteome, blocking/inhibiting specific signal transduction pathways, knocking out/down target genes and exploring the epigenetic involvement will also provide critical mechanisms in TCE-induced autoimmunity.

Acknowledgements This work was supported by Grant ES016302 from the National Institute of Environmental Health Sciences (NIEHS), National Institute of Health (NIH), and it contents are solely the responsibility of the authors and do not necessarily represent the official views of the NIEHS, NIH.

References

Albert DA, Albert AN, Vernace M, Sebastian JK, Hsia EC (2005) Analysis of a cluster of cases of Wegener granulomatosis. J Clin Rheumatol 11:188–193

Ali F, Sultana S (2012) Repeated short-term stress synergizes the ROS signaling through up regulation of NFkB and iNOS expression induced due to combined exposure of trichloroethylene and UVB rays. Mol Cell Biochem 360:133–145

Al-Shobaili HA, Rasheed Z (2012) Immunological studies of oxidized superoxide dismutase in patients with systemic lupus erythematosus. Correlation with disease induction and progression. Saudi Med J 33:1177–1184

Al-Shobaili HA, Robaee AA, Alzolibani AA, Rasheed Z (2013) Immunological studies of reactive oxygen species damaged catalase in patients with systemic lupus erythematosus: correlation with disease activity index. Immunol Invest 42:191–203

Ashley DL, Bonin MA, Cardinali FL, McCraw JM, Wooten JV (1994) Blood concentrations of volatile organic compounds in a nonoccupationally exposed US population and in groups with suspected exposure. Clin Chem 40(7 Pt 2):1401–1404

ATSDR (Agency for Toxic Substances and Disease Registry, Division of Toxicology) (2010) Public Health Statement for trichloroethylene. Toxicological Profile for Trichloroethylene (TCE), 1997. Available at http://www.atsdr.cdc.gov

Bakke B, Stewart PA, Waters MA (2007) Uses of and exposure to trichloroethylene in U.S. industry: a systematic literature review. J Occup Environ Hyg 4:375–390

Beal MF (2002) Oxidatively modified proteins in aging and disease. Free Radic Biol Med 32:797–803

Beaudreuil S, Lasfargues G, Lauériere L, El Ghoul Z, Fourquet F, Longuet C, Halimi JM, Nivet H, Büchler M (2005) Occupational exposure in ANCA-positive patients: a case-control study. Kidney Int 67:1961–1966

Ben Mansour R, Lassoued S, Elgaied A, Haddouk S, Marzouk S, Bahloul Z, Masmoudi H, Attia H, Aïfa MS, Fakhfakh F (2010) Enhanced reactivity to malondialdehyde-modified proteins by systemic lupus erythematosus autoantibodies. Scand J Rheumatol 39:247–253

Biemond P, Swaak AJ, Koster JF (1984) Protective factors against oxygen free radicals and hydrogen peroxide in rheumatoid arthritis synovial fluid. Arthritis Rheum 27:760–765

Blossom SJ, Gilbert KM (2006) Exposure to a metabolite of the environmental toxicant, trichloroethylene, attenuates CD4+ T cell activation-induced cell death by metalloproteinase-dependent FasL shedding. Toxicol Sci 92:103–114

Blossom SJ, Pumford NR, Gilbert KM (2004) Activation and attenuation of apoptosis of CD4+ T cells following in vivo exposure to two common environmental toxicants, trichloroacetaldehyde hydrate and trichloroacetic acid. J Autoimmun 23:211–220

Blossom SJ, Doss JC, Gilbert KM (2007) Chronic exposure to a trichloroethylene metabolite in autoimmune-prone MRL+/+ mice promotes immune modulation and alopecia. Toxicol Sci 95:401–411

Blossom SJ, Doss JC, Hennings LJ, Jernigan S, Melnyk S, James SJ (2008) Developmental exposure to trichloroethylene promotes CD4+ T cell differentiation and hyperactivity in association with oxidative stress and neurobehavioral deficits in MRL+/+ mice. Toxicol Appl Pharmacol 231:344–353

Blossom SJ, Melnyk S, Cooney CA, Gilbert KM, James SJ (2012) Postnatal exposure to trichloroethylene alters glutathione redox homeostasis, methylation potential, and neurotrophin expression in the mouse hippocampus. Neurotoxicology 33:1518–1527

Bourg ACM, Mouvet C, Lemer DN (1992) A review of the attenuation of trichloroethylene in soils and aquifers. Q J Eng Geol Hydrogeol 25:359–370

Boyer AS, Finch WT, Runyan RB (2000) Trichloroethylene inhibits development of embryonic heart valve precursors in vitro. Toxicol Sci 53:109–117

Byers VS, Levin AS, Ozonoff DM, Baldwin RW (1988) Association between clinical symptoms and lymphocyte abnormalities in a population with chronic domestic exposure to industrial solvent-contaminated domestic water supply and a high incidence of leukaemia. Cancer Immunol Immunother 27:77–81

Cai P, König R, Khan MF, Qiu S, Kaphalia BS, Ansari GA (2006) Autoimmune response in MRL+/+ mice following treatment with dichloroacetyl chloride or dichloroacetic anhydride. Toxicol Appl Pharmacol 216:248–255

Cai P, König R, Boor PJ, Kondraganti S, Kaphalia BS, Khan MF, Ansari GA (2008) Chronic exposure to trichloroethene causes early onset of SLE-like disease in female MRL +/+ mice. Toxicol Appl Pharmacol 228:68–75

Caldwell JC, Keshava N (2006) Key issues in the modes of action and effects of trichloroethylene metabolites for liver and kidney tumorigenesis. Environ Health Perspect 114:1457–1463

Channel SR, Latendresse JR, Kidneym JK, Grabaum JH, Lanem JW, Steel-Goodwinm L, Gothausm MC (1998) A subchronic exposure to trichloroethylene causes lipid peroxidation and hepatocellular proliferation in male B6C3F1 mouse liver. Toxicol Sci 43:145–154

Chen SJ, Wang JL, Chen JH, Huang RN (2002) Possible involvement of glutathione and p53 in trichloroethylene- and perchloroethylene-induced lipid peroxidation and apoptosis in human lung cancer cells. Free Radic Biol Med 33:464–472

Cojocel C, Beuter W, Muller W, Mayer D (1989) Lipid peroxidation: a possible mechanism of trichloroethylene-induced nephrotoxicity. Toxicology 55:131–141

Cooper GS, Makris SL, Nietert PJ, Jinot J (2009) Evidence of autoimmune-related effects of trichloroethylene exposure from studies in mice and humans. Environ Health Perspect 117:696–702

Cuzzocrea S (2006) Role of nitric oxide and reactive oxygen species in arthritis. Curr Pharm Des 12:3551–3570

D'souza A, Kurien BT, Rodgers R, Shenoi J, Kurono S, Matsumoto H, Hensley K, Nath SK, Scofield RH (2008) Detection of catalase as a major protein target of the lipid peroxidation product 4-HNE and the lack of its genetic association as a risk factor in SLE. BMC Med Genet 9:62–69

Diot E, Lesire V, Guilmot JL, Metzger MD, Pilore R, Rogier S, Stadler M, Diot P, Lemarie E, Lasfargues G (2002) Systemic sclerosis and occupational risk factors: a case-control study. Occup Environ Med 59:545–549

Drake VJ, Koprowski SL, Hu N, Smith SM, Lough J (2006) Cardiogenic effects of trichloroethylene and trichloroacetic acid following exposure during heart specification of avian development. Toxicol Sci 94:153–162

DuTeaux SB, Berger T, Hess RA, Sartini BL, Miller MG (2004) Male reproductive toxicity of trichloroethylene: sperm protein oxidation and decreased fertilizing ability. Biol Reprod 70:1518–1526

Fan X, Wang G, English RD, Khan MF (2012) Proteomic analysis of carbonylated protein in the kidney of Trichloroethene-exposed MRL+/+ mice. The Toxicologist (2012 SOT Annual Meeting), p 252

Finkel T (2011) Signal transduction by reactive oxygen species. J Cell Biol 194:7–15

Flindt-Hansen H, Isager H (1987) Scleroderma after occupational exposure to trichlorethylene and trichlorethane. Acta Derm Venereol 67:263–264

Frostegard J, Svenungsson E, Wu R, Gunnarsson I, Lundberg IE, Klareskog L, Horkko S, Witztum JL (2005) Lipid peroxidation is enhanced in patients with systemic lupus erythematosus and is associated with arterial and renal disease manifestations. Arthritis Rheum 52:192–200

Garabrant DH, Lacey JV Jr, Laing TJ, Gillespie BW, Mayes MD, Cooper BC, Schottenfeld D (2003) Scleroderma and solvent exposure among women. Am J Epidemiol 157:493–500

Gharib OA (2009) Effects of Kombucha on oxidative stress induced nephrotoxicity in rats. Chin Med 4:23

Gilbert KM, Griffin JM, Pumford NR (1999) Trichloroethylene activates $CD4^+$ T cells: potential role in an autoimmune response. Drug Metab Rev 31:901–916

Gilbert KM, Pumford NR, Blossom SJ (2006) Environmental contaminant trichloroethylene promotes autoimmune disease and inhibits T-cell apoptosis in MRL(+/+) mice. J Immunotoxicol 3:263–267

Gilkeson GS, Keil D, Peden-Adams MM (2004) Immune effects of trichloroethylene on autoimmune disease in mice. In: Mohr LC, Hoel DG, Jollow D (eds) Trichloroethylene: the scientific basis of risk assessment. Medical University of South Carolina, Charleston, pp 87–98

Griffin JM, Blossom SJ, Jackson SK, Gilbert KM, Pumford NR (2000a) Trichloroethylene accelerates an autoimmune response by Th1 T cell activation in MRL +/+ mice. Immunopharmacology 46:123–137

Griffin JM, Gilbert KM, Lamps LW, Pumford NR (2000b) $CD4^+$ T-cell activation and induction of autoimmune hepatitis following trichloroethylene treatment in MRL+/+ mice. Toxicol Sci 57:345–352

Griffin JM, Gilbert KM, Pumford NR (2000c) Inhibition of CYP2E1 reverses CD4+ T-cell alterations in trichloroethylene-treated MRL+/+ mice. Toxicol Sci 54:384–389

Grune T, Michel P, Sitte N, Eggert W, Albrecht-Nebe H, Esterbauer H, Siems WG (1997) Increased levels of 4-hydroxynonenal modified proteins in plasma of children with autoimmune diseases. Free Radic Biol Med 23:357–360

Hadjigogos K (2003) The role of free radicals in the pathogenesis of rheumatoid arthritis. Panminerva Med 45:7–13

Halliwell B, Gutteridge JM (1984) Lipid peroxidation, oxygen radicals, cell damage, and antioxidant therapy. Lancet 1:1396–1397

Hardin BD, Kelman BJ, Brent RL (2005) Trichloroethylene and dichloroethylene: a critical review of teratogenicity. Birth Defects Res A Clin Mol Teratol 73:931–955

Haustein UF, Ziegler V (1985) Environmentally induced systemic sclerosis-like disorders. Int J Dermatol 24:147–151

Hayashi N, Igarashi A, Matsuyama T, Harada S (2000) Eosinophilic fasciitis following exposure to trichloroethylene: successful treatment with cyclosporin. Br J Dermatol 142:830–832

Higdon A, Diers AR, Oh JY, Landar A, Darley-Usmar VM (2012) Cell signaling by reactive lipid species: new concepts and molecular mechanisms. Biochem J 442:453–464

Hill BG, Dranka BP, Bailey SM, Lancaster JR Jr, Darley-Usmar VM (2010) What part of NO don't you understand? Some answers to the cardinal questions in nitric oxide biology. J Biol Chem 285:19699–19704

Hu C, Jiang L, Geng C, Zhang X, Cao J, Zhong L (2008) Possible involvement of oxidative stress in trichloroethylene-induced genotoxicity in human HepG2 cells. Mutat Res 652:88–94

Huang H, Kamijima M, Wang H, Li S, Yoshikawa T, Lai G, Huang Z, Liu H, Chen J, Takeuchi Y, Nakajima T, Li L (2006) Human herpesvirus 6 reactivation in trichloroethylene-exposed workers suffering from generalized skin disorders accompanied by hepatic dysfunction. J Occup Health 48:417–423

Huang Z, Yue F, Yang X, Xia L, Chen C, Qiu X, Huang J, Li L, Kamijima M, Nakajima T, Huang H (2012) Upregulation of calprotectin and downregulation of retinol binding protein in the serum of workers with trichloroethylene-induced hypersensitivity dermatitis. J Occup Health 54:299–309

IARC (1995) IARC monographs on the evaluation of carcinogenic risks to humans. In: Dry cleaning. Some chlorinated solvents and other industrial chemicals, vol 63. International Agency for Research on Cancer, Lyon

Iuchi Y, Kibe N, Tsunoda S, Suzuki S, Mikami T, Okada F, Uchida K, Fujii J (2010) Implication of oxidative stress as a cause of autoimmune hemolytic anemia in NZB mice. Free Radic Biol Med 48:935–944

Jacobson DL, Gange SJ, Rose NR, Graham NM (1997) Epidemiology and estimated population burden of selected autoimmune diseases in the United States. Clin Immunol Immunopathol 84:223–243

Januszewski AS, Alderson NL, Jenkins AJ, Thorpe SR, Baynes JW (2005) Chemical modification of proteins during peroxidation of phospholipids. J Lipid Res 46:1440–1449

Jia Q, Zang D, Yi J, Dong H, Niu Y, Zhai Q, Teng Y, Bin P, Zhou W, Huang X, Li H, Zheng Y, Dai Y (2012) Cytokine expression in trichloroethylene-induced hypersensitivity dermatitis: an in vivo and in vitro study. Toxicol Lett 215:31–39

Kamijima M, Wang H, Huang H, Li L, Shibata E, Lin B, Sakai K, Liu H, Tsuchiyama F, Chen J, Okamura A, Huang X, Hisanaga N, Huang Z, Ito Y, Takeuchi Y, Nakajima T (2008) Trichloroethylene causes generalized hypersensitivity skin disorders complicated by hepatitis. J Occup Health 50:328–338

Karpuzoglu E, Ahmed SA (2006) Estrogen regulation of nitric oxide and inducible nitric oxide synthase (iNOS) in immune cells: implications for immunity, autoimmune diseases, and apoptosis. Nitric Oxide 15:177–186

Khan F, Ali R (2006) Antibodies against nitric oxide damaged poly L-tyrosine and 3-nitrotyrosine levels in systemic lupus erythematosus. J Biochem Mol Biol 39:189–196

Khan MF, Kaphalia BS, Prabhakar BS, Kanz MF, Ansari GAS (1995) Trichloroethene-induced autoimmune response in female MRL +/+ mice. Toxicol Appl Pharmacol 134:155–160

Khan MF, Kaphalia BS, Ansari GAS (1997) Time-dependent autoimmune response of dichloroacetyl chloride in female MRL +/+ mice. Immunopharmacol Immunotoxicol 19:265–277

Khan MF, Wu X, Boor PJ, Ansari GAS (1999) Oxidative modification of lipids and proteins in aniline-induced splenic toxicity. Toxicol Sci 48:134–140

Khan MF, Wu X, Ansari GAS (2001) Anti-malondialdehyde antibodies in MRL+/+ mice treated with trichloroethene and dichloroacetyl chloride: possible role of lipid peroxidation in autoimmunity. Toxicol Appl Pharmacol 170:88–92

Khan MF, Wu X, Kaphalia BS, Boor PJ, Ansari GA (2003) Nitrotyrosine formation in splenic toxicity of aniline. Toxicology 194:95–102

Khan MF, Kannan S, Wang J (2006) Activation of transcription factor AP-1 and mitogen-activated protein kinases in aniline-induced splenic toxicity. Toxicol Appl Pharmacol 210:86–93

Khan S, Priyamvada S, Khan SA, Khan W, Farooq N, Khan F, Yusufi AN (2009) Effect of trichloroethylene (TCE) toxicity on the enzymes of carbohydrate metabolism, brush border membrane and oxidative stress in kidney and other rat tissues. Food Chem Toxicol 47:1562–1568

Kilburn KH, Warshaw RH (1992) Prevalence of symptoms of systemic lupus erythematosus (SLE) and of fluorescent antinuclear antibodies associated with chronic exposure to trichloroethylene and other chemicals in well water. Environ Res 57:1–9

Kurien BT, Scofield RH (2003) Free radical mediated peroxidative damage in systemic lupus erythematosus. Life Sci 73:1655–1666

Kurien BT, Scofield RH (2008) Autoimmunity and oxidatively modified autoantigens. Autoimmun Rev 7:567–573

Kurien BT, Hensley K, Bachmann M, Scofield RH (2006) Oxidatively modified autoantigens in autoimmune diseases. Free Radic Biol Med 41:549–556

Lacey JV Jr, Garabrant DH, Laing TJ, Gillespie BW, Mayes MD, Cooper BC, Schottenfeld D (1999) Petroleum distillate solvents as risk factors for undifferentiated connective tissue disease (UCTD). Am J Epidemiol 149:761–770

Lockey JE, Kelly CR, Cannon GW, Colby TV, Aldrich V, Livingston GK (1987) Progressive systemic sclerosis associated with exposure to trichloroethylene. J Occup Med 29:493–496

Martín-Gallán P, Carrascosa A, Gussinyé M, Domínguez C (2003) Biomarkers of diabetes-associated oxidative stress and antioxidant status in young diabetic patients with or without subclinical complications. Free Radic Biol Med 34:1563–1574

Moran MJ, Zogorski JS, Squillace PJ (2007) Chlorinated solvents in groundwater of the United States. Environ Sci Technol 41:74–81

Morgan PE, Sturgess AD, Davies MJ (2005) Increased levels of serum protein oxidation and correlation with disease activity in systemic lupus erythematosus. Arthritis Rheum 52:2069–2079

Morgan PE, Sturgess AD, Davies MJ (2009) Evidence for chronically elevated serum protein oxidation in systemic lupus erythematosus patients. Free Radic Res 43:117–127

Nagy G, Koncz A, Fernandez D, Perl A (2007a) Nitric oxide, mitochondrial hyperpolarization, and T cell activation. Free Radic Biol Med 42:1625–1631

Nagy G, Clark JM, Buzás EI, Gorman CL, Cope AP (2007b) Nitric oxide, chronic inflammation and autoimmunity. Immunol Lett 111:1–5

Nietert PJ, Sutherland SE, Silver RM, Pandey JP, Knapp RG, Hoel DG, Dosemeci M (1998) Is occupational organic solvent exposure a risk factor for scleroderma? Arthritis Rheum 41:1111–1118

Oates JC, Christensen EF, Reilly CM, Self SE, Gilkeson GS (1999) Prospective measure of serum 3-nitrotyrosine levels in systemic lupus erythematosus: correlation with disease activity. Proc Assoc Am Physicians 111:611–621

Ogino K, Hobara T, Kobayashi H, Ishiyama H, Gotoh M, Imamura A, Egami N (1991) Lipid peroxidation induced by trichloroethylene in rat liver. Bull Environ Contam Toxicol 46:417–421

Ohmori H, Kanayama N (2005) Immunogenicity of an inflammation-associated product, tyrosine nitrated self-proteins. Autoimmun Rev 4:224–229

Phoon WH, Chan MO, Rajan VS, Tan K, Thirumoorthy T, Goh CL (1984) Stevens-Johnson syndrome associated with occupational exposure to trichloroethylene. Contact Dermatitis 10:270–276

Pralong P, Cavailhes A, Balme B, Cottin V, Skowron F (2009) Diffuse systemic sclerosis after occupational exposure to trichloroethylene and perchloroethylene. Ann Dermatol Venereol 136:713–717

Purdue MP, Bakke B, Stewart P, De Roos AJ, Schenk M, Lynch CF, Bernstein L, Morton LM, Cerhan JR, Severson RK, Cozen W, Davis S, Rothman N, Hartge P, Colt JS (2011)

A case-control study of occupational exposure to trichloroethylene and non-Hodgkin lymphoma. Environ Health Perspect 119:232–238

Reed TT, Pierce WM, Markesbery WR, Butterfield DA (2009) Proteomic identification of HNE-bound proteins in early Alzheimer disease: insights into the role of lipid peroxidation in the progression of AD. Brain Res 1274:66–76

Reinl W (1957) Scleroderma caused by trichloroethylene in workers. Bull Hyg 32:678–679

Renke J, Szlagatys A, Hansdorfer-Korzon R, Szumera M, Kamińska B, Knap N, Popadiuk S, Szarszewski A, Woźniak M (2007) Persistence of protein oxidation products and plasma antioxidants in juvenile idiopathic arthritis. A one-year follow-up study. Clin Exp Rheumatol 25:112–114

Rhomberg LR (2000) Dose-response analyses of the carcinogenic effects of trichloroethylene in experimental animals. Environ Health Perspect 108(Suppl 2):343–358

Shah D, Kiran R, Wanchu A, Bhatnagar A (2010) Oxidative stress in systemic lupus erythematosus: relationship to Th1 cytokine and disease activity. Immunol Lett 129:7–12

Sheikh Z, Ahmad R, Sheikh N, Ali R (2007) Enhanced recognition of reactive oxygen species damaged human serum albumin by circulating systemic lupus erythematosus autoantibodies. Autoimmunity 40:512–520

Shen T, Zhu QX, Yang S, Ding R, Ma T, Ye LP, Wang LJ, Liang ZZ, Zhang XJ (2007) Trichloroethylene induce nitric oxide production and nitric oxide synthase mRNA expression in cultured normal human epidermal keratinocytes. Toxicology 239:186–194

Shen T, Zhu QX, Yang S, Wu CH, Zhang HF, Zhou CF, Zhang XJ (2008) Trichloroethylene induced cutaneous irritation in BALB/c hairless mice: histopathological changes and oxidative damage. Toxicology 248:113–120

Stadtman ER, Berlett BS (1998) Reactive oxygen-mediated protein oxidation in aging and disease. Drug Metab Rev 30:225–243

Tabrez S, Ahmad M (2011) Some enzymatic/nonenzymatic antioxidants as potential stress biomarkers of trichloroethylene, heavy metal mixture, and ethyl alcohol in rat tissues. Environ Toxicol 26:207–216

Tam LS, Li EK, Leung VY, Griffith JF, Benzie IF, Lim PL, Whitney B, Lee VW, Lee KK, Thomas GN, Tomlinson B (2005) Effects of vitamins C and E on oxidative stress markers and endothelial function in patients with systemic lupus erythematosus: a double blind, placebo controlled pilot study. J Rheumatol 32:275–282

Toraason M, Clark J, Dankovic D, Mathias P, Skaggs S, Walker C, Werren D (1999) Oxidative stress and DNA damage in Fischer rats following acute exposure to trichloroethylene or perchloroethylene. Toxicology 138:43–53

NTP (National Toxocology Program) (1990) Carcinogenesis studies of trichloroethylene (without epichlorohydrin) in F344/N rats and BC63F1 mice (gavage study), NTP Technical Report 243, CAS No. 79-01-06. NTP, Research Triangle Park

Vasanthi P, Nalini G, Rajasekhar G (2009) Status of oxidative stress in rheumatoid arthritis. Int J Rheum Dis 12:29–33

Waller PA, Clauw D, Cupps T, Metcalf JS, Silver RM, Leroy EC (1994) Fasciitis (not scleroderma) following prolonged exposure to an organic solvent (trichloroethylene). J Rheumatol 21: 1567–1570

Walsh SJ, Rau LM (2000) Autoimmune diseases: a leading cause of death among young and middle-aged women in the United States. Am J Public Health 90:1463–1466

Wanchu A, Khullar M, Deodhar SD, Bambery P, Sud A (1998) Nitric oxide synthesis is increased in patients with systemic lupus erythematosus. Rheumatol Int 18:41–43

Wang G, Cai P, Ansari GA, Khan MF (2007a) Oxidative and nitrosative stress in trichloroethene-mediated autoimmune response. Toxicology 229:186–193

Wang G, Ansari GA, Khan MF (2007b) Involvement of lipid peroxidation-derived aldehyde-protein adducts in autoimmunity mediated by trichloroethene. J Toxicol Environ Health A 70:1977–1985

Wang G, König R, Ansari GAS, Khan MF (2008) Lipid peroxidation-derived aldehyde-protein adducts contribute to trichloroethene-mediated autoimmunity via activation of CD4+ T cells. Free Radic Biol Med 44:1475–1482

Wang G, Wang J, Ma H, Khan MF (2009) Increased nitration and carbonylation of proteins in MRL+/+ mice exposed to trichloroethene: potential role of protein oxidation in autoimmunity. Toxicol Appl Pharmacol 237:188–195

Wang G, Pierangeli SS, Papalardo E, Ansari GA, Khan MF (2010a) Markers of oxidative and nitrosative stress in systemic lupus erythematosus: correlation with disease activity. Arthritis Rheum 62:2064–2072

Wang G, Ma H, Fan X, Wang J, Khan MF (2010b) N-acetylcysteine supplementation protests against trichloroethene-induced autoimmunity: role of oxidative stress. The Toxicologist (Annual Meeting of Society of Toxicology), p 67

Wang G, Wang J, Fan X, Ansari GA, Khan MF (2012a) Protein adducts of malondialdehyde and 4-hydroxynonenal contribute to trichloroethene-mediated autoimmunity via activating Th17 cells: dose- and time-response studies in female MRL+/+ mice. Toxicology 292:113–122

Wang G, Li H, Khan MF (2012b) Differential oxidative modification of proteins in MRL+/+ and MRL/lpr mice: increased formation of lipid peroxidation-derived aldehyde-protein adducts may contribute to accelerated onset of autoimmune response. Free Radic Res 46:1472–1481

Wang G, Wakamiya M, Wang J, Ansari GA, Khan MF (2013) Attenuation of trichloroethene-mediated autoimmune response in iNOS-null MRL+/+ mice. The Toxicologist (Annual Meeting of Society of Toxicology), p 306

Weinberg JB, Granger DL, Pisetsky DS, Seldin MF, Misukonis MA, Mason SN, Pippen AM, Ruiz P, Wood ER, Gilkeson GS (1994) The role of nitric oxide in the pathogenesis of spontaneous murine autoimmune disease: increased nitric oxide production and nitric oxide synthase expression in MRL-lpr/lpr mice, and reduction of spontaneous glomerulonephritis and arthritis by orally administered NG-monomethyl-L-arginine. J Exp Med 179:651–660

Wu KL, Berger T (2007) Trichloroethylene metabolism in the rat ovary reduces oocyte fertilizability. Chem Biol Interact 170:20–30

Wu C, Schaum J (2000) Exposure assessment of trichloroethylene. Environ Health Perspect 108(Suppl 2):359–363

Xia Y, Zweier JL (1997) Superoxide and peroxynitrite generation from inducible nitric oxide synthase in macrophages. Proc Natl Acad Sci U S A 94:6954–6958

Yáñez Díaz S, Morán M, Unamuno P, Armijo M (1992) Silica and trichloroethylene-induced progressive systemic sclerosis. Dermatology 184:98–102

Zhu QX, Shen T, Ding R, Liang ZZ, Zhang XJ (2005) Cytotoxicity of trichloroethylene and perchloroethylene on normal human epidermal keratinocytes and protective role of vitamin E. Toxicology 209:55–67

Chapter 5
Brain and Behavioral Changes in Rodent Models

Ambuja S. Bale

Abstract Trichloroethylene (TCE) exposure results in central nervous system (CNS) effects in experimental animals that can result from acute, subchronic, or chronic exposure. This chapter focuses on the behavioral and neurological changes that have been observed, to date, in rodent models (primarily rat and mice). Some of the neurological changes such as hearing impairment (ototoxicity) and nerve degeneration are classified as persistent and/or non-reversible. Other impairments such as vision and motor movement have been classified as primarily reversible or non-permanent neurological changes. The neurological changes that are observed include nerve conduction changes, sensory effects, cognitive deficits, changes in psychomotor function, and changes in mood and sleep behaviors. In addition, general pathological and neurotransmitter changes in the brain are discussed and may be relevant to the observed behavioral changes.

Disclaimer: The views expressed in this book chapter are those of the author and do not represent the policy of the US EPA.

Keywords Neurotoxicity • Rodents • Trichloroethylene • Brain • CNS depressant

5.1 TCE Exposure to Rodents and the Nervous System

Several studies with TCE exposure in rodents have been conducted to better understand the neurological effects that have been observed in humans. Neurotoxicological effects have been studied in humans and, briefly, the major neurological impairments are changes in nerve function, some observations of

A.S. Bale, PhD, DABT
National Center For Environmental Assessment,
Washington, DC, USA
e-mail: bale.ambuja@epa.gov

K.M. Gilbert, S.J. Blossom (eds.), *Trichloroethylene: Toxicity and Health Risks*,
Molecular and Integrative Toxicology, DOI 10.1007/978-1-4471-6311-4_5,

hearing and vision impairment, decreased cognitive function and impaired vestibular function. Many studies found that humans exposed to TCE present abnormalities in trigeminal nerve function (Feldman et al. 1988, 1992; Kilburn and Warshaw 1993; Kilburn 2002; Ruijten et al. 1991). Auditory impairments in humans, with children under 9 years of age being particularly susceptible, were reported in three studies that were conducted on the population in the TCE Subregistry from the National Exposure Registry (NER) developed by the Agency for Toxic Substances Disease Registry (ATSDR) (Burg and Gist 1995, 1999; ATSDR 2003). Human studies have also reported visual impairment resulting from TCE exposure including color discrimination (Kilburn 2002), contrast sensitivity (Reif et al. 2003) and visual depth perception (Vernon and Ruff 1969). TCE exposure on psychomotor response in humans have been studied primarily as the impact on reaction times (RT), and generally it has been found the RT is increased with exposure (Kilburn 2002; Kilburn and Warshaw 1993; Reif et al. 2003; Kilburn and Thornton 1996). However, there are significant limitations in the human neurotoxicity studies with TCE. Including lack of exposure information and duration, inability to determine effect levels.

To circumvent to the inherent limitations of studies involving TCE-mediated neurotoxicity in humans, rodent models allow investigators to better characterize the potential neurological targets that may be associated with TCE exposure. For most of the neurological effects associated with TCE, there has been a high degree of concordance between the rodent models and the observed effects in humans. In vivo studies in rodents and in vitro models have demonstrated that TCE produces functional and physiological neurological changes. Documenting changes in brain pathology and neurotransmitter signaling in rodents exposed to TCE may help verify that the observed behavior changes are due to TCE exposure. Overall, these effects collectively indicate that TCE has CNS depressant-like effects at low level exposures and causes anesthetic-like effects at high exposures.

5.2 Persistent Neurotoxic Effects of TCE

5.2.1 Auditory Effects

Perhaps the most carefully assessed persistent and/or permanent neurotoxic effect of adult TCE exposure in laboratory animals is a loss in auditory sensitivity that occurs preferentially for mid-frequency sound. The ability of TCE to permanently disrupt auditory function and produce abnormalities in inner ear histopathology has been demonstrated in several rodent studies using a variety of test methods. At least four independent research groups have replicated the finding that subacute TCE inhalation exposures in the range of 2,000–4,000 ppm produce a preferential impairment of mid-frequency hearing in several strains of rats (see Table 5.1 for study details and findings). The effects have been well-characterized using a broad range

Table 5.1 Summary of animal auditory function studies – inhalation exposure

Reference	Species/strain/sex/number	Dose level/exposure duration	Effects
Rebert et al. (1991)	Rat, Long Evans, male, 10/group Rat, F344, male, 4–5/group	Long Evans: 0, 1,600, 3,200 ppm; 12 h/day, 12 weeks F344: 0, 2,000, 3,200 ppm; 12 h/day, 3 weeks	Brainstem auditory evoked responses (BAERs) were measured. Significant decreases in BAER amplitude and an increase in latency of appearance of the initial peak (P1). For Long Evans rats, exposures to 3,200 ppm resulted in significant decreases in BAER amplitude, whereas for the F344 rats the same effect was observed at 2,000 ppm TCE
Rebert et al. (1993)	Rat, Long Evans, male, 9/group	0, 2,500, 3,000, 3,500 ppm; 8 h/day, 5 days	BAERs were measured 1–2 weeks post-exposure to assess auditory function. Significant decreases in BAERs were noted with TCE exposures of 3,000 ppm and higher
Rebert et al. (1995)	Rat, Long Evans, male, 9/group	0, 2,800 ppm; 8 h/day, 5 days	BAER measured 2–14 days post-exposure at a 16 kHz tone. Hearing loss ranged from 55 to 85 dB in rats exposed to 2,800 ppm TCE
Crofton et al. (1994)	Rat, Long Evans, male, 7–8/group	0, 3,500 ppm TCE; 8 h/day, 5 days	BAER measured and auditory thresholds determined 5–8 weeks post-exposure. Selective impairment of auditory function for mid-frequency tones (8 and 16 kHz) in rats exposed to 3,500 ppm TCE
Crofton and Zhao (1997), Boyes et al. (2000)	Rat, Long Evans, male, 9–12/group Rat, Long Evans, male, 8–10/group Rat, Long Evans, male, 8–10/group Rat, Long Evans, male, 8–10/group	0, 4,000, 6,000, 8,000 ppm; 6 h 0, 1,600, 2,400, 3,200 ppm; 6 h/day, 5 days 0, 800, 1,600, 2,400, 3,200 ppm; 6 h/day, 5 days/week, 4 weeks 0, 800, 1,600, 2,400, 3,200 ppm; 6 h/day, 5 days/week, 13 weeks	Auditory thresholds as measured by BAERs for the 16 kHz tone increased with TCE exposure. The effect levels were dependent on exposure duration 6 h exposure – 8,000 ppm 5 day exposure – 3,200 ppm 4 week exposure – 3,200 ppm 13 week exposure – 2,400 ppm No significant changes in BAERs were observed at lower exposures for the individual exposure durations

(continued)

Table 5.1 (continued)

Reference	Species/strain/sex/number	Dose level/exposure duration	Effects
Fechter et al. (1998)	Rat, Long Evans, male, 12/group	0, 4,000 ppm; 6 h/day, 5 days	Cochlear function measured 5–7 weeks after exposure. Loss of spiral ganglion cells noted. Auditory function was significantly decreased as measured by compound action potentials in the animal exposed to 4,000 ppm
Jaspers et al. (1993)	Rat, Wistar derived WAG-Rii/MBL, male, 12/group	0, 1,500, 3,000 ppm; 18 h/day, 5 days/week, 3 weeks	Auditory function assessed repeatedly 1–5 weeks post-exposure for 5, 20, and 35 kHz tones; No effect at 5 or 35 kHz; Decreased auditory sensitivity at 20 kHz in animals exposed to 1,500 ppm and higher
Muijser et al. (2000)	Rat, Wistar derived WAG-Rii/MBL, male, 8	0, 3,000 ppm	Auditory sensitivity decreased with TCE exposure at 4, 8, 16, and 20 kHz tones for the animals exposed to 3,000 ppm
Albee et al. (2006)	Rat, Fischer 344, male and female, 10/sex/group	0, 250, 800, 2,500 ppm	Mild frequency specific hearing deficits; Focal loss of hair cells and cochlear lesions in only the animals exposed to 2,500 ppm
Yamamura et al. (1983)	Guinea Pig, albino Hartley, male, 7–10/group	0, 6,000, 12,000, 17,000 ppm; 4 h/day, 5 days	No change in auditory sensitivity at any exposure level as measured by cochlear action potentials and microphonics at any TCE exposure levels

of well validated test methods including brainstem auditory evoked response (BAER) (Rebert et al. 1993, 1995; Albee et al. 2006), a behavioral method known as reflex modification audiometry (Jaspers et al. 1993; Muijser et al. 2000; Crofton et al. 1994; Crofton and Zhao 1997; Boyes et al. 2000), and electrophysiological assessment of cochlear potentials (compound action potential threshold, compound action potential growth functions, and cochlear microphonic isopotential curves) (Fechter et al. 1998). In one study (Fechter et al. 1998), multiple methods were used to document qualitatively similar auditory impairments within the same subjects through collaborative investigations among laboratories. In addition, there are limited histopathological data to support a loss of sensory receptor cells and neuronal cells in the cochlea that are consistent with the functional impairment and document an irreversible effect (Albee et al. 2006; Fechter et al. 1998). Collectively, TCE has been shown to produce ototoxicity in rodents at mid-frequency tones (4–24 kHz) and no observed changes in auditory function were observed at either the low (<4 kHz) or high (>24 kHz) frequency tones. Additionally, deficits in auditory effects were found to persist for at least 7 weeks after the cessation of TCE exposure (Rebert et al. 1991; Jaspers et al. 1993; Crofton and Zhao 1997; Fechter et al. 1998; Boyes et al. 2000). For example, Jaspers et al. (1993) exposed Wistar rats to 1,500 and 3,000 ppm TCE for 18 h/day, 5 days/week, for 3 weeks. Selective hearing loss at the 20 kHz tone in the rats exposed to 3,000 ppm TCE, but not in the 1,500 ppm TCE group, was reported for up to 6 weeks after cessation of exposure. Similarly, in male Long-Evans hooded rats, animals had increased hearing thresholds at a 16 kHz tone for up to 5 weeks following a acute, short-term, or subchronic exposures (up to 13 weeks) (Crofton and Zhao 1997).

Decreased amplitude and latency were noted in the BAERs (Rebert et al. 1991, 1993, 1995) suggesting that TCE exposure affects central auditory processes. Decrements in auditory function following reflex modification audiometry (Jaspers et al. 1993; Crofton et al. 1994; Crofton and Zhao 1997; Muijser et al. 2000) combined with changes observed in cochlear histopathology (Fechter et al. 1998; Albee et al. 2006) suggest that ototoxicity is occurring at the level of the cochlea and/or brainstem.

5.2.1.1 Reflex Modification

Reflex modification was used in several studies to evaluate the auditory function in TCE-exposed animals (Jaspers et al. 1993; Muijser et al. 2000; Fechter et al. 1998; Crofton and Zhao 1993; Crofton et al. 1994; Crofton and Zhao 1997; Boyes et al. 2000; Yamamura et al. 1983). In these studies, tones of different kHz were presented at increasing decibels (sound intensity). The decibel level at which the rodent became startled (e.g. acoustic startle response) was recorded. A decrease in auditory function is noted when an increased decibel level is needed to elicit a startle response. These studies collectively demonstrate significant decreases in auditory function at mid- frequency tones (8–20 kHz tones) for TCE exposures greater than 1,500 ppm after acute, short-term, and chronic durations. Only one study in guinea pigs (Yamamura et al. 1983) did not demonstrate impairment in auditory function from TCE exposures as high as 17,000 ppm for 4 h/day over 5 days. However, auditory testing was not performed in an audiometric sound attenuating chamber and extraneous noise could have influenced the outcome, and the guinea pig has been reported to be far less sensitive than the rat to the effects of ototoxic aromatic hydrocarbons such as toluene.

5.2.1.2 Brainstem Auditory Evoked Responses (BAERs)

Brainstem auditory-evoked potentials (BAERs) were also measured in several studies (Rebert et al. 1991, 1993, 1995; Albee et al. 2006) following TCE exposures ranging from 3 to 13 weeks. BAERs were generally measured by presenting tones to the rodents and measuring the responses using an electrode that was placed over the brainstem area. Rebert et al. (1991) measured BAERs in male Long Evans rats (n = 10) and F344 rats (n = 4 – 5) following stimulation with 4, 8, and 16 kHz sounds. The Long-Evans rats were exposed to 0, 1,600, or 3,200 ppm TCE, 12 h/day for 12 weeks and the F344 rats were exposed to 0, 2,000, or 3,200 ppm TCE, 12 h/day for 3 weeks. BAER amplitudes were significantly decreased at all frequencies for F344 rats exposed to 2,000 and 3,000 ppm TCE and for Long Evans rats exposed to 3,200 ppm TCE. In subsequent studies Rebert et al. (1993, 1995) again demonstrated TCE significantly decreases BAER amplitudes and also significantly increases the latency of appearance. Similar results were obtained by Albee et al. (2006) for male and female F344 rats exposed to TCE for 13 weeks.

5.2.1.3 Pathology Changes in the Auditory System

Notable pathology changes were also reported in a few auditory studies. Histological data from cochleas in Long-Evans rats exposed to 4,000 ppm TCE indicated that there was a loss in spiral ganglion cells (Fechter et al. 1998). Similarly, there was an observed loss in hair cells in the upper basal turn of the cochlea in F344 rats exposed to 2,500-ppm TCE (Albee et al. 2006).

5.2.2 *Neuronal Degeneration and Neuronal Impairment*

There is evidence of persistent and/ or permanent neurological effects from TCE exposure on neuronal degeneration. The available studies primarily demonstrate selective neuronal injury following TCE administration.

5.2.2.1 Dopaminergic Neurons

In two separate animal studies, subchronic administration of TCE has resulted in a decrease of dopaminergic (DA) cells in both rats and mice.

Gash et al. (2008) assessed the effects of subchronic TCE administration on dopaminergic neurons in the central nervous system. Fischer 344 male rats were given 1,000 mg/kg TCE in olive oil by gavage, 5 days/week for 6 weeks. Degenerative changes in DA containing neurons in the substantia nigra were reported as indexed by a 45 % decrease in the number of tyrosine hydroxylase positive cells. Additionally, there was a decrease in the ratio of 3,4-dihydroxyphenylacetic acid (DOPAC), a metabolite of DA, to DA levels in the striatum. This shift in ratio, on the order of 35 %, was significant by Students t test, suggesting a decrease in release and utilization of this neurotransmitter. While it is possible that long-term adaptation might occur with regard to release rates for DA, the loss of DA cells in the substantia nigra is viewed as a permanent adverse effect. The exposure level used in this study was limited to one high dose and more confidence in the outcome will depend upon replication and development of a dose-response relationship. If the results are replicated, they might be important in understanding mechanisms by which TCE produces neurotoxicity in the central nervous system. The functional significance of such cellular loss has not yet been determined through behavioral testing.

Guehl et al. (1999) also reported persistent effects of TCE exposure on DA neurons. In this study, OF1 male mice (n = 10) were injected ip daily for 5 days/week for 4 weeks with TCE (400 mg/kg/day). Following a 7 day period when the subjects did not receive TCE, the mice were euthanized and tyrosine hydroxylase immunoreactivity was used to measure neuronal death in the substantia nigra pars compacta. Treated mice presented significant dopaminergic neuronal death (50 %) in comparison with control mice based upon total cell counts conducted by an examiner blinded as to treatment group in six samples per subject.

5.2.2.2 Gamma-Aminobutyric Acid (GABA) and Glutamatergic Neurons

Disruption of GABAergic and glutamatergic neurons by environmental agents can represent serious impairment as GABA serves as a key inhibitory neurotransmitter while glutamate is equally important as an excitatory neurotoxicant. Moreover, elevations in glutamatergic release have been identified as an important process by which more general neurotoxicity can occur through a process identified as excitotoxicity. Consequently, GABA and glutamatergic neurons represent potentially important targets of TCE neurotoxicity.

Briving et al. (1986) exposed Mongolian gerbils to 50 and 150 ppm TCE continuously for 12 months via inhalation and reported changes in amino acids levels in the hippocampus and cerebellar vermis, and on high affinity uptake of GABA and glutamate in those same structures. An elevation of glutamine in the hippocampus of approximately 20 % at 150 ppm was reported, but no other reliable changes in amino acids in either of these two structures. With regard to high affinity uptake of glutamate and GABA, there were no differences in the hippocampal uptake between control and treated gerbils although in the cerebellar vermis there was a dose related elevation in the high affinity uptake for both of these neurotransmitters. Glutamate uptake was increased about 50 % at 50 ppm and 100 % at 150 ppm. The corresponding increases for GABA were 69 and 74 %. It is unclear if this finding in cerebellar vermis is also present in other brain tissues and should be studied further.

Shih et al. (2001) provided indirect evidence in male Mf1 mice that TCE exposure by injection might alter GABAergic function. The mice were injected ip with 250, 500, 1,000 and 2,000 mg/kg TCE in corn oil and the effect of these treatments on susceptibility to seizure induced by a variety of drugs was observed. Shih et al. reported that doses of TCE as low as 250 mg/kg reduced signs of seizure induced by picrotoxin, bicuculline and pentylenetetrazol, all GABAergic antagonist drugs. TCE treatment had a more limited effect on seizure threshold induced by non-GABAergic convulsant drugs such as strychnine (glycine receptor antagonist), 4-aminopyridine (alcohol dehydrogenase inhibitor) and N-methyl-d-aspartate (glutamatergic agonist) than was observed with the GABAergic antagonists. While these data suggest the possibility that TCE could act at least acutely on GABAergic neurons, there are no direct measurements of such an effect.

5.2.3 Demyelination Following TCE Exposure

Because of its anaesthetic properties and lipophilicity, it is hypothesized that TCE may disrupt the lipid-rich sheaths that cover many central and peripheral nerves. This issue has also been studied both in specific cranial nerves known to be targets of TCE neurotoxicity (namely the trigeminal nerve) and in the central nervous system including the cerebral cortex, hippocampus and cerebellum in particular. For peripheral and cranial nerves, there are limited nerve conduction velocity studies that are relevant as a functional measure. For central pathways, the most common outcomes studied include histological endpoints and lipid profiles.

A significant difficulty in assessing these studies concerns the permanence or persistence of effect. There is a very large literature unrelated to TCE which demonstrates the potential for repair of the myelin sheath, and at least partial if not full recovery of function. In the studies reviewed, where nerve myelin markers were assessed, it was not possible to determine if the effects were transient or persistent.

5.2.3.1 Trigeminal Nerve Demyelination

Rodent studies that have examined nerve function focused on the trigeminal nerve, or the fifth cranial nerve, that mediates facial sensations and motor functions including chewing and biting. Rodent studies have focused on impaired trigeminal nerve function because there are several human studies that have associated TCE exposure to decreased functionality in this nerve. Recent findings have also suggested that dichloroacetylene (DCA), an ex vivo TCE degradation product, may also be responsible for the impairment of trigeminal nerve function. The overall published information from the rodent studies suggest that the breakdown product of TCE, DCA, may be responsible for the trigeminal nerve effects.

Morphological Changes

Rodent studies have found that exposure to TCE results in morphological changes following a 3 day or a 10 week oral dose of 2,500 mg/kg-day (Barret et al. 1991, 1992). Examination of the pathology of the nerve revealed that TCE exposed animals had thinner trigeminal nerve fibers. Specifically, the thickness of the myelin sheath was significantly decreased. Also, it was observed that the internodal length was decreased. Effects were also evaluated with DCA (17 mg/kg) and it was found that the morphological changes were more severe than with TCE alone. TCE-dosed animals only exhibited changes in the smaller Class A fibers where internode length increased marginally (<2 %) and fiber diameter increased by 6 %. Conversely, DCA-treated rats exhibited significant and more robust decreases in internode length and fiber diameter in both fiber classes A (decreased 8 %) and B (decreased 4 %). Although the changes were noted, the administered doses in these studies were significantly higher than an incidental human exposure to TCE. Thus, DCA, a degradation product of TCE, was found to produce more severe morphological changes, but in animals exposed to TCE only the smaller nerve fibers (class A) were impacted.

Functionality

In order to verify the observed morphological changes in the trigeminal nerve, evaluations of trigeminal nerve functionality were conducted in rodents following inhalation exposures to TCE. Rats were exposed to various concentrations (250–2,500 ppm) of TCE for 13 weeks, and the trigeminal sensory evoked potentials (TSEPs) were measured (Albee et al. 2006). Stimulation of the trigeminal nerve was accomplished by sending electrical signals at the vibrissae pads (whiskers of the rat), and the evoked potentials were measured with electrodes placed over the somatosensory cortex where responses from the trigeminal nerve would traverse. TSEPs were not changed with TCE exposure, but when rodents were exposed to DCA, there were significant disruptions in the TSEP (Albee et al. 1997, 2006).

Kulig (1987) also measured peripheral (caudal nerve) nerve conduction time in male Wistar rats and failed to show an effect of TCE with exposures as high as 1,500 ppm for 16 h/day, 5 days/week for 18 weeks.

5.2.3.2 Demyelination in Central Nervous System

There are two studies (Isaacson and Taylor 1989; Isaacson et al. 1990) that document selective hippocampal histopathology when Sprague Dawley rats are exposed to TCE within a developmental model. Both of these studies employed oral TCE administration via the drinking water.

Isaacson and Taylor (1989) examined the development of the hippocampus in neonatal rats that were exposed in utero and in the preweaning period to TCE via their dam. TCE was added to the drinking water of the dam, and daily maternal doses were estimated based upon water intake of the dam as being 4 and 8.1 mg/day. Based upon body weight norms for 70 day old female Sprague Dawley rats, which would predict body weights of about 250 g at that age, such a dose might approach 16–32 mg/kg/day initially during pregnancy. Even if these assumptions hold true, it was not possible to determine how much TCE was received by the pups although the authors did provide an estimate of fetal exposure expressed as μg/ml of TCE, trichloroethanol, and trichloroacetic acid. The authors reported a 40 % decline in myelinated fibers in the CA1 region of the hippocampus of the weanling rats where the dams were exposed to a daily dose of 4.0 or 8.1 mg/day TCE. Since there was no effect of TCE treatment on myelination in several other brain regions including the internal capsule, optic tract or fornix, this effect appeared to be restricted to the CA1 region of the hippocampus at the tested exposures.

In a second publication by that group (Isaacson et al. 1990), weanling rats were exposed to TCE via their drinking water at doses of 5.5 mg/day for 4 weeks or 5.5 mg/day for 4 weeks, followed by a 2 week period with no TCE, and then a final 2 weeks of exposure to 8.5 mg/day TCE. Spatial learning was studied using the Morris water maze and hippocampal myelination was examined histologically starting 1 day post exposure. The authors reported that the subjects receiving a total of 6 weeks exposure to TCE showed *better* performance in the Morris swim test ($p<.05$) than did controls while the 4 week exposed subjects performed at the same level as did controls. Despite this apparent improvement in performance, histological examination of the hippocampus demonstrated the hippocampal myelin was significantly reduced in the TCE exposed groups, while normal myelin patterns were found in the internal capsule, optic tract and fornix. The authors did not evaluate the signs of gross toxicity in treated animals such as growth rate which might have influenced hippocampal development.

Ohta et al. (2001) administered 300 or 1,000 mg/kg TCE, i.p., to male ddY mice. Twenty-four hours after TCE administration, the mice were sacrificed and hippocampal sections were prepared from the excised brains and long term potentiation was measured in the slices. A dose related reduction in the population spike was observed following a tetanic stimulation relative to the size of the population spike elicited in the TCE mice prior to tetany. The spike amplitude was reduced 14 % in the 300 mg/kg TCE group and 26 % in the 1,000 mg/kg group. Precisely how such a shift in excitability of hippocampal CA1 neurons relates to altered hippocampal function is not certain, but it does demonstrate that injection with 300 mg/kg TCE can have lingering consequences on the hippocampus at least 24 h following ip administration.

A critical area for future study is the potential that TCE might have to produce demyelination in the central nervous system. It is realistic to imagine that an anaesthetic and lipophilic agent such as TCE might interact with lipid membranes and produce alterations, for example, in membrane fluidity at least at anaesthetic levels. However, from the available data it appears that chronic lower doses of TCE (50 and

150 ppm for 12 months, 320 ppm for 90 days, 510 ppm 8 h/day for 5 months) might alter fatty acid metabolism in the brains of Sprague Dawley rats and Mongolian gerbils (Kyrklund et al. 1983, 1986, 2002; Kyrklund and Haglid 1990; Kyrklund 1992). High doses were not included in these studies. Because the lower doses produced only sporadic significant effects and those tended to be of small magnitude (5–10 %) it is not certain that they are truly observing events with biological significance, or whether they are observing random effects. It could be hypothesized that the alterations in fatty acid metabolism could be an underlying mechanism for demyelination. However, it is not apparent that one brain region is more vulnerable to the effects of TCE than is another region. Significant changes in levels of cholesterol, neutral and acidic phospholipids or total lipid phospholipids were reported throughout the brain regions that have been measured and suggested a shift in lipid profiles between treated and untreated subjects.

5.3 Nonpersistent Neurotoxic Effects of TCE

5.3.1 *Vestibular Function*

The effect of TCE on vestibular function in animals has been evaluated by either (1) promoting nystagmus (vestibular system dysfunction) and comparing the level of effort required to achieve nystagmus in the presence and absence of TCE, or (2) using an elevated beam apparatus and measuring the balance of subjects exposed to TCE. Overall, it was found that exposure to TCE disrupts vestibular function. Impairment of vestibular function in male and female pigmented rats was observed after an acute inhalation exposure to TCE (2,700–7,200 ppm; Niklasson et al. 1993). An increased ability to promote nystagmus (fast and uncontrollable movement of the eyes related to vestibular function) was observed with acute TCE exposure. Complete recovery of the vestibular function in rats was reported within minutes of terminating a direct arterial infusion of TCE (Tham et al. 1979, 1984).

5.3.2 *Visual Effects*

Changes in visual function have been demonstrated in rodent studies during acute (Boyes et al. 2003, 2005) and subchronic exposure (Rebert et al. 1991; Blain et al. 1994) to TCE. In these studies, the effect of TCE on visual evoked responses to patterns (Boyes et al. 2003, 2005; Rebert et al. 1991) or a flash stimulus (Rebert et al. 1991; Blain et al. 1994) were evaluated. Overall, the studies demonstrated that exposure to TCE results in significant changes in the visual evoked response, which is reversible once TCE exposure is stopped. All of the rodent studies evaluated central visual function by measuring changes in evoked potential response following a visual stimulus that was presented to the animal. Two acute exposure inhalation studies (Boyes et al. 2003, 2005) exposed Long Evans rats to TCE based on a concentration × time schedule (Haber's law), and reported decreases in visual evoked potential amplitude and indicated decreased visual function. Additionally, Boyes et al. (2003, 2005) found that brain TCE concentration was best correlated with changes in visual function as measured by evoked potentials under acute exposure conditions. Two subchronic exposure studies (Rebert et al. 1991; Blain et al. 1994) demonstrated visual function changes as measured by pattern reversal evoked potentials (Rebert et al. 1991) or electroretinograms/oscillatory potentials (Blain et al. 1994). In one of these studies changes in ERGs and oscillatory potentials were noted following a 12-week exposure at 350 ppm (LOAEL) in rabbits (Blain et al. 1994). In the second study rats exposed to 3,200-ppm TCE for 12 weeks showed decreases in pattern reversal evoked potentials but no effect was noted in the 1,600-ppm exposure group (Rebert et al. 1991). Both subchronic studies examined visual function following an exposure-free period of either 2 weeks (Rebert et al. 1991) or 6 weeks (Blain et al. 1994) and found that visual function returned to pre-exposure levels, thus demonstrating that the TCE effects were reversible.

5.3.3 Cognitive Function

Many rodent studies have demonstrated significant differences in performance of learning tasks such as the speed to complete the task following TCE exposure. Impairment in operant-conditioning cognitive tasks has been reported following TCE exposure in both rats and mice (Kulig (1987); Umezu et al. 1997; Bushnell and Oshiro 2000). Wistar rats exposed to 250–4,000 ppm TCE and higher showed a significant decrease both in the total number of lever presses and in avoidance responses compared with controls. The rats did not recover their pre-exposure performance until about 2 h after exposure (Kulig (1987)). Likewise, a depressed rate of operant responding in male ICR strain mice intraperitoneally injected with 1,000 mg/kg TCE was observed in a conditioned avoidance task. Increased responding during the signal avoidance period at lower doses (250 and 500 mg/kg) suggested an impairment in ability to inhibit responding or failure to recognize the signal.

Rats trained in an operant visual signal detection task and then exposed to TCE (2,000 or 2,400 ppm, inhalation, 70 min daily for 9 days) had significant decrements in the accuracy of signal detection and response time (Bushnell and Oshiro 2000). In a follow-up, repeated exposure study with the operant visual signal detection task rats were inhalationally exposed to TCE (0, 1,600, 2,400) for 6 h/day for 20 days (Oshiro et al. 2004). No significant differences were observed among the exposure groups with respect to acquisition of the visual discrimination response or in the reaction times. Therefore, it was suggested that TCE exposure with this cognitive task does not result in persistent effects since repeated inhalational exposure to TCE (Oshiro et al. 2004) did not result in the same impairments that was observed in the acute exposure study (Bushnell and Oshiro 2000).

Although cognitive impairments were noted in some studies, other studies indicated no change, or even improvement, in cognitive tasks with continuous TCE exposure. No decrements in cognitive function as measured by the radial arm maze were observed in Mongolian gerbils exposed continuously by inhalation to 320 ppm TCE for 9 months (Kjellstrand et al. 1980). Improved performance, despite a loss in hippocampal myelination, was noted in a Morris swim test for weanling rats orally dosed with 5.5 mg/day for 4 weeks followed by 2 weeks of no exposure and an additional 2 weeks of 8.5 mg/day (Isaacson et al. 1990). Overall, cognitive function is impaired by TCE exposure primarily in tests that measure working memory (e.g., avoidance, operant responding, visual signal detection task). For spatial learning and memory tasks, such as the radial arm maze and the Morris swim test, TCE exposure in animals does not result in impaired performance.

5.3.4 Psychomotor Effects

Several animal studies have demonstrated that TCE exposure produces changes in psychomotor function. At high doses (≥2,000 mg/kg) TCE causes mice to lose their righting reflex when the compound is injected intraperitoneally (Shih et al. 2001; Umezu et al. 1997). At lower exposures (inhalation and oral), TCE produces alterations in neurobehavioral measures including locomotor activity, gait, operant responding, and reactivity. However, these effects may also be due to anesthetic properties associated with TCE exposure.

5.3.4.1 Loss of Righting Reflex

Impaired righting reflexes have been primarily tested in mice. Acute intraperitoneal (i.p.) injections of TCE in male ICR mice resulted in a dose-dependent disrupted righting reflex at doses of 2,000 mg/kg and higher (Umezu et al. 1997). Similarly, impaired righting reflexes at exposure doses of 5,000 mg/kg (i.p.) in male Mf1 mice were observed (Shih et al. 2001). When mice were pretreated with a CYP2E1 inhibitor, dimethyl sulfoxide or disulfiram, the TCE-induced loss of the righting reflex was delayed in a dose related manner. In contrast, the alcohol dehydrogenase inhibitor, 4-metylpyradine, did not delay the loss of the righting reflex following TCE (5,000 mg/kg) treatment. These data suggest that the anesthetic properties of TCE involve its oxidation via CYP2E1 to an active metabolite.

5.3.4.2 Activity, Sensory-Motor and Neuromuscular Function

TCE exposure in animals has resulted in changes in sensory-motor and neuromuscular function primarily following acute or short-term exposure (Kishi et al. 1993; Moser et al. 1995, 2003). Male Wistar rats inhalationally exposed to 250–4,000 ppm TCE for 4 h showed a significant decrease both in the total number of lever presses and in avoidance responses at 140 min of exposure compared with controls (Kishi et al. 1993). In adult female Fischer 344 rats acute and short-term (14 day) administration of TCE resulted in decreased performance in the neuromuscular and sensorimotor function tests which were conducted as part of a functional observational battery (general battery of behavioral observations and tests; Moser et al. 1995). Acute exposure to TCE produced the most significant effects in motor activity (activity domain), gait (neuromuscular domain), and click response (sensorimotor domain). In the 14-day study, only the activity domain (rearing) and neuromuscular domain (forelimb grip strength) were significantly different from control animals. In a separate 10-day study, TCE administration significantly reduced motor activity, tail pinch responsiveness, reactivity to handling, hind limb grip strength and body weight (Moser et al. 2003).

Although significant changes in neuromuscular and sensorimotor function have been observed following a shorter term exposure to TCE, longer term exposures (13–18 weeks) have not been able to establish an association between TCE exposure and impairment in this neurological domain. Male and female Fischer 344 rats inhalationally exposed to TCE (250–2,500 ppm) for 13 weeks did not have any impairments in the neuromuscular and sensorimotor tests conducted as part of the functional observational battery (Albee et al. 2006). No treatment related differences in grip strength or landing foot splay were demonstrated in this study. Similarly, in an 18 week exposure study with TCE (500–1,500 ppm), no changes in spontaneous activity, grip strength, or coordinated hind limb movement were reported (Kulig (1987)). Measurements were made every 3 weeks during the exposure period and occurred between 45 and 180 min following the previous TCE inhalation exposure. Therefore, it appears that acute or short-term exposures to TCE result in neuromuscular and sensorimotor function deficits and there may be some development of tolerance in longer term exposures since these neurological functions were not impaired for the longer exposure durations even though the exposure concentrations were comparable.

5.3.4.3 Locomotor Activity

The observed effects of TCE on locomotor activity in rodents are inconsistent. Several studies showed that TCE exposure can decrease locomotor activity in mice and rats (Wolff and Siegmund 1978; Moser et al. 1995, 2003). Reduced locomotor activity was reported in including AB mice ($n=18$) treated acutely 182 mg/kg TCE, i.p. at one of four time points during a 24-h day (Wolff and Siegmund 1978) and in female Fischer 344 rats ($n=8-10$) gavaged with TCE over an acute

(LOAEL = 5,000 mg/kg TCE) or subacute period (LOAEL = 500 but no effect at 5,000 mg/kg) (Moser et al. 1995, 2003). Rats were also reported to have an increased response latency (potentially an indication of decreased locomotor activity) to a two choice visual discrimination following 1,000- and 1,500-ppm TCE exposures for 18 weeks. However, no significant changes in grip strength, hindlimb movement, or any other motor activity measurements were noted (Kulig (1987)).

There are also a few studies (Fredriksson et al. 1993; Waseem et al. 2001) generally conducted using lower exposure doses that failed to demonstrate impairment of motor activity or ability following TCE exposure. Male Wistar rats dosed with TCE (350, 700, and 1,400 ppm) in drinking water for 90 days or exposed to 500, 1,000, and 1,500 ppm for 16 h/day, 5 days/week, for 18 weeks did not have any changes in locomotor activity (Waseem et al. 2001). No changes in locomotor activity were observed for 17-day-old male NMRI mice that were dosed postnatally with 50 or 290 mg/kg/day from Day 10 to 16 (Fredriksson et al. 1993). However, rearing activity was significantly decreased in the NMRI mice at Day 60.

5.3.4.4 Mood Effects and Sleep Disorders

Evaluating mood changes in rodents is difficult, but some investigators have reported in rats that exposure to TCE results in increased handling reactivity. The increased handling reactivity or hostility in rats was reported in male and female Fischer 344 rats following either an oral gavage of TCE (Moser et al. 2003) or an inhalation exposure (Albee et al. 2006).

Sleep disturbances with TCE exposure have been demonstrated in male Wistar rats exposed to 50–300 ppm TCE inhalation for 8 h/day, 5 days/week, for 6 weeks. Electroencephalographic (EEG) responses were measured and used to determine the number of awake (wakefulness hours) and sleep hours. TCE exposure significantly decreased amount of time spent in wakefulness (W) during the exposure period (Arito et al. 1994). The sleep changes in rodents are a highly sensitive effect in comparison to other observed neurological changes.

5.4 Summary

Exposure to TCE results in several neurological effects in rodent models. The most studied TCE-mediated neurotoxicological effect in the rodents was hearing impairment. The major findings are that hearing loss occurs at mid frequency tones (8–20 kHz), at inhalation exposures starting at 2,000 ppm. These effects persist as measured by loss in spiral ganglion and focal hair cells in the cochlea in addition to the lack of recovery in the hearing function tests. Neuronal degeneration and neuronal impairment was also reported in rodent models and these findings may be extended to numerous human reports of TCE exposure and nerve (specifically trigeminal nerve) impairment. Changes in vestibular function, decrements in visual, cognitive, and psychomotor effects were also observed in the rodent models and consistent with human findings. Sleeping disorders were also observed with TCE exposure in rodents, but have not been studied, to date, in humans and may represent a neurological hazard for humans. Overall, the rodent models help to strengthen the association between TCE exposure and neurotoxicity in humans. In addition, further behavioral studies with these rodent models may help to uncover more neurological changes and mechanistic events that might be occurring in humans with TCE exposure.

References

Albee RR, Nitschke KD, Mattsson JL et al (1997) Dichloroacetylene: effects on the rat trigeminal nerve somatosensory evoked potential. Neurotoxicol Teratol 19(1):27–37

Albee RR, Spencer PJ, Johnson KA et al (2006) Lack of trigeminal nerve toxicity in rats exposed to trichloroethylene vapor for 13 weeks. Int J Toxicol 25(6):531–540

Arito H, Takahashi M, Ishikawa T (1994) Effect of subchronic inhalation exposure to low-level TCEchloroethylene on heart rate and wakefulness-sleep in freely moving rats. Jpn J Ind Health (Sangyo Igaku) 36(1):1–8

ATSDR (Agency for Toxic Substances and Disease Registry) (2003) Impact of trichloroethylene exposure on oral motor, speech, and hearing in children. U.S. Department of Health and Human Services, Atlanta

Barret L, Torch S, Usson Y, Barret L, Torch S, Usson Y et al (1991) A morphometric evaluation of the effects of trichloroethylene and dichloroacetylene on the rat mental nerve. Preliminary results. Neurosci Lett 131(2):141–144

Barret L, Torch S, Leray C, Sarliève L, Saxod R. (1992). Morphometric and biochemical studies in trigeminal nerve of rat after trichloroethylene or dichloroacetylene oral administration. Neurotoxicology 13: 601–614

Blain L, Lachapelle P, Molotchnikoff S (1994) Electroretinal responses are modified by chronic exposure to trichloroethylene. Neurotoxicology 15(3):627–631

Boyes WK, Bushnell PJ, Crofton KM et al (2000) Neurotoxic and pharmacokinetic responses to trichloroethylene as a function of exposure scenario. Environ Health Perspect 108(Suppl 2):317–322

Boyes WK, Bercegeay M, Ali JS et al (2003) Dose-based duration adjustments for the effects of inhaled trichloroethylene on rat visual function. Toxicol Sci 76(1):121–130

Boyes WK, Bercegeay M, Krantz T et al (2005) Momentary brain concentration of trichloroethylene predicts the effects on rat visual function. Toxicol Sci 87(1):187–196

Briving C, Jacobson I, Hamberger A et al (1986) Chronic effects of perchloroethylene and trichloroethylene on the gerbil brain amino acids and glutathione. Neurotoxicology 7(1):101–108

Burg JR, Gist GL (1999) Health effects of environmental contaminant exposure: an intrafile comparison of the Trichloroethylene Subregistry. Arch Environ Health 54(4):231–241

Burg JR, Gist G, Allred SL et al (1995) The national exposure registry – morbidity analyses of noncancer outcomes from the trichloroethylene subregistry baseline data. Int J Occup Med Toxicol 4(2):237–257

Bushnell PJ, Oshiro WM (2000) Behavioral components of tolerance to repeated inhalation of trichloroethylene (TCE) in rats. Neurotoxicol Teratol 22(2):221–229

Crofton KM, Zhao X (1993). Mid-frequency hearing loss in rats following inhalation exposure to trichloroethylene: evidence from reflex modification audiometry. Neurotoxicol Teratol. 15(6):413–23

Crofton KM, Zhao X (1997) The ototoxicity of trichloroethylene: extrapolation and relevance of high-concentration, short-duration animal exposure data. Fundam Appl Toxicol 38(1): 101–106

Crofton KM, Lassiter TL, Rebert CS (1994) Solvent-induced ototoxicity in rats: an atypical selective mid-frequency hearing deficit. Hear Res 80:25–30
Fechter LD, Liu Y, Herr DW, Crofton KM (1998) Trichloroethylene ototoxicity: evidence for a cochlear origin. Toxicol Sci 42:28–35
Feldman RG, Chirico-Post J, Proctor SP (1988) Blink reflex latency after exposure to trichloroethylene in well water. Arch Environ Health 43(2):143–148
Feldman RG, Niles C, Proctor SP, Jabre J (1992). Blink reflex measurement of effects of trichloroethylene exposure on the trigeminal nerve. Muscle Nerve. 15(4):490–5
Fredriksson A et al (1993) Altered behavior in adult mice orally exposed to tri- and tetrachloroethylene as neonates. Toxicol Lett 66:13–19
Gash DM, Rutland K, Hudson NL, Sullivan PG, Bing G, Cass WA, Pandya JD, Liu M, Choi DY, Hunter RL, Gerhardt GA, Smith CD, Slevin JT, Prince TS (2008). Trichloroethylene: Parkinsonism and complex 1 mitochondrial neurotoxicity. Ann Neurol. 63(2):184–92
Guehl, D; Bezard, E; Dovero, S; Boraud, T; Bioulac, B; Gross, C. (1999). Trichloroethylene and parkinsonism: A human and experimental observation. Eur J Neurol 6: 609–611
Isaacson LG, Taylor DH (1989) Maternal exposure to 1,1,2-trichloroethylene affects myelin in the hippocampal formation of the developing rat. Brain Res 488(1–2):403–407
Isaacson LG, Spohler SA, Taylor DH (1990) Trichloroethylene affects learning and decreases myelin in the rat hippocampus. Neurotoxicol Teratol 12(4):375–381
Jaspers RMA, Muijser H et al (1993) Mid-frequency hearing loss and reduction of acoustic startle responding in rats following trichloroethylene exposure. Neurotoxicol Teratol 15:407–412
Kilburn KH (2002) Is neurotoxicity associated with environmental trichloroethylene (TCE)? Arch Environ Health 57(2):113–120
Kilburn KH, Thornton JC (1996) Prediction equations for simple and visual two-choice reactions times in environmental neurotoxicology. Arch Environ Health 51(6):439–444
Kilburn KH, Warshaw RH (1993) Effects on neurobehavioral performance of chronic exposure to chemically contaminated well water. Toxicol Ind Health 9(3):391–404
Kishi R, Harabuchi I, Ikeda T et al (1993) Acute effects of trichloroethylene on blood concentrations and performance decrements in rats and their relevance to humans. Br J Ind Med 50(5): 470–480
Kjellstrand P, Lanke J, Bjerkemo M et al (1980) Irreversible effects of trichloroethylene exposure on the central nervous system. Scand J Work Environ Health 6(1):40–47
Kulig BM (1987) The effects of chronic trichloroethylene exposure on neurobehavioral functioning in the rat. Neurotoxicol Teratol 9(2):171–178
Kyrklund T (1992) The use of experimental studies to reveal suspected neurotoxic chemicals as occupational hazards: acute and chronic exposures to organic solvents. Am J Ind Med 21(1):15–24
Kyrklund T, Alling C, Haglid K et al (1983) Chronic exposure to trichloroethylene: lipid and acyl group composition in gerbil cerebral cortex and hippocampus. Neurotoxicology 4(4):35–42
Kyrklund T, Kjellstrand P, Haglid KG (1986) Fatty acid changes in rat brain ethanolamine phosphoglycerides during and following chronic exposure to trichloroethylene. Toxicol Appl Pharmacol 85:145–153
Kyrklund T, Kjellstrand P, Haglid KG (2002) Effects of exposure to Freon 11, 1,1,1-trichloroethane or perchloroethylene on the lipid and fatty-acid composition of rat cerebral cortex. Scand J Work 14–94
Kyrklung T, Haglid KG (1990) Exposure of rats to high concentrations of 1,1,1-trichloroethane and its effects on brain lipid and fatty acid composition. Pharmacol Toxicol 67(5):384–386
Moser VC, Cheek BM, MacPhail RC (1995) A multidisciplinary approach to toxicological screening III. Neurobehavioral toxicity. J Toxicol Environ Health 45:173–210
Moser VC, MacPhail RC, Gennings C (2003) Neurobehavioral evaluations of mixtures of trichloroethylene, heptachlor, and di(2-ethylhexyl)phthlate in a full-factorial design. Toxicology 188(2–3):125–137
Muijser H, Lammers J, Kulig BM (2000) Effects of exposure to trichloroethylene and noise on hearing in rats. Noise Health 6:57–66

Niklasson M, Tham R, Larsby B, Eriksson B (1993) Effects of toluene, styrene, trycholoeethylene and trichloroethane on the vestibule and opto-oculo motor system in rats. Neurotoxicol Teratol 15:327–334

Ohta M, Saito T, Saito K et al (2001) Effect of trichloroethylene on spatiotemporal pattern of LTP in mouse hippocampal slices. Int J Neurosci 111(3–4):257–271

Oshiro WM, Krantz QT, Bushnell PJ (2004) A search for residual behavioral effects of trichloroethylene (TCE) in rats exposed as young adults. Neurotoxicol Teratol 26(2):239–251

Rebert CS, Day VL, Matteucci MJ et al (1991) Sensory-evoked potentials in rats chronically exposed to trichloroethylene: predominant auditory dysfunction. Neurotoxicol Teratol 13(1):83–90

Rebert CS, Boyes WK, Pryor GT et al (1993) Combined effects of solvents on the rat's auditory system: styrene and trichloroethylene. Int J Psychophysiol 14(1):49–59

Rebert CS, Schwartz RW, Svendsgaard DJ et al (1995) Combined effects of paired solvents on the rat's auditory system. Toxicology 105(2–3):345–354

Reif JS, Burch JB, Nuckols JR et al (2003) Neurobehavioral effects of exposure to trichloroethylene through a municipal water supply. Environ Res 93(3):248–258

Ruijten M, Verberk, M, Sallé, H. (1991). Nerve function in workers with long term exposure to trichloroethene. Br J Ind Med 48: 87–92

Shih CL, Chen HH, Chiu TH (2001) Acute exposure to trichloroethylene differentially alters the susceptibility to chemoconvulsants in mice. Toxicology 162:35–42

Tham R, Larsby B, Odkvist LM et al (1979) The influence of trichloroethylene and related drugs on the vestibular system. Acta Pharmacol Toxicol (Copenh) 44(5):336–342

Tham R, BUNNFORS I, Eriksson B et al (1984) Vestibulo ocular disturbances in rats exposed to organic solvents. Acta Pharmacol Toxicol (Copenh) 54(1):58–63

Umezu T, Yonemoto J, Soma Y et al (1997) Behavioral effects of trichloroethylene and tetrachloroethylene in mice. Pharmacol Biochem Behav 58(3):665–671

Vernon RK, Ruff JF (1969) Effects of trichloroethylene on visual-motor performance. Arch Environ Health 18(6):894–900

Waseem M, Ali M, Dogra S et al (2001) Toxicity of trichloroethylene following inhalation and drinking contaminated water. J Appl Toxicol 21(6):441–444

Wolff DL, Siegmund R (1978) The effect of trichloroethylene on the spontaneous locomotor activity of mice and of tetrachloroethane on the mortality of mice as a function of the time of day. Biol Zbl 97:345–351

Yamamura K, Ikeda I, Sadamoto T et al (1983) Effects of trichloroethylene exposure on hearing. An investigation of cochlear microphonics and action potential of the guinea pig. Eur J Appl Physiol Occup Physiol 52:47–50

Chapter 6
Role of Trichloroethylene in Parkinson's Disease

Samuel M. Goldman and Stephanie Whisnant Cash

Abstract Parkinson's disease (PD) is the second most common neurodegenerative disorder, primarily affecting older adults. Given the projected growth and aging trend of the population, the prevalence of PD is expected to double by the year 2040, highlighting the imperative need for a better understanding of the disease. It is generally accepted that the development of idiopathic PD is multifactorial, involving both genetic and environmental factors; however, relatively little is known about specific exposures. An association between TCE exposure and increased PD risk has been suggested by several case reports/clusters, and recently, the first epidemiologic study found a significant six-fold increased risk of PD associated with occupational TCE exposure. Data from animal studies also support a plausible mechanistic pathway, with rodent models demonstrating TCE-induced neurotoxic effects similar to those occurring in PD. Given that TCE is ubiquitous, even a modest increased risk could have enormous population-level neurological health implications.

Keywords Parkinson's disease • Movement disorder • Neurodegenerative disease • Epidemiology • Dopamine • TaClo

S.M. Goldman, MD, MPH (✉)
Department of Clinical Research,
The Parkinson's Institute and Clinical Center,
675 Almanor Ave, Sunnyvale, CA 94085-2934, USA
e-mail: sgoldman@thepi.org

S.W. Cash, PhD, MPH
Department of Clinical Research,
The Parkinson's Institute, Sunnyvale, CA, USA

K.M. Gilbert, S.J. Blossom (eds.), *Trichloroethylene: Toxicity and Health Risks*, Molecular and Integrative Toxicology, DOI 10.1007/978-1-4471-6311-4_6,

6.1 Introduction

This chapter summarizes the potential role of trichloroethylene (TCE) in Parkinson's disease (PD), a debilitating progressive neurodegenerative disorder affecting approximately ten million people worldwide, including one million in the U.S. (Parkinson's Disease 2013). The sections that follow provide an introduction to the etiology, clinical features, and known and suspected risk factors for PD, focusing on the role of trichloroethylene (TCE). Specific topics include PD case reports and clusters thought to be associated with TCE exposure, epidemiologic evidence supporting an association between PD and TCE, animal models of TCE and parkinsonism, potential biologic mechanisms, and limitations of current research. By the end of this chapter, the reader should have an understanding of PD epidemiology, and how exposure to TCE may increase risk of PD, including likely disease processes.

6.2 Parkinson's Disease: Background and Epidemiology

6.2.1 History, Clinical Features, and Etiology of PD

Parkinson's disease (PD) is a disabling and progressive neurodegenerative disorder of aging. Features suggestive of PD were described as early as several thousand years ago in Indian texts (Manyam 1990; Gourie-Devi et al. 1991), and by twelfth century Spanish Jewish philosopher Maimonides (Rosner and Muntner 1970), but the full syndrome was not described until 1817, during the height of the industrial revolution. In "An Essay on the Shaking Palsy" (Parkinson 2002), British apothecary surgeon James Parkinson documented the constellation of features that now bear his name (Goetz 1986) — coined several decades later by French neurologist Jean-Martin Charcot (Goetz 1986; Charcot 1892). The cardinal clinical features of PD include a classic low frequency "pill rolling" tremor at rest, slowness and paucity of movements and gait (bradykinesia, akinesia), muscular rigidity, and impaired balance. Over the last few decades, it has become increasingly recognized that PD also includes many other "non-motor" features such as hyposmia (decreased sense of smell) and autonomic and cognitive dysfunction (Bonnet et al. 2012; Stern et al. 2012; Postuma et al. 2012; Chaudhuri et al. 2006; Lang 2011). In contrast to PD, the term "parkinsonism" more generally describes the presence of some of the motor impairments associated with PD, and may result from a number of causes (e.g. medications, vascular insults), of which PD is the most common (Baldereschi et al. 2000).

Pathologically, PD is marked by the loss of pigmented dopaminergic neurons in the substantia nigra pars compacta (SNpc), with distinctive intracytoplasmic proteinaceous inclusions known as Lewy bodies, as well as Lewy neurites, seen in neuronal processes. In the late 1990s, Spillantini and colleagues identified the protein α-synuclein as the primary component of the Lewy body and Lewy neurite (Spillantini et al. 1998), and specific stains were developed that showed that synuclein pathology was widely distributed throughout the nervous system. Although most of the motor symptoms of PD result from degeneration of nigral dopaminergic neurons, synuclein pathology can be found in the olfactory bulb, in the enteric and autonomic nervous systems, and throughout the brainstem and cortex, and this pathology correlates with the non-motor features of the disease (Braak et al. 2003; Langston 2006).

Diagnosis of PD is based on clinical examination and recognition of motor deficits (Gelb et al. 1999), as there is currently no diagnostic test. Post-mortem studies have found that diagnostic accuracy by neurologists with specialty training in movement disorders is high, but non-specialist accuracy may be much lower (Hughes et al. 1992). Unfortunately, by the time clinical signs and symptoms are sufficient to meet diagnostic criteria, substantial neurological damage has already occurred, with an estimated 60–80 % loss of striatal dopamine (Bernheimer et al. 1973). Because putative disease-slowing therapies are much less likely to be effective at this point (Chen 2010), intensive efforts are focused on identifying biomarkers for detecting early disease.

PD is thought to have a long pre-clinical "prodromal" period estimated at years to decades (Koller et al. 1991), and early recognition of associated non-motor features may provide a window of opportunity to intervene with treatments to slow or halt disease progression. Seminal work by Braak et al. developed a model of PD progression based on a large autopsy series (Braak et al. 2003). In this model, PD begins in lower brainstem autonomic nuclei and olfactory bulb, or even in the gut, and slowly advances through susceptible regions upward to the midbrain (motor PD) and cortex (cognitive impairment, dementia). They propose that this process could potentially be triggered by a toxin or infection gaining access via the gastrointestinal tract.

This staging model is supported by prospective epidemiologic studies, which have found that reduced bowel movement frequency may occur decades before diagnosis, and correlates with Lewy pathology (Abbott et al. 2001, 2007; Ross et al. 2006, 2012). Similarly, impaired olfaction may develop at least 5 years before the onset of motor symptoms (Ross et al. 2008a, 2012). Other features associated with prodromal PD include cardiac autonomic dysfunction (CAD) (Kallio et al. 2000; Devos et al. 2003; Miyamoto et al. 2008), and idiopathic REM sleep behavior disorder (iRBD) (Schenck et al. 1996; Boeve et al. 2003; Iranzo et al. 2006; Postuma et al. 2009)—a condition associated with an approximately 50 % risk of PD. Research is currently underway to identify the most predictive combinations of tests and measures, with population-level screening as an ultimate goal.

6.3 Disease Causation: Theories and Mechanisms

Theories regarding the causes of PD have been debated for well over 150 years (Goetz 1986; Factor et al. 1988; Keppel Hesselink 1989), but like other chronic diseases of aging, it is now generally accepted that most PD results from a combination of genetic and environmental factors (Warner and Schapira 2003; Tanner 2003; Chade et al. 2006).

6.3.1 Genetics

Despite the long-held suspicion that PD occurred more commonly in family members of affected individuals (Charcot 1892; Mjones 1949; Gowers 1900; Bell and Clark 1926), the first causal genetic mutation wasn't identified until 1997. Polymeropolous et al. identified the highly penetrant autosomal dominant A53T mutation in the *SNCA* gene encoding α-synuclein protein (Polymeropoulos et al. 1997). Two other *SNCA* mutations have since been identified, but these are exceedingly rare, occurring in only a few families worldwide. Individuals with multiple copies of the *SNCA* gene (i.e., duplications or triplications) have also been identified, and they tend to have young onset rapidly progressive disease (Ross et al. 2008b). This, and observations that common variants that increase α-synuclein expression are associated with modestly increased risk (Maraganore et al. 2006; Mata et al. 2010) suggests that over-production of α-synuclein is likely to be an important mechanism. Additional research has associated at least 11 genes and 16 genetic loci with increased risk of PD (Corti et al. 2011). The most common mutation is the G2019S variant in the *LRRK2* (leucine-rich repeat kinase 2) gene, which accounts for 1–2 % of typical PD in the general population, with the frequency being much higher among certain subgroups (e.g. North Africans, Ashkenazi Jewish) (Correia Guedes et al. 2010; Hulihan et al. 2008; Ozelius et al. 2006).

Despite the evidence for the role of genetics in PD risk, Mendelian mutations explain only a few percent of PD cases (Martin et al. 2011; Klein and Ziegler 2011). Although some of the very rare genetic variants are highly penetrant, the more common *LRRK2* G2019S mutation (Goldwurm et al. 2005) is only 30–40 % penetrant (Bonifati 2007), suggesting an important role for environmental factors. Perhaps even more striking is the observation that monozygotic twins (who share 100 % of their genetic makeup) are no more likely to be concordant for PD than are dizygotic twins (who share only 50 % of their genetic makeup) (Tanner et al. 1999; Wirdefeldt et al. 1923, 2004). Together these observations argue for a major etiologic role for environment, a notion that was first supported by anecdotal evidence and later explored more systematically in animal and human studies.

6.3.2 Environment

6.3.2.1 The MPTP Discovery

Although there had been reports of acute parkinsonism resulting from large toxicant exposures (Rodier 1955; Tetrud et al. 1994; Pezzoli et al. 1989; Klawans et al. 1982), the PD environmental hypothesis was propelled by the discovery of a cluster of acute parkinsonism among intravenous drug users. Unlike the previous cases,

MPP+:

Paraquat:

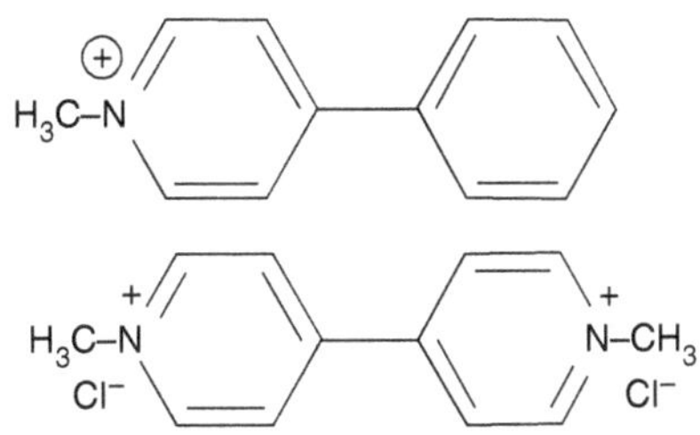

Fig. 6.1 Molecular structures of MPP+ and paraquat. *Abbreviations*: *MPP+* 1-methyl-4-phenylpyridinium; $C_{12}H_{12}N^+$, *Paraquat* N,N'-dimethyl-4,4'-bipyridinium dichloride; $C_{12}H_{14}Cl_2N_2$

which were highly atypical, these individuals presented with a syndrome nearly identical to typical PD. They showed classic PD motor signs including rest tremor, rigidity, impaired balance and severe bradykinesia, and were highly responsive to treatment with L-dopa, the standard PD therapy. The toxicant was identified as MPTP (N-methyl-4-phenyl-1,2,3,6 tetrahydropyridine), which had been inadvertently created during the synthesis of a meperidine analog (Langston et al. 1983). Toxicologically, MPTP is lipophilic and readily crosses the blood brain barrier (BBB). Once in the brain, it is oxidized by glial monoamine oxidase B (MAO-B) to the proximate toxin MPP^+ (1-methyl-4-phenylpyridinium; Fig. 6.1). The organic cation transporter-3 (Oct-3) then shuttles the MPP^+ out of the astrocytes (Cui et al. 2009). MPP+ is taken up by the dopamine transporter, and that which is not sequestered in vesicles accumulates in mitochondria where it inhibits Complex 1 of the electron transport chain. This causes oxidative stress, reduced ATP synthesis, perturbed calcium homeostasis and subsequent cell death (Singer et al. 1988; Przedborski and Jackson-Lewis 1998; Heikkila et al. 1985), and like PD, in primate models, toxicity is highly specific for nigral dopaminergic neurons (Langston et al. 1984). The discovery of MPTP propelled investigations into both disease mechanisms and other environmental agents, particularly those with similar molecular structures.

6.3.2.2 Disease Mechanisms

Dopaminergic neurons may be particularly susceptible to certain types of toxic insults. This is likely due to their high energy needs, dopamine oxidation that results in highly reactive oxygen radicals, and the presence of neuromelanin—which binds pro-oxidant metals (Halliday et al. 2005; Uversky et al. 2001a; Fahn and Cohen 1992; Carlsson and Fornstedt 1991; Guzman et al. 2010). Based on observations in PD patients, the discovery of genetic variants, and the MPTP experience, important pathogenic mechanisms have been identified; these include mitochondrial dysfunction, oxidative stress, inflammation, aggregation of α-synuclein protein, and impaired autophagy. The sequence of initiation versus propagation of the disease process is not understood.

Impaired mitochondrial function is believed to play an important role in PD pathogenesis (Di Monte 1991; Sherer et al. 2007). Patients have reduced mitochondrial function in brain and in peripheral tissues, and impaired mitochondrial transport (Perier and Vila 2012). MPTP is toxic to mitochondria, and several of the rare genetic forms of PD involve proteins essential to mitochondrial function (Martin et al. 2011). These include Parkin, an E3 ubiquitin ligase, PINK1, a mitochondrially-targeted kinase, and DJ-1, a mitochondrial peroxidase (Bonifati et al. 2003; Abbas et al. 1999; Kitada et al. 1998).

Mitochondrial dysfunction leads to oxidative damage, which in turn damages mitochondria. Patients manifest high levels of oxidative damage to membranes and proteins (Jenner 2003; Dexter et al. 1989), as well as depletion of anti-oxidants such as reduced glutathione (GSH) (Sian et al. 1994). Oxidative stress also increases α-synuclein aggregation (Souza et al. 2000). Enzymatic and non-enzymatic metabolism of dopamine generates hydrogen peroxide, and neuromelanin, a breakdown product of dopamine metabolism, binds pro-oxidant metals such as iron, producing highly reactive hydroxyl radical via the Fenton reaction (Lloyd and Phillips 1999). PD patients also manifest microglial activation early in the course of disease (Gerhard et al. 2006). Activated microglia produce inflammatory cytokines, which are increased in PD brain (Whitton 2007), trigger immune effectors and further increase oxidative stress (Collins et al. 2012).

α-synuclein aggregation plays a fundamental role in PD pathogenesis (Lashuel et al. 2013). It is the major component of Lewy bodies, and genetic variants that increase its expression are associated with increased PD risk (Ross et al. 2008b; Linnertz et al. 2009; Kay et al. 2008; Mata et al. 2010). Factors that increase its production or that enhance its propensity to misfold and aggregate are likely to increase PD risk (Freundt et al. 2012; Luk et al. 2012). Chaperone-mediated autophagy and lysosomal degradation are required to clear protein aggregates and damaged organelles from cells. PD patients manifest impairments in these systems, and genetic forms of PD may also cause impairment of autophagy (Alvarez-Erviti et al. 2010; Mak et al. 2010; Chu et al. 2009; Orenstein et al. 2013; Dehay et al. 2010).

The discovery of MPTP toxicity and the elucidation of potential etiologic and pathogenic mechanisms has sparked research into environmental factors that may act via similar pathways, including lifestyle factors, metals, polychlorinated biphenyls (PCBs), pesticides, and solvents (TCE being one example).

6.3.3 Epidemiology of PD

6.3.3.1 Challenges

PD has no diagnostic test, is relatively rare, has a long pathologic evolution, and important environmental exposures may occur decades before motor symptoms manifest. Thus, misdiagnosis is common (Schrag et al. 2002) and inconsistent over time and place, patients with underlying early disease may die from other causes before they are diagnosed, and accurate characterization of potential causative factors is very difficult (Ross et al. 2004, 2006). Furthermore, self-reporting of environmental exposures is subject to recall bias—whereby individuals with disease are prone to report exposures more frequently than are control subjects. Despite these barriers, the past few decades have seen great progress in identifying potential environmental causes of PD.

6.3.3.2 Descriptive Epidemiology

PD is the second most common neurodegenerative disorder behind Alzheimer's disease, affecting approximately 1 % of the population over age 60 (Tanner and Goldman 1996). In the United States, there were approximately 630,000 people living with *diagnosed* PD in 2010 (Kowal et al. 2013); however, many cases go undetected, and prevalence may be closer to one million, with approximately 60,000 incident (new) cases per year (Parkinson's Disease 2013). Enumerating the true burden of the illness in the population is challenging given the lack of a comprehensive disease registry. With the aging of the population (Census US 2011), the prevalence of PD is expected to increase dramatically (Parkinson's Disease 2013; Kowal et al. 2013), possibly even doubling by 2040 (Kowal et al. 2013). As a result, PD-related costs in the U.S., currently estimated at $23 billion annually, will soar to $50 billion by 2040 (Huse et al. 2005).

Age, Gender, Race

Age is the strongest risk factor for PD. Incidence rates increase drastically, from approximately 2.5 per 100,000 person-years at age 50 to more than 100 per 100,000 person-years by age 80 (Van Den Eeden et al. 2003). Incidence in men is nearly twice that in women (Tanner and Goldman 1996). Possible explanations include the role of hormones, differential exposure to occupational environmental toxicants, and higher stores of fat-soluble toxicants. Rates may be highest in Caucasians, intermediate in Asians, and lowest in persons of African descent (Van Den Eeden et al. 2003).

Geographic and Temporal Trends

Prevalence and incidence estimates vary widely (Tanner and Goldman 1996). Several studies reported higher incidence in more industrially developed nations compared to Africa and pre-industrial China (Okubadejo et al. 2006; Chen et al. 2001), suggesting a role of industrial toxicants. This is further supported by ecological evidence of higher than expected PD rates in farming and industrial areas (Rybicki et al. 1993; Barbeau 1986; Willis et al. 2010). In addition, PD incidence may be higher among people living at the northernmost latitudes, which may reflect exposure to environmental toxicants bioaccumulated in fish and marine mammals (Bjerregaard and Hansen 2000; Chiu et al. 2004; Wermuth et al. 2008).

If environmental factors related to industrialization affect PD risk, age-specific incidence should increase over time. However, because of changing diagnostic criteria and improved diagnostic accuracy, temporal changes are difficult to assess. Two studies using data from a county-wide medical database in Rochester, Minnesota, found that incidence did not change from 1967 to 1990 (Rajput et al. 1984; Rocca et al. 2001); however, these studies may have failed to detect a change over time due to overestimation of rates earlier in the period, since exclusions were not made for parkinsonism resulting from causes other than PD. A Finnish study found that rates increased by 50 % in men but decreased by 30 % in women between 1971 and 1992; (Kuopio et al. 1999). One possible explanation is that, as for the Minnesota studies, earlier rates were falsely elevated, and true rates may have increased substantially among men.

The totality of temporal and geographic evidence suggests that industrial-associated exposures may increase PD risk. Nonetheless, geographic and temporal data must be interpreted with caution given that comparisons may be compromised by a number of factors, including access to medical care, inconsistent reporting, and differences in study design.

6.3.3.3 Analytic Epidemiology: Assessing Risk Factors for PD

In order to understand the possible role of TCE as a risk factor for PD, it is important to briefly review other environmental associations with PD, since xenobiotics are likely to act via common or related mechanisms.

Lifestyle Factors

Providing further evidence for the role of environment, many lifestyle and behavioral factors are associated with PD risk. Cigarette smoking is consistently associated with a *reduced* risk of PD. In a meta-analysis of 44 case–control studies and four cohort studies, risk reductions were 61 % in current-smokers, 41 % in "ever" smokers, and 20 % in former smokers, compared to never-smokers (Hernan et al. 2002). Dose–response effects have also been observed (Ritz et al. 2007). Nicotine is neuroprotective in some animal models (Ryan et al. 2001; Park et al. 2007), although potential mechanisms are not well-understood, and cigarette smoke contains thousands of other compounds that could directly or indirectly reduce risk (Ross and Petrovitch 2001). Caffeine and coffee consumption are also consistently associated with *reduced risk* of PD, with magnitudes and dose–response effects similar to those of smoking (Hernan et al. 2002). Antagonism of the adenosine a2a receptor by caffeine is a proposed mechanism (Ross and Petrovitch 2001).

Evidence for other lifestyle factors is more limited and less consistent. Non-steroidal anti-inflammatory medications (Gao et al. 2011; Samii et al. 2009), dihydropyridine calcium channel blocking drugs (Marras et al. 2012), and cholesterol-lowering statins (Huang et al. 2007; Wolozin et al. 2007) have been associated with reduced risk. Diets high in fruits and vegetables (Gao et al. 2007; Sofi et al. 2008), low in animal fat (Gaenslen et al. 2008; Kyrozis et al. 2013), high in polyunsaturated fat (Kyrozis et al. 2013), and high in antioxidants (Miyake et al. 2011) and flavonoids (Gao et al. 2012) may be associated with reduced risk. Mild traumatic brain injury has been associated with increased risk with some consistency (Goldman et al. 2006).

Metals and PCBs

Mechanistically, several metals have been linked to PD pathogenesis. Dysregulation of iron homeostasis in PD is well described (Sofic et al. 1988; Griffiths and Crossman 1993; Dexter et al. 1991; Martin 2009; Zhang et al. 2010). Individuals with PD have increased nigral free iron, which binds to neuromelanin and can generate large amounts of highly reactive hydroxyl radical through the Fenton reaction (see Sect. 6.3.2.2). This in turn damages lipid membranes and mitochondria (Huang et al. 2006; Mehta et al. 2006) and leads to increased oxidative stress (Berg et al. 2004; Stohs and Bagchi 1995). Copper, manganese, nickel and other metals may act via similar mechanisms (Lloyd and Phillips 1999; Cavallo et al. 2003; Leonard et al. 2004). Lead and mercury are well-recognized neurotoxins (ATSDR 2007). They alter dopaminergic neuronal function in cellular and animal models, impair mitochondrial electron transport and autophagy, and generate reactive oxygen species (Farina et al. 2011; Yamin et al. 2003). In addition, most metal ions can catalyze the fibrillization and aggregation of α-synuclein protein (Yamin et al. 2003; Uversky et al. 2001a, b; Santner and Uversky 2010). Despite the mechanistic evidence, there is limited epidemiologic data to support an association with environmental exposure to these metals and PD risk (Logroscino et al. 2008; Tanner et al. 2009; Firestone et al. 2010; Feldman et al. 2011; Miyake et al. 2011; Ngim and Devathasan 1989; Petersen et al. 2008; Ohlson and Hogstedt 1981; Seidler et al. 1996; Semchuk et al. 1993; Gorell et al. 1999; Thygesen et al. 2011; Park et al. 2005). Of note, although exposure to very high levels of manganese can induce a parkinsonian syndrome that shares some clinical characteristics with PD ("manganism"), a recent systematic review concludes that manganese exposure is not associated with risk of PD (Mortimer et al. 2012). Data are strongest for lead; two studies found that individuals in the highest quartile of bone lead levels, a cumulative measure of decades of exposure, had twice the risk of PD (Coon et al. 2006; Weisskopf et al. 2010a).

Polychlorinated biphenyls (PCBs) were extensively used as lubricants and coolants. Despite their banning in 1977, these lipophilic compounds persist in the environment, particularly in fatty tissues of circumpolar fish and marine mammals (Welfinger-Smith et al. 2011; CDC 2009; Carpenter 2006; Hardell et al. 2010; ATSDR 2000; Chiu et al. 2004). Rates of PD may be higher in Inuit peoples who consume large amounts of these foods (Petersen et al. 2008; Wermuth et al. 2000). Post-mortem studies have found higher levels of PCBs in PD striatum and nigra compared to controls (Corrigan et al. 1998; Corrigan et al. 2000), but studies of serum levels are not consistent (Weisskopf et al. 2012; Petersen et al. 2008). Mechanistically, cellular and animal models have shown that PCBs damage dopaminergic neurons (Wolff et al. 1982; Richardson and Miller 2004; Seegal et al. 1991, 1994) and increase oxidative stress (Lee et al. 2012).

Pesticides

Epidemiologic evidence for environmental toxicants and PD risk is strongest for pesticides. Rural living has been associated with increased risk of PD for decades (Tanner and Goldman 1996), and farming occupation is also consistently associated with increased risk of PD (Priyadarshi et al. 2000; Goldman et al. 2005; Dick et al. 2007a; Kamel et al. 2007). Exposure to pesticides is a likely explanation, and epidemiologic and animal studies provide strong support for a direct causal relationship between pesticide exposure and increased risk of PD (Petrovitch et al. 2002; Ascherio et al. 2006; Dick et al. 2007a; Kamel et al. 2007; Hancock et al. 2008; Tanner et al. 2009, 2011). A meta-analysis of 19 studies estimated that a history of pesticide exposure is associated with an approximately two-fold increased risk (Priyadarshi et al. 2000), and even greater risk (2.5-fold) in studies that used job-based exposure classification rather than self-reported exposure (1.5-fold risk) (van der Mark et al. 2012).

Few studies have examined associations with specific pesticides, but evidence is most convincing for rotenone (Drolet et al. 2009; Greene et al. 2009; Pan-Montojo et al. 2010a, b; Tanner et al. 2011), paraquat (Tanner et al. 2011; Kamel et al. 2007; Hertzman et al. 1990; Liou et al. 1997), and the organochlorines (Hancock et al. 2008; Elbaz et al. 2009; Seidler et al. 1996). Specific organochlorines associated with increased PD risk include dieldrin, γ-HCH (hexachlorohexane; lindane) (Corrigan et al. 2000; Richardson et al. 2009, 2011) and β-HCH (Richardson et al. 2009, 2011; Weisskopf et al. 2010b; Petersen et al. 2008). Like MPP^+, rotenone (Sherer et al. 2007; Greenamyre et al. 1999) is a potent inhibitor of mitochondrial Complex 1 activity (Di Monte 1991; Sherer et al. 2007). Paraquat (Fig. 6.1) is a structural analog of MPTP that potently induces oxidative stress through redox cycling (Dinis-Oliveira et al. 2006; Kuter et al. 2010; McCormack et al. 2002). Both rotenone and paraquat produce animal models of PD, recapitulating PD features and pathology, with selective loss of nigrostriatal dopaminergic neurons and aggregation of α-synuclein (Cannon et al. 2009; McCormack et al. 2002). In addition, several organochlorine pesticides are known dopaminergic toxins (Kanthasamy et al. 2005).

6.4 TCE and Parkinson's Disease (PD) Risk: Human Studies

6.4.1 Solvents

Over the past several decades, there have been a handful of case reports of acute parkinsonism resulting from very large exposures to hydrocarbon solvents. However, these cases differed clinically and pathologically from PD (Guggenheim et al. 1971; Pezzoli et al. 1989; McCrank and Rabheru 1989; Tetrud et al. 1994; Uitti et al. 1994; Melamed and Lavy 1977). The epidemiologic literature examining solvents and PD is inconsistent, with some studies reporting no association, and others finding increased risk or younger disease onset (Hertzman et al. 1994; Pals et al. 2003; McDonnell et al. 2003; Tanner et al. 2009; Dick et al. 2007b; Seidler et al. 1996; Firestone et al. 2010; Pezzoli et al. 2000; Rango et al. 2006). This inconsistency may result for several reasons. First, most historical epidemiologic studies did not investigate specific compounds, instead considering solvents as a single exposure variable. The term "solvent" is exceedingly broad, encompassing a huge range of compounds whose only common characteristic is the ability to dissolve other substances. Although most solvents are oxidative stressors (Lam et al. 1995; Mattia et al. 1993), there is little reason to assume these disparate compounds share a common neuronal toxicity (Mutti and Franchini 1987). This approach may obscure disease relationships with any particular etiologic agent. Second, most studies relied on self-reporting of past solvent exposures, and many individuals do not know to which specific solvents they were exposed. Furthermore, self-reporting of exposures is subject to recall bias. Third, exposures were typically characterized as ever/never, without considering dose or duration in analyses (Hertzman et al. 1994; Pals et al. 2003; McDonnell et al. 2003; Tanner et al. 2009; Dick et al. 2007b; Seidler et al. 1996; Firestone et al. 2010; Pezzoli et al. 2000). Fourth, etiologically important exposures may have occurred years or even decades prior to disease onset.

6.4.2 TCE Overview

In contrast to the very general approach of most epidemiologic studies of solvent exposure and PD, recent work has implicated a specific association with TCE. Observations in humans, though limited, are supported by in vitro and animal models, and given what is known about PD pathogenesis, the association has considerable biological plausibility. In the following sections we review and critically evaluate the literature on TCE and PD, parkinsonism, and related pathogenetic pathways, and synthesize the evidence to date.

TCE is a ubiquitous hydrocarbon that has been used globally in a wide range of industrial and household applications since the 1920s (EPA 2009; IARC 1995); NIOSH 1978; WHO 1997; ATSDR 1997a). It is still commonly used today, most importantly as a vapor degreasing of metal parts in the automotive and metals industries, but also in adhesives, cleaning products, lubricants, paints and pesticides, and in the manufacture of plastics, automobiles, textiles, and paper and glass. As many as 3.5 million workers may have been exposed to TCE on a full or part-time basis (NIOSH 1978), and currently an estimated 400,000 workers are regularly exposed (IARC 1995); ATSDR 1997a). Occupational TCE exposure, which may be 100–1,000-fold higher than non-occupational exposure, is through both inhalation and dermal routes (Bogen et al. 1992, 1998). Although TCE is not thought to bioaccumulate in the food chain, it is detected in air, soil, and food, and is the common organic contaminant in groundwater, found in 30 % of U.S. drinking water supplies (EPA 2009; ATSDR 1997a; Wu and Schaum 2000; EPA United States Environmental

Protection Agency). Additional details of TCE use and human exposure are described in Chap. 1 of this textbook.

Tetrachloroethylene (perchloroethylene, PERC), a chemical with a very similar molecular structure and shared industrial uses, will be briefly discussed below. PERC was introduced in the 1930s, and by the 1960s was the predominant dry cleaning solvent (Andrasik 1990; ATSDR 1997b; EPA and Office of Compliance 1995). Like TCE, PERC is also widely used for metal cleaning and vapor degreasing, textile processing and as a chemical intermediate. At least 600,000 workers are likely exposed to PERC on a daily basis (ATSDR 1997b; National Center for Environmental Assessment and EPA 2001), primarily through inhalation. Secondary exposure to workers' family members can occur given that absorbed PERC is slowly exhaled (Aggazzotti et al. 1994a, b). Similar to TCE, PERC is not thought to bioaccumulate in the food chain, but it can persist in subsurface groundwater for many years.

6.4.3 Case Reports

The first report of a potential association between occupational TCE exposure and PD was published by Guehl and colleagues in 1999 (Guehl et al. 1999). The report described a single case, a woman who had been heavily exposed to TCE (among other solvents) for an unknown period of time before being diagnosed with PD. The woman was first exposed to TCE around age 27, while working as a house cleaner in small rooms with poor ventilation for several months. Immediately following this acute period of exposure, she worked for 6 years in the plastics industry in a small unventilated office, where she was exposed to TCE and other volatile compounds. Three years after leaving this position, she was diagnosed with PD (around age 37). Presenting signs and symptoms included gait disturbance and asymmetric rigidity, and consistent with typical PD, although disease features became bilateral, they remained asymmetric after 10 years of disease. She had fairly typical progression of her symptoms, but developed a pronounced "on-off" response to dopaminergic therapy and severe dyskinesias. While this was a compelling seminal case report, the potential effects of TCE cannot be isolated from those of the other volatile compounds to which she was exposed. Nonetheless, the complementary rodent experiment described in the same report, and discussed in further detail below (see Sect. 6.5), provides additional supportive evidence of an association specifically with TCE.

The next case reports were published by Kochen and colleagues several years later, in 2003 (Kochen et al. 2003). They described three unrelated patients with PD being seen for various reasons in their clinic. The first, "patient M", presented with elevated blood cholesterol, and possibly hepatitis. The authors were developing a method to detect exhaled volatile compounds associated with lipid peroxidation, and were studying patients with "hepatic lesions". They were surprised to find a TCE level of 128 ng/L. A review of the patient's history revealed that he had been occupationally exposed to TCE for 20 years, and was ultimately forced to retire due to the onset of unspecified neurological symptoms and hepatitis. He developed parkinsonian features at age 67, 6 years after terminating TCE exposure. Details regarding the nature or intensity of his occupational exposure or his clinical features are not provided. The authors note that the first exhalation measurement was made at least 3 years after cessation of exposure, and TCE was detectable in exhaled air up to 6 years after cessation of exposure. They concluded that TCE must persist in adipose tissue, and estimated a total body burden of 24 g at the time of exposure cessation—an extraordinary figure given the time-weight average (TWA) permissible exposure level (PEL) of 270 mg/m^3 in effect in Germany.

The second case reported by Kochen et al. (2003), "patient T", was identified in an attempt to elucidate potential mechanisms using blood samples, which were not available for "patient M". "Patient T" had PD onset approximately 11 years after initial exposure to TCE, and experienced two major unspecified solvent accidents during his/her last occupational year (at age 58). PD onset age or features were not reported, but he was retired due to "encephalopathy type IIB". TCE was detected in exhaled air at a concentration of 70 ng/L 2 months after termination of exposure, and was detectable up to 30 months after cessation. Importantly, blood levels of the TCE metabolite 1-trichloromethyl-1,2,3,4-tetrahydro-beta-carboline (TaClo) were also detectable up to 18 months post-exposure (see Sect. 6.6).

The third and final case reported by Kochen et al. (2003), "patient 11", was identified among 12 workers with high TCE exposures. No recruitment details are provided. Patient #11 developed PD 2 years after being forced to retire due to "encephalopathy grade IIB". He/she had worked as a painter with long-term, chronic exposure to TCE and PERC, including several "TCE accidents." One month after retiring, he had high blood levels of TaClo.

Although these four case reports are intriguing, they have some major shortcomings, and provide only anecdotal support for an association of TCE exposure and PD. First, exposure data and occupational conditions are not provided. Second, with the partial exception of the report by Guehl et al., there is little or no information regarding clinical features of parkinsonism or the criteria used to diagnose PD. Third, most cases had diagnoses of "encephalopathy," implying very broad neurological damage more consistent with some of the other reports of severe acute parkinsonism secondary to large solvent exposures. The relevance of this non-specific neurological damage to idiopathic PD is unclear.

6.4.4 A PD Cluster

Temporal and geographic clusters of PD are extremely rare (Kim et al. 2008). A major development occurred in 2008. Gash and colleagues were conducting a clinical trial for a PD medication (Gash et al. 2008), and one of their patients (the "index case") reported that he had worked with TCE for many years, suspecting this may have played a role in his PD. He informed them that two of his coworkers from a small manufacturing plant also had developed PD, prompting the authors to investigate a possible cluster of TCE-related PD. They mailed questionnaires regarding movement abnormalities to 134 former employees of the factory. Twenty-one of the 65 respondents reported three or more signs of parkinsonism (e.g. stooped posture, trouble with balance, rigidity or stiffness), and 23 additional individuals reported one or two symptoms.

Thorough clinical and occupational histories were collected for the three individuals with suspect PD. PD diagnosis was confirmed using standard diagnostic criteria, including at least two of: rest tremor, rigidity and bradykinesia, as well as a clear response to dopaminergic therapy. Fourteen individuals who self-reported three or more parkinsonian symptoms, and 13 who reported no symptoms agreed to undergo a clinical evaluation. The three patients with confirmed PD were diagnosed at age 42 (the index case), 68 and 53; they had worked directly with TCE for durations of 25, 25, and 29 years, respectively. The 14 individuals reporting parkinsonian symptoms ranged from 31 to 66 years of age and had been exposed for durations of 11–35 years.

The index case had worked in a degreasing area with large vats containing TCE. He regularly fully submersed his arms in the vats without using any gloves or other personal protective equipment. Although he met diagnostic criteria for PD, he had some atypical features including tics and L-dopa-responsive dystonic movements of his face. The second PD case had worked alongside the index case for 25 years, performing the same degreasing tasks without any protective equipment. His clinical course was also somewhat atypical, with disturbed cognition, impaired balance and oral dyskinesias as presenting symptoms. The third case worked near the other two adjacent to the TCE vat, and handled wet parts provided to her by the other two cases immediately after degreasing. Her clinical features were also somewhat atypical, with rapid progression, early freezing, disequilibrium and tongue movements.

Although they manifested parkinsonian signs or symptoms, none of the other 14 symptomatic employees who were evaluated met formal diagnostic criteria for PD. Workstation mapping revealed that they had also worked near the vat containing TCE for many years (11–35 years), and the severity of their symptoms appeared to be related to proximity to the vat. Interestingly, even those individuals who reported no symptoms, who ranged in age from 46 to 63 years and had exposure durations ranging from 8 to 33 years, had significantly slower fine motor hand movement times than age-matched healthy controls ($p<0.001$)—although the control population is not described in the paper.

Rodent experiments reported in the same publication are summarized below (see Sect. 6.5).

The strengths of this paper are the in-person movement evaluations, the application of standard diagnostic criteria for PD, and more detailed occupational histories. However, subjects manifested an array of atypical parkinsonian features, it is unclear if a true control population was used, and subjects were likely aware of study hypotheses, and thus might have some motivation for possible secondary gain. Most importantly, this is not an analytic study, i.e., subjects were not systematically enrolled to test a pre-specified hypothesis, and thus causal inferences are limited. Nonetheless, this report and the four prior case reports suggest a possible association of TCE exposure and PD, and data from animal and in vitro studies provide further support.

6.4.5 *Analytic Epidemiological Studies*

Several studies that examined specific occupations have reported increased risk of PD associated with occupations that might expose individuals to TCE or PERC (Schulte et al. 1996). Dry cleaning workers more frequently reported neurological symptoms, had slower reaction times and slower cognitive processing (Seeber 1989; Ferroni et al. 1992; NIOSH 1976). Working in cleaning and textiles occupations has been associated with an increased risk of PD (Fall et al. 1999; Tuchsen and Jensen 2000). Others have reported increased frequency of subtle neurological features suggestive of possible parkinsonism associated with TCE or PERC exposures, including hand tremor (Murata et al. 2010; Bruning et al. 1998; Liu et al. 1988; McCunney 1988).

Despite these observations, several case reports, and mechanistic plausibility (reviewed below), only a single analytic epidemiologic study of TCE exposure and PD risk has been published. Goldman and colleagues (Goldman et al. 2012a) studied PD risk associated with exposure to six solvents specified a priori: n-hexane, xylene, toluene, CCl_4, TCE, and PERC. These solvents were selected because they had either been associated with parkinsonism in case reports, or occupational studies had raised the possibility of an association with PD. By specifying the solvents a priori, they helped to mitigate recall bias to some extent.

The study was conducted in a 20,000 member cohort of elderly male twins, the National Academy of Sciences – National Research Council World War II-Veteran Twins Cohort (Jablon et al. 1967). A major strength of this truly population-based cohort is that it was established from military records in the 1960s when the twins were quite young, and subjects were unselected for diseases of old age or other factors that might bias results. In addition, twins constitute an ideal population in which to study environmental risk factors, because they are either genetically identical (monozygotic), or they share approximately 50 % of their nuclear genome (dizygotic). Thus, differences in health outcomes between twins are more likely to represent differential environment rather than differential genetics. The investigators identified 198 twin pairs discordant for PD using a multi-stage screening process and in-person examinations, and applying standard diagnostic criteria. Detailed methods are described by Tanner et al. (1999). Risk factor interviews were conducted by telephone, and included questionnaires on numerous lifestyle factors including residence, diet, smoking, head injury, and occupation.

Lifelong occupational and hobby histories were obtained by interview using job-task-specific structured questionnaires that collected detailed information on years of employment, company name and location, products made, job titles, and task, process and material-specific data for all jobs and hobbies held at least 6 months from age 10 onward. Fifty occupation-specific questionnaires were developed. Querying subjects with specific closed-ended responses for each job task minimized recall bias. Industrial hygienists blinded to disease status reviewed job histories and inferred exposure likelihood, intensity and duration for the 6 solvents of interest using probability databases and multiple reference sources. This provided a method for estimating exposures that avoided possible recall bias. Ninety-nine twin pairs discordant for PD completed all interviews.

Case and control exposure frequencies for TCE were 10 and 3 %, respectively, with 12 % of all pairs having at least one twin exposed. The most frequent occupations or hobbies with TCE exposure were electrician, dry cleaner, industrial machinery repairer, and health worker. Strikingly, after adjusting for potential confounding factors such as smoking, "ever" exposure to TCE was associated with a significant sixfold increased risk of PD (odds ratio (OR): 6.1; 95 % confidence interval (CI): 1.2–33.0; p=0.034). Risk was even greater for exposure to *either* TCE or PERC, which was associated with a significant nine-fold increased risk (OR: 8.9; 95 % CI: 1.7–47; p=0.010). Consistent with prior case reports, exposures occurred several decades prior to disease onset. With the exception of a borderline increased risk for carbon tetrachloride, none of the other solvents were associated with PD, suggesting that the association with TCE and PERC was specific and not due to recall bias or other types of bias. A dose–response was seen for exposure duration and cumulative exposure estimates.

This study has several important strengths. First, it was a population-based sample—selected decades before disease onset. Second, twins are a unique and powerful epidemiologic population. Because twins have very similar lifestyles and genes, observed differences are more likely to represent a true biological effect of differential environment. Third, exposures were estimated by industrial hygienists unaware of disease status, minimizing recall bias. The major weaknesses of this study are its relatively small size and the imprecision of exposure estimates. Nonetheless, this seminal analytic study strengthens evidence from the prior case reports, and suggests a possible causal association of PD and TCE (Table 6.1).

6.5 TCE and PD: Animal Studies

The case reports described above provided a basis for conducting animal studies, and two of these reports simultaneously presented results from complementary experimental rodent models (Guehl et al. 1999; Gash et al. 2008).

Guehl and colleagues modeled TCE exposure in rodents to observe neurological effects (Guehl et al. 1999). They intraperitoneally (i.p.) exposed ten 28-week-old OF1 male mice to 400 mg/kg of TCE per day for 5 days/week for a duration of 4 weeks. Although they observed no parkinsonian motor symptoms, the average number of tyrosine hydroxylase-positive (TH+) neurons in the substantia nigra pars compacta (SNpc) was 50 % lower in TCE-treated mice compared to controls

Table 6.1 Human studies of TCE exposure and PD or parkinsonism

Study [a]	Location	Design	Sample size	Exposure	Results
Epidemiologic evidence					
Goldman et al. (2012a)	USA	Discordant twin study; industrial hygienist inferred exposures	99 twin pairs discordant for PD	Occupational- and hobby-based TCE exposure (ever/never[b])	OR = 6.1 95 % CI: 1.2–33 p = 0.034
				Occupational- and hobby-based TCE *or* PERC exposure (ever/never[b])	OR = 8.9 95 % CI: 1.7–47 p = 0.010
Case reports/clusters					
Gash et al. (2008)	USA	Cluster of coworkers from small industrial plant	30 coworkers: 3 PD cases, 14 with parkinsonism, 13 self-reported asymptomatic	Occupational TCE exposure; 25–29 years for PD cases, 11–35 years for parkinsonism cases, 8–33 years for asymptomatics; Extensive dermal and respiratory exposures	Severity of parkinsonian symptoms appeared related to proximity to TCE vat; even asymptomatic individuals had significantly slower hand movements than age-matched controls
Kochen et al. (2003)	Germany	Case reports	3 unrelated PD case reports	Long-term occupational TCE exposures: 1. 128 ng/L TCE exhaled 3 years after 20-year exposure period 2. 70 ng/L TCE exhaled 2 months after exposure ended 3. 35 ng/5 ml in blood after long-term exposure and several "accidents" with TCE	No comparison group or formal statistical tests, but authors suggested an association
Guehl et al. (1999)	France	Case report	1 PD case (47-year-old female)	Occupational TCE exposure, approximately 7 years in duration	No comparison group or formal statistical tests

Abbreviations: *TCE* trichloroethylene, PD Parkinson's disease, *OR* odds ratio, *CI* confidence interval

[a]Studies included in this table are limited to those that examined TCE exposure specifically, and do not include those that only examined solvents as a larger category

[b]Exposures were derived from linking occupational and hobby histories reported in interviews to probability exposure databases. "Ever" exposure to TCE was considered to be exposure at least 2 % of work time or 1 h per week

($p<0.001$). As the authors noted, the neuronal degradation observed in their model was similar to that associated with a mid-range dose of MPTP in a similar model (Bezard et al. 1997).

Gash and colleagues (2008) also published a rodent experimental model in conjunction with their human case series. They exposed nine 5-month-old Fischer 344 male rats to 1 g/kg of TCE per day for 5 days/week for 6 weeks by oral gavage. Dopamine (DA) and its metabolites, which included DOPAC (dihydroxyphenylacetic acid) and HVA (homovanillic acid), were measured in the SNpc and striatum. Within the SNpc, TCE-treated rats had a 20 % lower DA concentration ($p<0.01$) and 25 % lower DOPAC concentration ($p<0.01$) compared to controls. HVA concentration in the SNpc did not vary between TCE-treated rats and controls. Within the striatum, significant differences were not observed for DA, but TCE-treated rats had 30 % lower DOPAC concentration compared to controls ($p<0.001$), and 10 % lower HVA ($p<0.01$). Of note, the DOPAC/DA ratio was 0.113 in TCE-treated rats, compared to 0.176 in controls, suggesting that the normal concentration of DA may have resulted from a compensatory reduction of DA metabolism. Similar to Guehl et al. (1999), they observed a 45 % decrease in SNpc TH+ positive neurons. They also investigated mitochondrial bioenergetics, and observed that the respiratory control ratio (state III/state IV oxygen consumption) was significantly lower for TCE-treated rats compared to controls in the SNpc ($p<0.01$), indicating selective inhibition of mitochondrial Complex 1. Finally, they observed α-synuclein-positive cytoplasmic inclusions in the SNpc and dorsal motor nucleus of the vagus nerve in the TCE-treated animals, mirroring pathological changes in PD patients according to Braak staging (Braak et al. 2003); similar inclusions were either absent or rarely observed in the control animals.

In a more detailed follow-up study that used a similar experimental protocol, the same research group exposed 5-month-old Fischer 344 male rats to 1 g/kg, 0.5 g/kg or 0.2 g/kg of TCE per day by oral gavage for 5 days/week for either 2 weeks or 6 weeks (Liu et al. 2010). They measured mitochondrial enzyme activity and oxidative damage in SN, levels of monoamines and their metabolites in SN and striatum (DA, DOPAC, HVA), and used silver staining and immunohistochemistry to detect dopaminergic neurons (TH+), cholinergic neurons (choline acetyltransferase), GABAergic neurons (dopamine and cAMP-regulated phosphoprotein 32; DARPP-32), microglia (OX-42), nitrative stress (3-nitrotyrosine), apoptosis (cleaved caspase 3), and α-synuclein. Neuronal counts were determined using unbiased stereology. They also used a Rotarod treadmill test to assess functional motor deficits. The Rotarod test requires animals to stay on a rotating cylinder as it accelerates.

At 6 weeks, there was a dose-dependent loss of SN TH+ neurons, with loss of 20, 25, and 40 % for 0.2, 0.5 and 1 mg/kg respectively. At the 1 mg/kg dose there was also a 34 % reduction in total neuron counts in SN. Remarkably, TH+ neuron counts were not reduced in the ventral tegmental area, indicating selectivity for SN. However, unlike typical PD, there were no differences in TH+ cell counts in the dorsal motor nucleus of the vagus nerve (a site of very early pathology during PD pathogenesis).(Braak et al. 2003) There were no differences in counts of striatal cholinergic or GABAergic neurons, nor of cerebellar GABAergic Purkinje cells, further demonstrating the specificity of TCE for dopaminergic neurons. Similar to their prior study (Gash et al. 2008), after 6 weeks dosing at 1 g/kg, striatal DOPAC levels decreased 33 %, HVA by 11 % and DA was unchanged. No changes were seen for serotonin or its metabolite, again highlighting the dopaminergic specificity of TCE.

Behavioral and mitochondrial functions were tested only at the 1 mg/kg treatment group. Rotarod treadmill duration, tested weekly, was significantly reduced relative to control animals at weeks 5 and 6, although there was no difference in spontaneous locomotor activity.

As measured by NADH substrate oxidation, mitochondrial Complex I activity was significantly reduced at both 2 and 6 weeks, and was correlated with increases in caspase 3-positivity, consistent with apoptosis. In addition, they observed increased lipid peroxidation, tyrosine nitration, and nigral microglial activation. In contrast, mitochondrial Complex II-driven respiration was unchanged, suggesting that TCE toxicity derives from specific inhibition of Complex I and an associated inflammatory cascade. Further recapitulating PD pathogenesis, they also observed marked intracellular α-synuclein accumulation in TH+ neurons in the dorsal motor nucleus of the vagus nerve, and modest increases in SNpc, but no cortical accumulation.

The final animal model discussed here examined the effects of TCE exposure alone and in combination with traumatic brain injury (TBI) (Sauerbeck et al. 2012). TBI is associated with increased risk of PD, and is known to perturb CNS function in numerous ways. It impairs the BBB, increases inflammatory cytokines, activates microglia, enhances oxidative and nitrative stress, inhibits proteosomal function, increases expression and aggregation of α-synuclein, and impairs mitochondrial function (Goldman et al. 2012b). Because TBI and TCE affect many of the same mechanistic pathways, both are associated with PD risk, and both are extremely common, Sauerbeck et al. (2012) hypothesized that they might have important synergistic effects.

The investigators exposed 5-month-old Fischer 344 male rats to 1 g/kg of TCE daily by oral gavage for either 1 or 2 weeks. Following exposure they were subjected to sham (craniotomy, but no impact), mild or moderate unilateral cortical impact injury, and sacrificed 6 h afterward. They examined effects on mitochondrial function (Complex I dependent oxygen consumption), TH+ and total cell counts in SN and striatum, and motor function.

After 2 weeks of TCE exposure, in contrast to findings of Gash et al. (2008) and Liu et al. (2010), there were no reductions in SN mitochondrial function in TCE only, TBI only, or in the TCE and TBI groups. However, in striatum, moderate TBI alone caused a non-significant 30 % decrease, TCE alone caused a significant 75 % decrease, but TBI and TCE together did not cause any additional reduction in mitochondrial function. The authors concluded that the failure to see synergism might be because of the severity of the deficit, and repeated the experiment after only 1 week of TCE exposure. In this model, TCE caused minimal reduction in striatal mitochondrial function, but TCE and TBI together caused a significant 50 % reduction.

The authors concluded that this shows evidence of a synergism between TCE and TBE-induced injury.

Motor impairment was tested after 2 weeks of TCE exposure using a Rotarod model with accelerating rotation (12 days after TBI), and a cylinder test where animals were placed in a plastic cylinder and paw touches contralateral to the cortical impact were counted (30 days after TBI). In the Rotarod model, neither mild nor moderate TBI, or TCE alone had an effect. However, TCE and moderate TBI caused a significant 50 % reduction in performance. In the cylinder model, TBI or TCE alone did nothing, but TCE+ moderate TBI caused a significant 30 % reduction in paw touches.

Finally, they looked at neuronal cell counts in SN and striatum using unbiased stereological counting. TCE and mild or moderate TBI caused a significant 13–17 % loss of TH positivity in SN, while either alone did nothing. However, neuronal counts were not decreased, unlike the results of Gash et al. (2008) and Liu et al. (2010). No effects were observed in striatum in any models.

In total, these experiments suggest a biologically plausible synergism between TCE and TBI.

Synthesizing the evidence from animal studies, it is clear that TCE exposure can impair mitochondrial function and specifically damage dopaminergic neurons in the SN. However, these effects have been observed at relatively high doses (Lock et al. 2013), and although α-synuclein accumulation was observed selectively in SNpc and dorsal motor nucleus of the vagus, the authors did not measure aggregated phosphorylated synuclein—which is thought to more specifically represent the pathological form of α-synuclein (Cavallarin et al. 2010). As discussed in the limitations and future directions section below, it will be important to expand the current evidence to incorporate doses which would be encountered by occupationally-exposed workers, as well as those persons in the general population (Table 6.2).

6.6 TCE and Parkinson's Disease (PD): Potential Biological Mechanisms

Although many solvents increase oxidative and nitrative stress, TCE and PERC more specifically affect mechanisms of primary importance to PD, mirroring the actions of MPTP. Most importantly, three animal studies reviewed here found that TCE exposure was associated with inhibition of mitochondrial Complex I (Gash et al. 2008; Liu et al. 2010; Sauerbeck et al. 2012).

TCE metabolism is reviewed in detail elsewhere in this text, and by Lash et al. (2000). TCE is lipophilic and readily crosses the blood brain barrier. CYP2E1-mediated oxidation of TCE generates chloral, which spontaneously reacts with tryptamine to produce 1-trichloromethyl-1,2,3,4-tetrahydro-β-carboline (TaClo) (Bringmann et al. 1995; Riederer et al. 2002; Heim and Sontag 1997) (Fig. 6.2). PERC metabolism, reviewed by Lash and Parker (2001), can also generate chloral and thus TaClo, in a similar pathway. Tryptamine is a trace biogenic amine present

Table 6.2 Animal models of TCE exposure in relation to deficits in neurological functioning and motor abilities

Study	Design	Exposure	Results
Sauerbeck et al. (2012)	4-month-old male Fischer 344 rats; four exposure arms for each of striatum and SNpc experiments 5–6 rats in each of four arms: 1. TCE & TBI 2. TCE & sham 3. TBI & vehicle 4. Vehicle & sham	TCE 1 g/kg/day, 7 days/week for either 1-week or 2-week. Oral administration TBI followed TCE administration	Mitochondrial Complex I activity: SN: no effects any treatment arm Striatum: 1 week TCE no effect, TBI 30 %↓, TCE+TBI 50 % ↓ Motor impairment: Rotarod time (12 days post TBI): no effect 2 weeks TCE or TBI, TCE+TBI 50 % ↓ ($p<0.01$) Cylinder test (30 days post TBI): no effect 2 weeks TCE or TBI, TCE+TBI 34 % ↓decreased use contralateral forepaw ($p<0.05$) Neurons: SN TH+neurons: no effect TCE or TBI, TCE+TBI 15 % ↓ Striatum: no effect any arm
Liu et al. (2010)	5-month-old male Fischer 344 rats; two exposure arms for each of striatum and SNpc experiments 6 TCE-exposed, 6 controls	TCE: 1 g/kg/day, 5 days/week for either 2-week or 6-week duration. Oral administration	Mitochondrial Complex I activity: SNpc: 2-week TCE 25 % ↓ Motor impairment: Rotarod time ~50 % ↓ after 5 weeks TCE Neurons: SNpc 6-week TCE: 40 % ↓ TH+ neurons, 35 % ↓ total neurons, dose dependent Ventral tegmentum or dorsal motor nucleus vagus: no effect Cholinergic, GABAergic neurons in striatum: no effect DA & metabolites in striatum: 2-week TCE: DOPAC ↓ 30 %, DA no effect 6-week TCE: DOPAC ↓ 33 %, HVA ↓ 11 %, DA no effect

(continued)

Table 6.2 (continued)

Study	Design	Exposure	Results
Gash et al. (2008)	5-month-old male Fischer 344 rats; two exposure arms 17 TCE-exposed, 17 controls in histopathologic assays. Mitochondrial assays had 9 animals pooled into 3 samples	TCE: 1 g/kg/day, 5 days/week for 6-week duration. Oral administration	Mitochondrial Complex I activity: SNpc: ~ 40 % ↓ ($p<0.01$) Striatum or liver: no effect Neurons: SNpc: 45 % ↓ TH+ neurons α-synuclein inclusions in SNpc, dorsal motor nucleus of vagus DA & metabolites SN: DA 20 % ↓, DOPAC 25 % ↓ Striatum: DA unchanged, DOPAC 30 % ↓ ($p<0.01$)
Guehl et al. (1999)	28-week-old OF1 male mice; two exposure arms 10 TCE-exposed, 10 controls	TCE: 0.4 g/kg/day, 5 days/week for 4-week duration. Intraperitoneal administration	Neurons: SNpc 50 % ↓TH+ neurons ($p<0.001$) Motor impairment: none observed

Abbreviations: *DA* dopamine, *SNpc* substantia nigra pars compacta, *DOPAC* dihydroxyphenylacetic acid, a dopamine metabolite, *HVA* homovanillic acid, a dopamine metabolite, *TH+* tyrosine hydroxylase positive, *TBI* traumatic brain injury

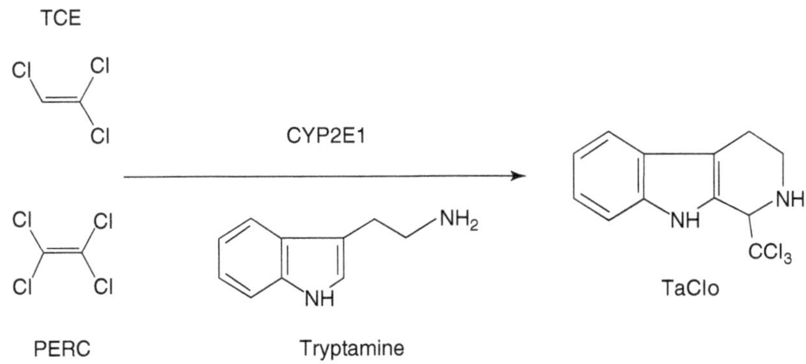

Fig. 6.2 Generation of TaClo from TCE and PERC. *Abbreviations*: *TCE* trichloroethylene; C_2HCL_3, *PERC* tetrachloroethylene; C_2Cl_4, *CYP2E1* cytochrome P450 2E1, *Tryptamine* 3-(2-aminoethyl)indole/2-(1H-indol-3-yl)ethanamine; $C_{10}H_{12}N_2$, *TaClo* 1-trichloromethyl-1,2,3,4-tetrahydro-beta-carboline; $C_{12}H_{11}Cl_3N_2$

throughout the brain (Berry 2004). It is formed from L-tyrosine by the enzyme aromatic L-amino acid decarboxylase (AADC). Although present at very low levels, its rate of synthesis is comparable to that of other biogenic amines such as dopamine, and levels may be highest in the striatum (Berry 2004).

TaClo is structurally similar to MPTP. Like MPTP, it freely crosses the BBB (Bringmann et al. 2006), and it specifically inhibits mitochondrial Complex 1—though TaClo and its metabolites are at least an order of magnitude more potent (Janetzky et al. 1995; Bringmann et al. 1995). TaClo is a dopaminergic toxin both in culture and in vivo (Rausch et al. 1995; Bringmann et al. 1995, 1996; Riederer et al. 2002; Heim and Sontag 1997; Grote et al. 1995). Injection of TaClo into the SNpc of rats resulted in a 15 % decrease in neuronal count (Bringmann et al. 1995). TaClo and its metabolites also inhibit the activity of tyrosine hydroxylase (Bringmann et al. 2002). However, there is some evidence that it is less specific for dopaminergic neurons than is MPP+ (Storch et al. 2006). The specificity of TaClo toxicity for SN dopaminergic neurons could result from the selective expression of CYP2E1 in SNpc TH+ neurons (Shahabi et al. 2008; Watts et al. 1998). However, unlike MPTP, which produces an acute parkinsonian syndrome, TaClo can initiate a slowly progressive motor deficit in a rat model, more closely recapitulating human PD (Sontag et al. 1995; Heim and Sontag 1997).

The TaClo hypothesis of TCE-associated PD is further strengthened by observations of Kochen et al. (2003), who were able to detect TaClo in PD "patient T", and recorded levels that decreased from 45 ng/5 ml blood at 2 months post-exposure to only trace amounts 30 months after cessation of exposure. They also were able to detect TaClo in 4 of 12 other workers exposed to high doses of TCE, one of whom later developed PD (Kochen et al. 2003).

Additional hypotheses of potential PD-related TCE toxicity have been proposed, including other major metabolites of TCE, such as trichloroethanol (TCE-OH) and trichloroacetic acid (TCA) (Lock et al. 2013; Zaheer and Slevin 2011). Because

CYP2E1 is expressed in SNpc TH+ neurons, Lock and colleagues (2013) suggest that SNpc-specific metabolism of TCE could produce regionally specific oxidative stress (Liu et al. 2010; Shahabi et al. 2008), without requiring the presence of TaClo.

Additional mechanisms may also contribute to oxidative stress. Consistent with the observation that PD patients have reduced nigral levels of the anti-oxidant glutathione (GSH) (Sian et al. 1994), Blossom et al. showed that TCE exposure in mice decreases cerebellar GSH (Blossom et al. 2013). Specifically, the concentration of GSH in cerebellar tissue of mice exposed to 0.01 and 0.1 mg/ml TCE in drinking water was significantly decreased by 12 and 20 %, respectively. Although the levels of GSSG (inactive oxidized disulfide form) were not significantly affected by TCE exposure, the high-dose TCE exposure resulted in a significant 27 % decrease in the GSH/GSSG ratio and a 25 % increase in oxidized GSH. The results of this study indicate that oxidative stress may be one mechanism by which TCE could increase risk of PD.

6.7 Limitations and Future Directions

There are a number of limitations that are important to consider when assessing the evidence to date regarding TCE and PD risk. Solvents have most often been examined epidemiologically as a single, aggregate category, and the effects of particular solvents have rarely been assessed; given that solvents represent a very large group of disparate compounds, any true association between PD and a particular agent (e.g. TCE) could be obscured (Goldman 2010; Goldman et al. 2012a). Exposure assessment methods are very limited, and are likely to misclassify exposed individuals. Recall bias is a major concern. Because PD manifests clinically after a long evolutionary process, there is likely a long latent period between exposure and disease, and many relevant exposures may occur years or even decades prior to disease onset. Exposure misclassification also presents a problem, as recognition and/or recollection of exposures may be poor. Estimation of dose can also be challenging (Goldman 2010), limiting our ability to detect any dose–response effects. In addition, solvent exposures are unlikely to occur in isolation, making it difficult to tease apart effects specifically attributable to TCE. Finally, current animal models do not fully reflect the complex motor, non-motor, and pathologic syndrome of PD (Martinez and Greenamyre 2012).

In light of these limitations, it will be important for future studies to focus on single solvents, and to measure exposures as accurately as possible. Specifically, future animal models should consider doses and exposure durations that reflect real-work human exposures. The rodent model used by Gash et al. (2008) exposed animals to doses that may have resulted in peak blood levels at least 35 times greater than that of typical industrial workers. Another major concern, as noted by Lock et al. (2013), is that most of the animal studies modeling TCE effects and parkinsonism originated from the same investigator group at a single institution (Gash et al. 2008; Liu et al. 2010; Sauerbeck et al. 2012), and it will be important for other groups to confirm their findings.

Future research should also consider potential interactions between TCE and other genetic and environmental risk factors. Environmental toxins, specifically those that inhibit mitochondrial Complex I, are good candidates. As shown by Sauerbeck et al., environmental insults that affect common pathways may interact synergistically (Sauerbeck et al. 2012). Genetic factors that affect TCE metabolism or PD-related pathogenic pathways should also be investigated, but will likely require very large epidemiologic studies, or pooling of data across studies.

As with all epidemiologic studies, results require replication across populations, research groups, and using a variety of study designs. Although animal and in vitro data are compelling, the human epidemiologic literature investigating TCE and PD risk is extremely small, consisting of a single retrospective analytic study (Goldman et al. 2012a), and case reports comprising a total of seven individuals—some of whom are not well characterized clinically (Guehl et al. 1999; Kochen et al. 2003; Gash et al. 2008). Parkinson's disease (PD) represents a major societal burden that is expected to increase rapidly in the coming decades. Identifying potentially modifiable risk factors is of the utmost importance.

References

Abbas N, Lucking CB, Ricard S, Durr A, Bonifati V, De Michele G, Bouley S, Vaughan JR, Gasser T, Marconi R, Broussolle E, Brefel-Courbon C, Harhangi BS, Oostra BA, Fabrizio E, Bohme GA, Pradier L, Wood NW, Filla A, Meco G, Denefle P, Agid Y, Brice A (1999) A wide variety of mutations in the parkin gene are responsible for autosomal recessive parkinsonism in Europe. French Parkinson's Disease Genetics Study Group and the European Consortium on Genetic Susceptibility in Parkinson's disease. Hum Mol Genet 8(4):567–574. doi:ddc081 [pii]

Abbott RD, Petrovitch H, White LR, Masaki KH, Tanner CM, Curb JD, Grandinetti A, Blanchette PL, Popper JS, Ross GW (2001) Frequency of bowel movements and the future risk of Parkinson's disease. Neurology 57(3):456–462

Abbott RD, Ross GW, Petrovitch H, Tanner CM, Davis DG, Masaki KH, Launer LJ, Curb JD, White LR (2007) Bowel movement frequency in late-life and incidental Lewy bodies. Mov Disord 22(11):1581–1586. doi:10.1002/mds.21560

Aggazzotti G, Fantuzzi G, Predieri G, Righi E, Moscardelli S (1994a) Indoor exposure to perchloroethylene (PCE) in individuals living with dry-cleaning workers. Sci Total Environ 156(2):133–137

Aggazzotti G, Fantuzzi G, Righi E, Predieri G, Gobba FM, Paltrinieri M, Cavalleri A (1994b) Occupational and environmental exposure to perchloroethylene (PCE) in dry cleaners and their family members. Arch Environ Health 49(6):487–493

Alvarez-Erviti L, Rodriguez-Oroz MC, Cooper JM, Caballero C, Ferrer I, Obeso JA, Schapira AH (2010) Chaperone-mediated autophagy markers in Parkinson disease brains. Arch Neurol 67(12):1464–1472. doi:10.1001/archneurol.2010.198

Andrasik F (1990) Psychologic and behavioral aspects of chronic headache. Neurol Clin 8(4):961–976

Ascherio A, Chen H, Weisskopf MG, O'Reilly E, McCullough ML, Calle EE, Schwarzschild MA, Thun MJ (2006) Pesticide exposure and risk for Parkinson's disease. Ann Neurol 60(2): 197–203. doi:10.1002/ana.20904

ATSDR (1997a) Agency for Toxic Substances and Disease Registry – Toxicological Profile for Trichlorethylene. United States Department of Health and Human Services, Atlanta

ATSDR (1997b) Agency for Toxic Substances and Disease Registry – Toxicological Profile for Tetrachloroethylene. United States Department of Health and Human Services, Atlanta

ATSDR (2000) Toxicological profile for polychlorinated biphenyls (PCBs). U.S. Department of Health and Human Services, Atlanta

ATSDR (2007) Toxicological profile for lead. Agency for Toxic Substances and Disease Registry Division of Toxicology and Environmental Medicine/Applied Toxicology Branch, Atlanta

Baldereschi M, Di Carlo A, Rocca WA, Vanni P, Maggi S, Perissinotto E, Grigoletto F, Amaducci L, Inzitari D (2000) Parkinson's disease and parkinsonism in a longitudinal study: two-fold higher incidence in men. ILSA Working Group. Italian Longitudinal Study on Aging. Neurology 55(9):1358–1363

Barbeau A (1986) At the frontiers of the brain – the neurologist and his literature. Union Med Can 115(12):884–890

Bell J, Clark AJ (1926) A pedigree of paralysis agitans. Ann Eugen 1:455–462

Berg D, Youdim MB, Riederer P (2004) Redox imbalance. Cell Tissue Res 318(1):201–213. doi:10.1007/s00441-004-0976-5

Bernheimer H, Birkmayer W, Hornykiewicz O, Jellinger K, Seitelberger F (1973) Brain dopamine and the syndromes of Parkinson and Huntington. Clinical, morphological and neurochemical correlations. J Neurol Sci 20(4):415–455

Berry MD (2004) Mammalian central nervous system trace amines. Pharmacologic amphetamines, physiologic neuromodulators. J Neurochem 90(2):257–271. doi:10.1111/j.1471-4159.2004.02501.x

Bezard E, Dovero S, Bioulac B, Gross C (1997) Effects of different schedules of MPTP administration on dopaminergic neurodegeneration in mice. Exp Neurol 128(1):288–292

Bjerregaard P, Hansen JC (2000) Organochlorines and heavy metals in pregnant women from the Disko Bay area in Greenland. Sci Total Environ 245(1–3):195–202. doi:S0048-9697(99)00444-1 [pii]

Blossom SJ, Cooney CA, Melnyk SB, Rau JL, Swearingen CJ, Wessinger WD (2013) Metabolic changes and DNA hypomethylation in cerebellum are associated with behavioral alterations in mice exposed to trichloroethylene postnatally. Toxicol Appl Pharmacol 269(3):263–269. doi:10.1016/j.taap.2013.03.025, S0041-008X(13)00132-4 [pii]

Boeve BF, Silber MH, Parisi JE, Dickson DW, Ferman TJ, Benarroch EE, Schmeichel AM, Smith GE, Petersen RC, Ahlskog JE, Matsumoto JY, Knopman DS, Schenck CH, Mahowald MW (2003) Synucleinopathy pathology and REM sleep behavior disorder plus dementia or parkinsonism. Neurology 61(1):40–45

Bogen KT, Colston BW Jr, Machicao LK (1992) Dermal absorption of dilute aqueous chloroform, trichloroethylene, and tetrachloroethylene in hairless guinea pigs. Fundam Appl Toxicol 18(1):30–39

Bogen KT, Keating GA, Meissner S, Vogel JS (1998) Initial uptake kinetics in human skin exposed to dilute aqueous trichloroethylene in vitro. J Expo Anal Environ Epidemiol 8(2):253–271

Bonifati V (2007) LRRK2 low-penetrance mutations (Gly2019Ser) and risk alleles (Gly2385Arg)-linking familial and sporadic Parkinson's disease. Neurochem Res 32(10):1700–1708. doi:10.1007/s11064-007-9324-y

Bonifati V, Rizzu P, van Baren MJ, Schaap O, Breedveld GJ, Krieger E, Dekker MC, Squitieri F, Ibanez P, Joosse M, van Dongen JW, Vanacore N, van Swieten JC, Brice A, Meco G, van Duijn CM, Oostra BA, Heutink P (2003) Mutations in the DJ-1 gene associated with autosomal recessive early-onset parkinsonism. Science 299(5604):256–259. doi:10.1126/science.1077209, 1077209 [pii]

Bonnet AM, Jutras MF, Czernecki V, Corvol JC, Vidailhet M (2012) Nonmotor symptoms in Parkinson's disease in 2012: relevant clinical aspects. Parkinsons Dis 2012:198316. doi:10.1155/2012/198316

Braak H, Del Tredici K, Rub U, de Vos RA, Jansen Steur EN, Braak E (2003) Staging of brain pathology related to sporadic Parkinson's disease. Neurobiol Aging 24(2):197–211. doi:S0197458002000659 [pii]

Bringmann G, God R, Feineis D, Wesemann W, Riederer P, Rausch WD, Reichmann H, Sontag KH (1995) The TaClo concept: 1-trichloromethyl-1,2,3,4-tetrahydro-beta-carboline (TaClo), a new toxin for dopaminergic neurons. J Neural Transm Suppl 46:235–244

Bringmann G, Friedrich H, Birner G, Koob M, Sontag KH, Heim C, Kolasiewicz W, Fahr S, Stablein M, God R, Feineis D (1996) Endogenous alkaloids in man. XXVI. Determination of the dopaminergic neurotoxin 1-trichloromethyl-1,2,3,4-tetrahydro-beta-carboline (TaClo) in biological samples using gas chromatography with selected ion monitoring. J Chromatogr B Biomed Appl 687(2):337–348

Bringmann G, Feineis D, God R, Peters K, Peters EM, Scholz J, Riederer F, Moser A (2002) 1-Trichloromethyl-1,2,3,4-tetrahydro-beta-carboline (TaClo) and related derivatives: chemistry and biochemical effects on catecholamine biosynthesis. Bioorg Med Chem 10(7): 2207–2214

Bringmann G, Feineis D, Bruckner R, God R, Grote C, Wesemann W (2006) Synthesis of radiolabelled 1-trichloromethyl-1,2,3,4-tetrahydro-beta-carboline (TaClo), a neurotoxic chloral-derived mammalian alkaloid, and its biodistribution in rats. Eur J Pharm Sci 28(5): 412–422. doi:10.1016/j.ejps.2006.03.009, S0928-0987(06)00104-7 [pii]

Bruning T, Vamvakas S, Makropoulos V, Birner G (1998) Acute intoxication with trichloroethene: clinical symptoms, toxicokinetics, metabolism, and development of biochemical parameters for renal damage. Toxicol Sci 41(2):157–165. doi:10.1006/toxs.1997.2401, S1096-6080(97)92401-X [pii]

Cannon JR, Tapias V, Na HM, Honick AS, Drolet RE, Greenamyre JT (2009) A highly reproducible rotenone model of Parkinson's disease. Neurobiol Dis 34(2):279–290

Carlsson A, Fornstedt B (1991) Possible mechanisms underlying the special vulnerability of dopaminergic neurons. Acta Neurol Scand Suppl 136:16–18

Carpenter DO (2006) Polychlorinated biphenyls (PCBs): routes of exposure and effects on human health. Rev Environ Health 21(1):1–23

Cavallarin N, Vicario M, Negro A (2010) The role of phosphorylation in synucleinopathies: focus on Parkinson's disease. CNS Neurol Disord Drug Targets 9(4):471–481

Cavallo D, Ursini CL, Setini A, Chianese C, Piegari P, Perniconi B, Iavicoli S (2003) Evaluation of oxidative damage and inhibition of DNA repair in an in vitro study of nickel exposure. Toxicolo In Vitro 17(5–6):603–607

Centers for Disease Control and Prevention (2009) Fourth report on human exposure to environmental chemicals. Atlanta, GA: U.S. Department of Health and Human Services, Centers for Disease Control and Prevention

Census US (2011) Projections of the population by selected age groups and sex for the United States: 2010 to 2050: national population projections (released 2008). US Census Bureau (Population Division), US Department of Commerce, Washington, DC

Chade AR, Kasten M, Tanner CM (2006) Nongenetic causes of Parkinson's disease. J Neural Transm Suppl 70:147–151

Charcot JM (1892) Lectures on the diseases of the nervous system. In: Delivered at La Salpetriere, Paris: Bureaux do Progres Medical, 1877. New Sydenham Society, London

Chaudhuri KR, Healy DG, Schapira AH (2006) Non-motor symptoms of Parkinson's disease: diagnosis and management. Lancet Neurol 5(3):235–245. doi:10.1016/S1474-4422(06)70373-8, S1474-4422(06)70373-8 [pii]

Chen JJ (2010) Parkinson's disease: health-related quality of life, economic cost, and implications of early treatment. Am J Manag Care 16(Suppl Implications):S87–S93. doi:12606 [pii]

Chen RC, Chang SF, Su CL, Chen TH, Yen MF, Wu HM, Chen ZY, Liou HH (2001) Prevalence, incidence, and mortality of PD: a door-to-door survey in Ilan county, Taiwan. Neurology 57(9):1679–1686

Chiu A, Beaubier J, Chiu J, Chan L, Gerstenberger S (2004) Epidemiologic studies of PCB congener profiles in North American fish consuming populations. J Environ Sci Health C Environ Carcinog Ecotoxicol Rev 22(1):13–36. doi:10.1081/GNC-120038004

Chu Y, Dodiya H, Aebischer P, Olanow CW, Kordower JH (2009) Alterations in lysosomal and proteasomal markers in Parkinson's disease: relationship to alpha-synuclein inclusions. Neurobiol Dis 35(3):385–398. doi:10.1016/j.nbd.2009.05.023

Collins LM, Toulouse A, Connor TJ, Nolan YM (2012) Contributions of central and systemic inflammation to the pathophysiology of Parkinson's disease. Neuropharmacology 62(7): 2154–2168. doi:10.1016/j.neuropharm.2012.01.028, S0028-3908(12)00048-2 [pii]

Coon S, Stark A, Peterson E, Gloi A, Kortsha G, Pounds J, Chettle D, Gorell J (2006) Whole-body lifetime occupational lead exposure and risk of Parkinson's disease. Environ Health Perspect 114(12):1872–1876

Correia Guedes L, Ferreira JJ, Rosa MM, Coelho M, Bonifati V, Sampaio C (2010) Worldwide frequency of G2019S LRRK2 mutation in Parkinson's disease: a systematic review. Parkinsonism Relat Disord 16(4):237–242. doi:10.1016/j.parkreldis.2009.11.004

Corrigan FM, Murray L, Wyatt CL, Shore RF (1998) Diorthosubstituted polychlorinated biphenyls in caudate nucleus in Parkinson's disease. Exp Neurol 150(2):339–342. doi:10.1006/exnr.1998.6776, S0014-4886(98)96776-0 [pii]

Corrigan FM, Wienburg CL, Shore RF, Daniel SE, Mann D (2000) Organochlorine insecticides in substantia nigra in Parkinson's disease. J Toxicol Environ Health A 59(4):229–234

Corti O, Lesage S, Brice A (2011) What genetics tells us about the causes and mechanisms of Parkinson's disease. Physiol Rev 91(4):1161–1218. doi:10.1152/physrev.00022.2010

Cui M, Aras R, Christian WV, Rappold PM, Hatwar M, Panza J, Jackson-Lewis V, Javitch JA, Ballatori N, Przedborski S, Tieu K (2009) The organic cation transporter-3 is a pivotal modulator of neurodegeneration in the nigrostriatal dopaminergic pathway. Proc Natl Acad Sci U S A 106(19):8043–8048. doi:10.1073/pnas.0900358106, 0900358106 [pii]

Dehay B, Bove J, Rodriguez-Muela N, Perier C, Recasens A, Boya P, Vila M (2010) Pathogenic lysosomal depletion in Parkinson's disease. J Neurosci 30(37):12535–12544. doi:10.1523/JNEUROSCI.1920-10.2010

Devos D, Kroumova M, Bordet R, Vodougnon H, Guieu JD, Libersa C, Destee A (2003) Heart rate variability and Parkinson's disease severity. J Neural Transm 110(9):997–1011. doi:10.1007/s00702-003-0016-8

Dexter DT, Carter CJ, Wells FR, Javoy-Agid F, Agid Y, Lees A, Jenner P, Marsden CD (1989) Basal lipid peroxidation in substantia nigra is increased in Parkinson's disease. J Neurochem 52(2):381–389

Dexter DT, Carayon A, Javoy-Agid F, Agid Y, Wells FR, Daniel SE, Lees AJ, Jenner P, Marsden CD (1991) Alterations in the levels of iron, ferritin and other trace metals in Parkinson's disease and other neurodegenerative diseases affecting the basal ganglia. Brain 114(Pt 4):1953–1975

Di Monte DA (1991) Mitochondrial DNA and Parkinson's disease. Neurology 41(5 Suppl 2):38–42; discussion 42–33

Dick FD, De Palma G, Ahmadi A, Scott NW, Prescott GJ, Bennett J, Semple S, Dick S, Counsell C, Mozzoni P, Haites N, Wettinger SB, Mutti A, Otelea M, Seaton A, Soderkvist P, Felice A, Geoparkinson Study Group (2007a) Environmental risk factors for Parkinson's disease and parkinsonism: the Geoparkinson study. Occup Environ Med 64(10):666–672. doi:10.1136/oem.2006.027003, oem.2006.027003 [pii]

Dick S, Semple S, Dick F, Seaton A (2007b) Occupational titles as risk factors for Parkinson's disease. Occup Med (Lond) 57(1):50–56. doi:10.1093/occmed/kql109, kql109 [pii]

Dinis-Oliveira RJ, Remiao F, Carmo H, Duarte JA, Navarro AS, Bastos ML, Carvalho F (2006) Paraquat exposure as an etiological factor of Parkinson's disease. Neurotoxicology 27(6): 1110–1122. doi:10.1016/j.neuro.2006.05.012, S0161-813X(06)00119-7 [pii]

Drolet RE, Cannon JR, Montero L, Greenamyre JT (2009) Chronic rotenone exposure reproduces Parkinson's disease gastrointestinal neuropathology. Neurobiol Dis 36(1):96–102. doi:10.1016/j.nbd.2009.06.017, S0969-9961(09)00168-5 [pii]

Elbaz A, Clavel J, Rathouz PJ, Moisan F, Galanaud JP, Delemotte B, Alperovitch A, Tzourio C (2009) Professional exposure to pesticides and Parkinson disease. Ann Neurol 66(4):494–504. doi:10.1002/ana.21717

EPA (2009) Iris toxicological review of trichloroethylene. EPA, Washington, DC

EPA United States Environmental Protection Agency (EPA). Toxic Release Inventory Program. http://www.epa.gov/tri/index.htm

EPA, Office of Compliance (1995) Profile of the Dry Cleaning Industry. EPA Office of Compliance Sector Notebook Project. EPA, Washington, DC

Factor SA, Sanchez-Ramos J, Weiner WJ (1988) Trauma as an etiology of parkinsonism: a historical review of the concept. Mov Disord 3(1):30–36. doi:10.1002/mds.870030105

Fahn S, Cohen G (1992) The oxidant stress hypothesis in Parkinson's disease: evidence supporting it. Ann Neurol 32(6):804–812. doi:10.1002/ana.410320616

Fall PA, Fredrikson M, Axelson O, Granerus AK (1999) Nutritional and occupational factors influencing the risk of Parkinson's disease: a case–control study in southeastern Sweden. Mov Disord 14(1):28–37

Farina M, Rocha JB, Aschner M (2011) Mechanisms of methylmercury-induced neurotoxicity: evidence from experimental studies. Life Sci 89(15–16):555–563. doi:10.1016/j.lfs.2011.05.019, S0024-3205(11)00265-7 [pii]

Feldman AL, Johansson AL, Nise G, Gatz M, Pedersen NL, Wirdefeldt K (2011) Occupational exposure in Parkinsonian disorders: a 43-year prospective cohort study in men. Parkinsonism Relat Disord 17(9):677–682. doi:10.1016/j.parkreldis.2011.06.009, S1353-8020(11)00185-4 [pii]

Ferroni C, Selis L, Mutti A, Folli D, Bergamaschi E, Franchini I (1992) Neurobehavioral and neuroendocrine effects of occupational exposure to perchloroethylene. Neurotoxicology 13(1):243–247

Firestone JA, Lundin JI, Powers KM, Smith-Weller T, Franklin GM, Swanson PD, Longstreth WT Jr, Checkoway H (2010) Occupational factors and risk of Parkinson's disease: a population-based case–control study. Am J Ind Med 53(3):217–223. doi:10.1002/ajim.20788

Freundt EC, Maynard N, Clancy EK, Roy S, Bousset L, Sourigues Y, Covert M, Melki R, Kirkegaard K, Brahic M (2012) Neuron-to-neuron transmission of alpha-synuclein fibrils through axonal transport. Ann Neurol 72(4):517–524. doi:10.1002/ana.23747

Gaenslen A, Gasser T, Berg D (2008) Nutrition and the risk for Parkinson's disease: review of the literature. J Neural Transm 115(5):703–713. doi:10.1007/s00702-007-0005-4

Gao X, Chen H, Fung TT, Logroscino G, Schwarzschild MA, Hu FB, Ascherio A (2007) Prospective study of dietary pattern and risk of Parkinson disease. Am J Clin Nutr 86(5):1486–1494

Gao X, Chen H, Schwarzschild MA, Ascherio A (2011) Use of ibuprofen and risk of Parkinson disease. Neurology 76(10):863–869. doi:10.1212/WNL.0b013e31820f2d79

Gao X, Cassidy A, Schwarzschild MA, Rimm EB, Ascherio A (2012) Habitual intake of dietary flavonoids and risk of Parkinson disease. Neurology 78(15):1138–1145. doi:10.1212/WNL.0b013e31824f7fc4, WNL.0b013e31824f7fc4 [pii]

Gash DM, Rutland K, Hudson NL, Sullivan PG, Bing G, Cass WA, Pandya JD, Liu M, Choi DY, Hunter RL, Gerhardt GA, Smith CD, Slevin JT, Prince TS (2008) Trichloroethylene: parkinsonism and complex 1 mitochondrial neurotoxicity. Ann Neurol 63(2):184–192. doi:10.1002/ana.21288

Gelb DJ, Oliver E, Gilman S (1999) Diagnostic criteria for Parkinson disease. Arch Neurol 56(1):33–39

Gerhard A, Pavese N, Hotton G, Turkheimer F, Es M, Hammers A, Eggert K, Oertel W, Banati RB, Brooks DJ (2006) In vivo imaging of microglial activation with [11C](R)-PK11195 PET in idiopathic Parkinson's disease. Neurobiol Dis 21(2):404–412. doi:10.1016/j.nbd.2005.08.002

Goetz CG (1986) Charcot on Parkinson's disease. Mov Disord 1(1):27–32. doi:10.1002/mds.870010104

Goldman SM (2010) Trichloroethylene and Parkinson's disease: dissolving the puzzle. Expert Rev Neurother 10(6):835–837. doi:10.1586/ern.10.61

Goldman SM, Tanner CM, Olanow CW, Watts RL, Field RD, Langston JW (2005) Occupation and parkinsonism in three movement disorders clinics. Neurology 65(9):1430–1435. doi:10.1212/01.wnl.0000180361.74060.70, 01.wnl.0000180361.74060.70 [pii]

Goldman SM, Tanner CM, Oakes D, Bhudhikanok GS, Gupta A, Langston JW (2006) Head injury and Parkinson's disease risk in twins. Ann Neurol 60(1):65–72. doi:10.1002/ana.20882

Goldman SM, Quinlan PJ, Ross GW, Marras C, Meng C, Bhudhikanok GS, Comyns K, Korell M, Chade AR, Kasten M, Priestley B, Chou KL, Fernandez HH, Cambi F, Langston JW, Tanner CM (2012a) Solvent exposures and Parkinson disease risk in twins. Ann Neurol 71(6): 776–784. doi:10.1002/ana.22629

Goldman SM, Kamel F, Ross GW, Jewell SA, Bhudhikanok GS, Umbach D, Marras C, Hauser RA, Jankovic J, Factor SA, Bressman S, Lyons KE, Meng C, Korell M, Roucoux DF, Hoppin JA, Sandler DP, Langston JW, Tanner CM (2012b) Head injury, alpha-synuclein Rep1, and Parkinson's disease. Ann Neurol 71(1):40–48. doi:10.1002/ana.22499

Goldwurm S, Di Fonzo A, Simons EJ, Rohe CF, Zini M, Canesi M, Tesei S, Zecchinelli A, Antonini A, Mariani C, Meucci N, Sacilotto G, Sironi F, Salani G, Ferreira J, Chien HF, Fabrizio E, Vanacore N, Dalla Libera A, Stocchi F, Diroma C, Lamberti P, Sampaio C, Meco G, Barbosa E, Bertoli-Avella AM, Breedveld GJ, Oostra BA, Pezzoli G, Bonifati V (2005) The G6055A (G2019S) mutation in LRRK2 is frequent in both early and late onset Parkinson's disease and originates from a common ancestor. J Med Genet 42(11):e65. doi:10.1136/jmg.2005.035568

Gorell JM, Johnson CC, Rybicki BA, Peterson EL, Kortsha GX, Brown GG, Richardson RJ (1999) Occupational exposure to manganese, copper, lead, iron, mercury and zinc and the risk of Parkinson's disease. Neurotoxicology 20(2–3):239–247

Gourie-Devi M, Ramu MG, Venkataram BS (1991) Treatment of Parkinson's disease in 'Ayurveda' (ancient Indian system of medicine): discussion paper. J R Soc Med 84(8):491–492

Gowers WR (1900) Diseases of the nerves and spinal cord. In: A manual of diseases of the nervous system, vol 1. Blakiston's Son, Philadelphia

Greenamyre JT, MacKenzie G, Peng TI, Stephans SE (1999) Mitochondrial dysfunction in Parkinson's disease. Biochem Soc Symp 66:85–97

Greene JG, Noorian AR, Srinivasan S (2009) Delayed gastric emptying and enteric nervous system dysfunction in the rotenone model of Parkinson's disease. Exp Neurol 218(1):154–161. doi:10.1016/j.expneurol.2009.04.023, S0014-4886(09)00158-7 [pii]

Griffiths PD, Crossman AR (1993) Distribution of iron in the basal ganglia and neocortex in postmortem tissue in Parkinson's disease and Alzheimer's disease. Dementia 4(2):61–65

Grote C, Clement HW, Wesemann W, Bringmann G, Feineis D, Riederer P, Sontag KH (1995) Biochemical lesions of the nigrostriatal system by TaClo (1-trichloromethyl-1,2,3,4-tetrahydro-beta-carboline) and derivatives. J Neural Transm Suppl 46:275–281

Guehl D, Bezard E, Dovero S, Boraud T, Bioulac B, Gross C (1999) Trichloroethylene and parkinsonism: a human and experimental observation. Eur J Neurol 6(5):609–611. doi:NE060513 [pii]

Guggenheim MA, Couch JR, Weinberg W (1971) Motor dysfunction as a permanent complication of methanol ingestion. Presentation of a case with a beneficial response to levodopa treatment. Arch Neurol 24(6):550–554

Guzman JN, Sanchez-Padilla J, Wokosin D, Kondapalli J, Ilijic E, Schumacker PT, Surmeier DJ (2010) Oxidant stress evoked by pacemaking in dopaminergic neurons is attenuated by DJ-1. Nature 468(7324):696–700. doi:10.1038/nature09536, nature09536 [pii]

Halliday GM, Ophof A, Broe M, Jensen PH, Kettle E, Fedorow H, Cartwright MI, Griffiths FM, Shepherd CE, Double KL (2005) Alpha-synuclein redistributes to neuromelanin lipid in the substantia nigra early in Parkinson's disease. Brain 128(Pt 11):2654–2664. doi:10.1093/brain/awh584

Hancock DB, Martin ER, Mayhew GM, Stajich JM, Jewett R, Stacy MA, Scott BL, Vance JM, Scott WK (2008) Pesticide exposure and risk of Parkinson's disease: a family-based case–control study. BMC Neurol 8:6. doi:10.1186/1471-2377-8-6, 1471-2377-8-6 [pii]

Hardell S, Tilander H, Welfinger-Smith G, Burger J, Carpenter DO (2010) Levels of polychlorinated biphenyls (PCBs) and three organochlorine pesticides in fish from the Aleutian Islands of Alaska. PLoS One 5(8):e12396. doi:10.1371/journal.pone.0012396

Heikkila RE, Nicklas WJ, Vyas I, Duvoisin RC (1985) Dopaminergic toxicity of rotenone and the 1-methyl-4-phenylpyridinium ion after their stereotaxic administration to rats: implication for the mechanism of 1-methyl-4-phenyl-1,2,3,6-tetrahydropyridine toxicity. Neurosci Lett 62(3):389–394. doi:0304-3940(85)90580-4 [pii]

Heim C, Sontag KH (1997) The halogenated tetrahydro-beta-carboline "TaClo": a progressively-acting neurotoxin. J Neural Transm Suppl 50:107–111

Hernan MA, Takkouche B, Caamano-Isorna F, Gestal-Otero JJ (2002) A meta-analysis of coffee drinking, cigarette smoking, and the risk of Parkinson's disease. Ann Neurol 52(3):276–284. doi:10.1002/ana.10277

Hertzman C, Wiens M, Bowering D, Snow B, Calne D (1990) Parkinson's disease: a case–control study of occupational and environmental risk factors. Am J Ind Med 17(3):349–355

Hertzman C, Wiens M, Snow B, Kelly S, Calne D (1994) A case–control study of Parkinson's disease in a horticultural region of British Columbia. Mov Disord 9(1):69–75. doi:10.1002/mds.870090111

Huang XP, O'Brien PJ, Templeton DM (2006) Mitochondrial involvement in genetically determined transition metal toxicity I. Iron toxicity. Chem Biol Interact 163(1–2):68–76. doi:10.1016/j.cbi.2006.05.007, S0009-2797(06)00113-X [pii]

Huang X, Chen H, Miller WC, Mailman RB, Woodard JL, Chen PC, Xiang D, Murrow RW, Wang YZ, Poole C (2007) Lower low-density lipoprotein cholesterol levels are associated with Parkinson's disease. Mov Disord 22(3):377–381. doi:10.1002/mds.21290

Hughes AJ, Daniel SE, Kilford L, Lees AJ (1992) Accuracy of clinical diagnosis of idiopathic Parkinson's disease: a clinico-pathological study of 100 cases. J Neurol Neurosurg Psychiatry 55(3):181–184

Hulihan MM, Ishihara-Paul L, Kachergus J, Warren L, Amouri R, Elango R, Prinjha RK, Upmanyu R, Kefi M, Zouari M, Sassi SB, Yahmed SB, El Euch-Fayeche G, Matthews PM, Middleton LT, Gibson RA, Hentati F, Farrer MJ (2008) LRRK2 Gly2019Ser penetrance in Arab-Berber patients from Tunisia: a case–control genetic study. Lancet Neurol 7(7):591–594. doi:10.1016/S1474-4422(08)70116-9, S1474-4422(08)70116-9 [pii]

Huse DM, Schulman K, Orsini L, Castelli-Haley J, Kennedy S, Lenhart G (2005) Burden of illness in Parkinson's disease. Mov Disord 20(11):1449–1454. doi:10.1002/mds.20609

IARC (International Agency for Research on Cancer) (1995) Dry cleaning, some chlorinated solvents and other industrial chemicals. Summary of data reported and evaluation. IARC monographs on the evaluation of carcinogenic risks to humans, vol 63. WHO, Lyon

Iranzo A, Molinuevo JL, Santamaria J, Serradell M, Marti MJ, Valldeoriola F, Tolosa E (2006) Rapid-eye-movement sleep behaviour disorder as an early marker for a neurodegenerative disorder: a descriptive study. Lancet Neurol 5(7):572–577. doi:10.1016/S1474-4422(06)70476-8, S1474-4422(06)70476-8 [pii]

Jablon S, Neel JV, Gershowitz H, Atkinson GF (1967) The NAS-NRC twin panel: methods of construction of the panel, zygosity diagnosis, and proposed use. Am J Hum Genet 19(2):133–161

Janetzky B, God R, Bringmann G, Reichmann H (1995) 1-Trichloromethyl-1,2,3,4-tetrahydro-beta-carboline, a new inhibitor of complex I. J Neural Transm Suppl 46:265–273

Jenner P (2003) Oxidative stress in Parkinson's disease. Ann Neurol 53(Suppl 3):S26–S36. doi:doi:10.1002/ana.10483; discussion S36–28

Kallio M, Haapaniemi T, Turkka J, Suominen K, Tolonen U, Sotaniemi K, Heikkila VP, Myllyla V (2000) Heart rate variability in patients with untreated Parkinson's disease. Eur J Neurol 7(6):667–672. doi:ene127 [pii]

Kamel F, Tanner C, Umbach D, Hoppin J, Alavanja M, Blair A, Comyns K, Goldman S, Korell M, Langston J, Ross G, Sandler D (2007) Pesticide exposure and self-reported Parkinson's disease in the agricultural health study. Am J Epidemiol 165(4):364–374. doi:10.1093/aje/kwk024, kwk024 [pii]

Kanthasamy AG, Kitazawa M, Kanthasamy A, Anantharam V (2005) Dieldrin-induced neurotoxicity: relevance to Parkinson's disease pathogenesis. Neurotoxicology 26(4):701–719. doi:10.1016/j.neuro.2004.07.010, S0161-813X(04)00113-5 [pii]

Kay DM, Factor SA, Samii A, Higgins DS, Griffith A, Roberts JW, Leis BC, Nutt JG, Montimurro JS, Keefe RG, Atkins AJ, Yearout D, Zabetian CP, Payami H (2008) Genetic association between alpha-synuclein and idiopathic Parkinson's disease. Am J Med Genet B Neuropsychiatr Genet 147B(7):1222–1230. doi:10.1002/ajmg.b.30758

Keppel Hesselink JM (1989) Trauma as an etiology of parkinsonism: opinions in the nineteenth century. Mov Disord 4(3):283–285. doi:10.1002/mds.870040312

Kim JH, Ryu SJ, Kim BG, Jhun HJ, Park JT, Kim HJ (2008) A case of trichloroethylene intoxication with neuropsychiatric symptoms. Korean J Occup Environ Med 20(1):54–61

Kitada T, Asakawa S, Hattori N, Matsumine H, Yamamura Y, Minoshima S, Yokochi M, Mizuno Y, Shimizu N (1998) Mutations in the parkin gene cause autosomal recessive juvenile parkinsonism. Nature 392(6676):605–608. doi:10.1038/33416

Klawans HL, Stein RW, Tanner CM, Goetz CG (1982) A pure parkinsonian syndrome following acute carbon monoxide intoxication. Arch Neurol 39(5):302–304

Klein C, Ziegler A (2011) From GWAS to clinical utility in Parkinson's disease. Lancet 377(9766):613–614. doi:10.1016/S0140-6736(11)60062-7

Kochen W, Kohlmuller D, De Biasi P, Ramsay R (2003) The endogeneous formation of highly chlorinated tetrahydro-beta-carbolines as a possible causative mechanism in idiopathic Parkinson's disease. Adv Exp Med Biol 527:253–263

Koller WC, Langston JW, Hubble JP, Irwin I, Zack M, Golbe L, Forno L, Ellenberg J, Kurland L, Ruttenber AJ et al (1991) Does a long preclinical period occur in Parkinson's disease? Neurology 41(5 Suppl 2):8–13

Kowal SL, Dall TM, Chakrabarti R, Storm MV, Jain A (2013) The current and projected economic burden of Parkinson's disease in the United States. Mov Disord. doi:10.1002/mds.25292

Kuopio AM, Marttila RJ, Helenius H, Rinne UK (1999) Environmental risk factors in Parkinson's disease. Mov Disord 14(6):928–939

Kuter K, Nowak P, Golembiowska K, Ossowska K (2010) Increased reactive oxygen species production in the brain after repeated low-dose pesticide paraquat exposure in rats. A comparison with peripheral tissues. Neurochem Res 35(8):1121–1130. doi:10.1007/s11064-010-0163-x

Kyrozis A, Ghika A, Stathopoulos P, Vassilopoulos D, Trichopoulos D, Trichopoulou A (2013) Dietary and lifestyle variables in relation to incidence of Parkinson's disease in Greece. Eur J Epidemiol 28(1):67–77. doi:10.1007/s10654-012-9760-0

Lam HR, Ostergaard G, Ladefoged O (1995) Three weeks' and six months' exposure to aromatic white spirit affect synaptosomal neurochemistry in rats. Toxicol Lett 80(1–3):39–48

Lang AE (2011) A critical appraisal of the premotor symptoms of Parkinson's disease: potential usefulness in early diagnosis and design of neuroprotective trials. Mov Disord 26(5):775–783. doi:10.1002/mds.23609

Langston JW (2006) The Parkinson's complex: parkinsonism is just the tip of the iceberg. Ann Neurol 59(4):591–596. doi:10.1002/ana.20834

Langston JW, Ballard P, Tetrud JW, Irwin I (1983) Chronic Parkinsonism in humans due to a product of meperidine-analog synthesis. Science 219(4587):979–980

Langston JW, Forno LS, Rebert CS, Irwin I (1984) Selective nigral toxicity after systemic administration of 1-methyl-4-phenyl-1,2,5,6-tetrahydropyrine (MPTP) in the squirrel monkey. Brain Res 292(2):390–394

Lash LH, Parker JC (2001) Hepatic and renal toxicities associated with perchloroethylene. Pharmacol Rev 53(2):177–208

Lash LH, Fisher JW, Lipscomb JC, Parker JC (2000) Metabolism of trichloroethylene. Environ Health Perspect 108(Suppl 2):177–200. doi:sc271_5_1835 [pii]

Lashuel HA, Overk CR, Oueslati A, Masliah E (2013) The many faces of alpha-synuclein: from structure and toxicity to therapeutic target. Nat Rev Neurosci 14(1):38–48. doi:10.1038/nrn3406

Lee DW, Notter SA, Thiruchelvam M, Dever DP, Fitzpatrick R, Kostyniak PJ, Cory-Slechta DA, Opanashuk LA (2012) Subchronic polychlorinated biphenyl (aroclor 1254) exposure produces oxidative damage and neuronal death of ventral midbrain dopaminergic systems. Toxicol Sci 125(2):496–508. doi:10.1093/toxsci/kfr313, kfr313 [pii]

Leonard SS, Harris GK, Shi X (2004) Metal-induced oxidative stress and signal transduction. Free Radic Biol Med 37(12):1921–1942. doi:10.1016/j.freeradbiomed.2004.09.010

Linnertz C, Saucier L, Ge D, Cronin KD, Burke JR, Browndyke JN, Hulette CM, Welsh-Bohmer KA, Chiba-Falek O (2009) Genetic regulation of alpha-synuclein mRNA expression in various human brain tissues. PLoS One 4(10):e7480. doi:10.1371/journal.pone.0007480

Liou HH, Tsai MC, Chen CJ, Jeng JS, Chang YC, Chen SY, Chen RC (1997) Environmental risk factors and Parkinson's disease: a case–control study in Taiwan. Neurology 48(6): 1583–1588

Liu YT, Jin C, Chen Z, Cai SX, Yin SN, Li GL, Watanabe T, Nakatsuka H, Seiji K, Inoue O et al (1988) Increased subjective symptom prevalence among workers exposed to trichloroethylene at sub-OEL levels. Tohoku J Exp Med 155(2):183–195

Liu M, Choi DY, Hunter RL, Pandya JD, Cass WA, Sullivan PG, Kim HC, Gash DM, Bing G (2010) Trichloroethylene induces dopaminergic neurodegeneration in Fisher 344 rats. J Neurochem 112(3):773–783. doi:10.1111/j.1471-4159.2009.06497.x, JNC6497 [pii]

Lloyd DR, Phillips DH (1999) Oxidative DNA damage mediated by copper(II), iron(II) and nickel(II) fenton reactions: evidence for site-specific mechanisms in the formation of double-strand breaks, 8-hydroxydeoxyguanosine and putative intrastrand cross-links. Mutat Res 424(1–2):23–36

Lock EA, Zhang J, Checkoway H (2013) Solvents and Parkinson disease: a systematic review of toxicological and epidemiological evidence. Toxicol Appl Pharmacol 266(3):345–355. doi:10.1016/j.taap.2012.11.016, S0041-008X(12)00497-8 [pii]

Logroscino G, Gao X, Chen H, Wing A, Ascherio A (2008) Dietary iron intake and risk of Parkinson's disease. Am J Epidemiol 168(12):1381–1388. doi:10.1093/aje/kwn273, kwn273 [pii]

Luk KC, Kehm VM, Zhang B, O'Brien P, Trojanowski JQ, Lee VM (2012) Intracerebral inoculation of pathological alpha-synuclein initiates a rapidly progressive neurodegenerative alpha-synucleinopathy in mice. J Exp Med 209(5):975–986. doi:10.1084/jem.20112457

Mak SK, McCormack AL, Manning-Bog AB, Cuervo AM, Di Monte DA (2010) Lysosomal degradation of alpha-synuclein in vivo. J Biol Chem 285(18):13621–13629. doi:10.1074/jbc.M109.074617

Manyam BV (1990) Paralysis agitans and levodopa in "Ayurveda": ancient Indian medical treatise. Mov Disord 5(1):47–48. doi:10.1002/mds.870050112

Maraganore DM, de Andrade M, Elbaz A, Farrer MJ, Ioannidis JP, Kruger R, Rocca WA, Schneider NK, Lesnick TG, Lincoln SJ, Hulihan MM, Aasly JO, Ashizawa T, Chartier-Harlin MC, Checkoway H, Ferrarese C, Hadjigeorgiou G, Hattori N, Kawakami H, Lambert JC, Lynch T, Mellick GD, Papapetropoulos S, Parsian A, Quattrone A, Riess O, Tan EK, Van Broeckhoven C (2006) Collaborative analysis of alpha-synuclein gene promoter variability and Parkinson disease. JAMA 296(6):661–670. doi:10.1001/jama.296.6.661, 296/6/661 [pii]

Marras C, Gruneir A, Rochon P, Wang X, Anderson G, Brotchie J, Bell CM, Fox S, Austin PC (2012) Dihydropyridine calcium channel blockers and the progression of parkinsonism. Ann Neurol 71(3):362–369. doi:10.1002/ana.22616

Martin WR (2009) Quantitative estimation of regional brain iron with magnetic resonance imaging. Parkinsonism Relat Disord 15(Suppl 3):S215–S218. doi:10.1016/S1353-8020(09)70818-1, S1353-8020(09)70818-1 [pii]

Martin I, Dawson VL, Dawson TM (2011) Recent advances in the genetics of Parkinson's disease. Annu Rev Genomics Hum Genet 12:301–325. doi:10.1146/annurev-genom-082410-101440

Martinez TN, Greenamyre JT (2012) Toxin models of mitochondrial dysfunction in Parkinson's disease. Antioxid Redox Signal 16(9):920–934. doi:10.1089/ars.2011.4033

Mata IF, Shi M, Agarwal P, Chung KA, Edwards KL, Factor SA, Galasko DR, Ginghina C, Griffith A, Higgins DS, Kay DM, Kim H, Leverenz JB, Quinn JF, Roberts JW, Samii A, Snapinn KW, Tsuang DW, Yearout D, Zhang J, Payami H, Zabetian CP (2010) SNCA variant associated with

Parkinson disease and plasma alpha-synuclein level. Arch Neurol 67(11):1350–1356. doi:10.1001/archneurol.2010.279

Mattia CJ, Ali SF, Bondy SC (1993) Toluene-induced oxidative stress in several brain regions and other organs. Mol Chem Neuropathol 18(3):313–328

McCormack AL, Thiruchelvam M, Manning-Bog AB, Thiffault C, Langston JW, Cory-Slechta DA, Di Monte DA (2002) Environmental risk factors and Parkinson's disease: selective degeneration of nigral dopaminergic neurons caused by the herbicide paraquat. Neurobiol Dis 10(2):119–127. doi:S0969996102905073 [pii]

McCrank E, Rabheru K (1989) Four cases of progressive supranuclear palsy in patients exposed to organic solvents. Can J Psychiatry 34(9):934–936

McCunney RJ (1988) Diverse manifestations of trichloroethylene. Br J Ind Med 45(2):122–126

McDonnell L, Maginnis C, Lewis S, Pickering N, Antoniak M, Hubbard R, Lawson I, Britton J (2003) Occupational exposure to solvents and metals and Parkinson's disease. Neurology 61(5):716–717

Mehta R, Templeton DM, O'Brien PJ (2006) Mitochondrial involvement in genetically determined transition metal toxicity II. Copper toxicity. Chem Biol Interact 163(1–2):77–85. doi:10.1016/j.cbi.2006.05.011, S0009-2797(06)00134-7 [pii]

Melamed E, Lavy S (1977) Parkinsonism associated with chronic inhalation of carbon tetrachloride. Lancet 1(8019):1015. doi:S0140-6736(77)92325-X [pii]

Miyake Y, Tanaka K, Fukushima W, Sasaki S, Kiyohara C, Tsuboi Y, Yamada T, Oeda T, Miki T, Kawamura N, Sakae N, Fukuyama H, Hirota Y, Nagai M (2011) Dietary intake of metals and risk of Parkinson's disease: a case–control study in Japan. J Neurol Sci 306(1–2):98–102. doi:10.1016/j.jns.2011.03.035, S0022-510X(11)00165-1 [pii]

Miyamoto T, Miyamoto M, Suzuki K, Nishibayashi M, Iwanami M, Hirata K (2008) 123I-MIBG cardiac scintigraphy provides clues to the underlying neurodegenerative disorder in idiopathic REM sleep behavior disorder. Sleep 31(5):717–723

Mjones H (1949) Paralysis agitans. A clinical genetic study. Acta Psychiatr Neurol 25(Suppl 54):1–195

Mortimer JA, Borenstein AR, Nelson LM (2012) Associations of welding and manganese exposure with Parkinson disease: review and meta-analysis. Neurology 79(11):1174–1180. doi:10.1212/WNL.0b013e3182698ced, 79/11/1174 [pii]

Murata K, Inoue O, Akutsu M, Iwata T (2010) Neuromotor effects of short-term and long-term exposures to trichloroethylene in workers. Am J Ind Med 53(9):915–921. doi:10.1002/ajim.20850

Mutti A, Franchini I (1987) Toxicity of metabolites to dopaminergic systems and the behavioural effects of organic solvents. Br J Ind Med 44(11):721–723

National Center for Environmental Assessment, EPA (2001) Sources, emission and exposure for trichloroethylene (TCE) and related chemicals. EPA, Washington, DC

Ngim CH, Devathasan G (1989) Epidemiologic study on the association between body burden mercury level and idiopathic Parkinson's disease. Neuroepidemiology 8(3):128–141

NIOSH, Division of Criteria Documentation and Standards Development (1976) Criteria for occupational exposure to tetrachloroethylene. NIOSH criteria for a recommended standard. NIOSH, CDC, HHS, Washington, DC

NIOSH, Division of Criteria Documentation and Standards Development (1978) Special occupational hazard review of trichloroethylene. Special occupational hazard review with control recommendations. NIOSH, CDC, DHHS, Rockville

Ohlson CG, Hogstedt C (1981) Parkinson's disease and occupational exposure to organic solvents, agricultural chemicals and mercury–a case-referent study. Scand J Work Environ Health 7(4):252–256. doi:2549 [pii]

Okubadejo NU, Bower JH, Rocca WA, Maraganore DM (2006) Parkinson's disease in Africa: a systematic review of epidemiologic and genetic studies. Mov Disord 21(12):2150–2156. doi:10.1002/mds.21153

Orenstein SJ, Kuo SH, Tasset I, Arias E, Koga H, Fernandez-Carasa I, Cortes E, Honig LS, Dauer W, Consiglio A, Raya A, Sulzer D, Cuervo AM (2013) Interplay of LRRK2 with chaperone-mediated autophagy. Nat Neurosci 16(4):394–406. doi:10.1038/nn.3350

Ozelius LJ, Senthil G, Saunders-Pullman R, Ohmann E, Deligtisch A, Tagliati M, Hunt AL, Klein C, Henick B, Hailpern SM, Lipton RB, Soto-Valencia J, Risch N, Bressman SB (2006) LRRK2 G2019S as a cause of Parkinson's disease in Ashkenazi Jews. N Engl J Med 354(4):424–425. doi:10.1056/NEJMc055509, 354/4/424 [pii]

Pals P, Van Everbroeck B, Grubben B, Viaene MK, Dom R, van der Linden C, Santens P, Martin JJ, Cras P (2003) Case–control study of environmental risk factors for Parkinson's disease in Belgium. Eur J Epidemiol 18(12):1133–1142

Pan-Montojo F, Schwarz M, Winkler C, Arnhold M, O'Sullivan GA, Pal A, Said J, Marsico G, Verbavatz JM, Rodrigo-Angulo M, Gille G, Funk RH, Reichmann H (2010a) Environmental toxins trigger PD-like progression via increased alpha-synuclein release from enteric neurons in mice. Sci Rep 2:898. doi:10.1038/srep00898

Pan-Montojo F, Anichtchik O, Dening Y, Knels L, Pursche S, Jung R, Jackson S, Gille G, Spillantini MG, Reichmann H, Funk RH (2010b) Progression of Parkinson's disease pathology is reproduced by intragastric administration of rotenone in mice. PLoS One 5(1):e8762. doi:10.1371/journal.pone.0008762

Park RM, Schulte PA, Bowman JD, Walker JT, Bondy SC, Yost MG, Touchstone JA, Dosemeci M (2005) Potential occupational risks for neurodegenerative diseases. Am J Ind Med 48(1):63–77. doi:10.1002/ajim.20178

Park HJ, Lee PH, Ahn YW, Choi YJ, Lee G, Lee DY, Chung ES, Jin BK (2007) Neuroprotective effect of nicotine on dopaminergic neurons by anti-inflammatory action. Eur J Neurosci 26(1):79–89. doi:10.1111/j.1460-9568.2007.05636.x, EJN5636 [pii]

Parkinson J (2002) An essay on the shaking palsy. 1817. J Neuropsychiatry Clin Neurosci 14(2):223–236; discussion 222

Parkinson's Disease Foundation (2013) Statistics on Parkinson's disease. http://www.pdf.org/en/parkinson_statistics

Perier C, Vila M (2012) Mitochondrial biology and Parkinson's disease. Cold Spring Harb Perspect Med 2(2):a009332. doi:10.1101/cshperspect.a009332, a009332 [pii]

Petersen MS, Halling J, Bech S, Wermuth L, Weihe P, Nielsen F, Jorgensen PJ, Budtz-Jorgensen E, Grandjean P (2008) Impact of dietary exposure to food contaminants on the risk of Parkinson's disease. Neurotoxicology 29(4):584–590. doi:10.1016/j.neuro.2008.03.001, S0161-813X(08)00040-5 [pii]

Petrovitch H, Ross GW, Abbott RD, Sanderson WT, Sharp DS, Tanner CM, Masaki KH, Blanchette PL, Popper JS, Foley D, Launer L, White LR (2002) Plantation work and risk of Parkinson disease in a population-based longitudinal study. Arch Neurol 59(11):1787–1792. doi:noc20055 [pii]

Pezzoli G, Barbieri S, Ferrante C, Zecchinelli A, Foa V (1989) Parkinsonism due to n-hexane exposure. Lancet 2(8667):874. doi:S0140-6736(89)93050-X [pii]

Pezzoli G, Canesi M, Antonini A, Righini A, Perbellini L, Barichella M, Mariani CB, Tenconi F, Tesei S, Zecchinelli A, Leenders KL (2000) Hydrocarbon exposure and Parkinson's disease. Neurology 55(5):667–673

Polymeropoulos MH, Lavedan C, Leroy E, Ide SE, Dehejia A, Dutra A, Pike B, Root H, Rubenstein J, Boyer R, Stenroos ES, Chandrasekharappa S, Athanassiadou A, Papapetropoulos T, Johnson WG, Lazzarini AM, Duvoisin RC, Di Iorio G, Golbe LI, Nussbaum RL (1997) Mutation in the alpha-synuclein gene identified in families with Parkinson's disease. Science 276(5321):2045–2047

Postuma RB, Gagnon JF, Vendette M, Fantini ML, Massicotte-Marquez J, Montplaisir J (2009) Quantifying the risk of neurodegenerative disease in idiopathic REM sleep behavior disorder. Neurology 72(15):1296–1300. doi:10.1212/01.wnl.0000340980.19702.6e, 01.wnl.0000340980.19702.6e [pii]

Postuma RB, Aarsland D, Barone P, Burn DJ, Hawkes CH, Oertel W, Ziemssen T (2012) Identifying prodromal Parkinson's disease: pre-motor disorders in Parkinson's disease. Mov Disord 27(5):617–626. doi:10.1002/mds.24996

Priyadarshi A, Khuder SA, Schaub EA, Shrivastava S (2000) A meta-analysis of Parkinson's disease and exposure to pesticides. Neurotoxicology 21(4):435–440

Przedborski S, Jackson-Lewis V (1998) Mechanisms of MPTP toxicity. Mov Disord 13(Suppl 1):35–38
Rajput AH, Offord KP, Beard CM, Kurland LT (1984) Epidemiology of parkinsonism: incidence, classification, and mortality. Ann Neurol 16(3):278–282. doi:10.1002/ana.410160303
Rango M, Canesi M, Ghione I, Farabola M, Righini A, Bresolin N, Antonini A, Pezzoli G (2006) Parkinson's disease, chronic hydrocarbon exposure and striatal neuronal damage: a 1-H MRS study. Neurotoxicology 27(2):164–168. doi:10.1016/j.neuro.2005.08.006, S0161-813X(05)00138-5 [pii]
Rausch WD, Abdel-mohsen M, Koutsilieri E, Chan WW, Bringmann G (1995) Studies of the potentially endogenous toxin TaClo (1-trichloromethyl-1,2,3,4-tetrahydro-beta-carboline) in neuronal and glial cell cultures. J Neural Transm Suppl 46:255–263
Richardson JR, Miller GW (2004) Acute exposure to aroclor 1016 or 1260 differentially affects dopamine transporter and vesicular monoamine transporter 2 levels. Toxicol Lett 148(1–2):29–40. doi:10.1016/j.toxlet.2003.12.006, S0378427403004636 [pii]
Richardson JR, Shalat SL, Buckley B, Winnik B, O''Suilleabhain P, Diaz-Arrastia R, Reisch J, German DC (2009) Elevated serum pesticide levels and risk of Parkinson disease. Arch Neurol 66(7):870–875. doi:10.1001/archneurol.2009.89, 66/7/870 [pii]
Richardson JR, Roy A, Shalat SL, Buckley B, Winnik B, Gearing M, Levey AI, Factor SA, O''Suilleabhain P, German DC (2011) beta-Hexachlorocyclohexane levels in serum and risk of Parkinson's disease. Neurotoxicology 32(5):640–645. doi:10.1016/j.neuro.2011.04.002, S0161-813X(11)00074-X [pii]
Riederer P, Foley P, Bringmann G, Feineis D, Bruckner R, Gerlach M (2002) Biochemical and pharmacological characterization of 1-trichloromethyl-1,2,3,4-tetrahydro-beta-carboline: a biologically relevant neurotoxin? Eur J Pharmacol 442(1–2):1–16. doi:S0014299902013080 [pii]
Ritz B, Ascherio A, Checkoway H, Marder KS, Nelson LM, Rocca WA, Ross GW, Strickland D, Van Den Eeden SK, Gorell J (2007) Pooled analysis of tobacco use and risk of Parkinson disease. Arch Neurol 64(7):990–997. doi:10.1001/archneur.64.7.990, 64/7/990 [pii]
Rocca WA, Bower JH, McDonnell SK, Peterson BJ, Maraganore DM (2001) Time trends in the incidence of parkinsonism in Olmsted County, Minnesota. Neurology 57(3):462–467
Rodier J (1955) Manganese poisoning in Moroccan miners. Br J Ind Med 12(1):21–35
Rosner F, Muntner S (1970) The medical aphorisms of Moses Maimonides, vol I and II. Yeshiva University Press, New York
Ross GW, Petrovitch H (2001) Current evidence for neuroprotective effects of nicotine and caffeine against Parkinson's disease. Drugs Aging 18(11):797–806. doi:181101 [pii]
Ross GW, Petrovitch H, Abbott RD, Nelson J, Markesbery W, Davis D, Hardman J, Launer L, Masaki K, Tanner CM, White LR (2004) Parkinsonian signs and substantia nigra neuron density in decendents elders without PD. Ann Neurol 56(4):532–539. doi:10.1002/ana.20226
Ross GW, Petrovitch H, Abbott RD, Tanner CM, White LR (2006) Pre-clinical indicators of Parkinson's disease: recent findings from the Honolulu-Asia Aging Study. Mov Disord 21(Suppl 13):S2
Ross GW, Petrovitch H, Abbott RD, Tanner CM, Popper J, Masaki K, Launer L, White LR (2008a) Association of olfactory dysfunction with risk for future Parkinson's disease. Ann Neurol 63(2):167–173. doi:10.1002/ana.21291
Ross OA, Braithwaite AT, Skipper LM, Kachergus J, Hulihan MM, Middleton FA, Nishioka K, Fuchs J, Gasser T, Maraganore DM, Adler CH, Larvor L, Chartier-Harlin MC, Nilsson C, Langston JW, Gwinn K, Hattori N, Farrer MJ (2008b) Genomic investigation of alpha-synuclein multiplication and parkinsonism. Ann Neurol 63(6):743–750. doi:10.1002/ana.21380
Ross GW, Abbott RD, Petrovitch H, Tanner CM, White LR (2012) Pre-motor features of Parkinson's disease: the Honolulu-Asia Aging Study experience. Parkinsonism Relat Disord 18(Suppl 1):S199–S202. doi:10.1016/S1353-8020(11)70062-1, S1353-8020(11)70062-1 [pii]
Ryan RE, Ross SA, Drago J, Loiacono RE (2001) Dose-related neuroprotective effects of chronic nicotine in 6-hydroxydopamine treated rats, and loss of neuroprotection in alpha4 nicotinic

receptor subunit knockout mice. Br J Pharmacol 132(8):1650–1656. doi:10.1038/sj.bjp.0703989

Rybicki BA, Johnson CC, Uman J, Gorell JM (1993) Parkinson's disease mortality and the industrial use of heavy metals in Michigan. Mov Disord 8(1):87–92. doi:10.1002/mds.870080116

Samii A, Etminan M, Wiens MO, Jafari S (2009) NSAID use and the risk of Parkinson's disease: systematic review and meta-analysis of observational studies. Drugs Aging 26(9):769–779. doi:10.2165/11316780-000000000-00000

Santner A, Uversky VN (2010) Metalloproteomics and metal toxicology of alpha-synuclein. Metallomics 2(6):378–392. doi:10.1039/b926659c

Sauerbeck A, Hunter R, Bing G, Sullivan PG (2012) Traumatic brain injury and trichloroethylene exposure interact and produce functional, histological, and mitochondrial deficits. Exp Neurol 234(1):85–94. doi:10.1016/j.expneurol.2011.12.012, S0014-4886(11)00464-X [pii]

Schenck CH, Bundlie SR, Mahowald MW (1996) Delayed emergence of a parkinsonian disorder in 38 % of 29 older men initially diagnosed with idiopathic rapid eye movement sleep behaviour disorder. Neurology 46(2):388–393

Schrag A, Ben-Shlomo Y, Quinn N (2002) How valid is the clinical diagnosis of Parkinson's disease in the community? J Neurol Neurosurg Psychiatry 73(5):529–534

Schulte PA, Burnett CA, Boeniger MF, Johnson J (1996) Neurodegenerative diseases: occupational occurrence and potential risk factors, 1982 through 1991. Am J Public Health 86(9): 1281–1288

Seeber A (1989) Neurobehavioral toxicity of long-term exposure to tetrachloroethylene. Neurotoxicol Teratol 11(6):579–583

Seegal RF, Bush B, Brosch KO (1991) Sub-chronic exposure of the adult rat to Aroclor 1254 yields regionally-specific changes in central dopaminergic function. Neurotoxicology 12(1):55–65

Seegal RF, Bush B, Brosch KO (1994) Decreases in dopamine concentrations in adult, non-human primate brain persist following removal from polychlorinated biphenyls. Toxicology 86 (1–2):71–87. doi:0300-483X(94)90054-X [pii]

Seidler A, Hellenbrand W, Robra BP, Vieregge P, Nischan P, Joerg J, Oertel WH, Ulm G, Schneider E (1996) Possible environmental, occupational, and other etiologic factors for Parkinson's disease: a case–control study in Germany. Neurology 46(5):1275–1284

Semchuk KM, Love EJ, Lee RG (1993) Parkinson's disease: a test of the multifactorial etiologic hypothesis. Neurology 43(6):1173–1180

Shahabi HN, Andersson DR, Nissbrandt H (2008) Cytochrome P450 2E1 in the substantia nigra: relevance for dopaminergic neurotransmission and free radical production. Synapse 62(5):379–388. doi:10.1002/syn.20505

Sherer TB, Richardson JR, Testa CM, Seo BB, Panov AV, Yagi T, Matsuno-Yagi A, Miller GW, Greenamyre JT (2007) Mechanism of toxicity of pesticides acting at complex I: relevance to environmental etiologies of Parkinson's disease. J Neurochem 100(6):1469–1479. doi:10.1111/j.1471-4159.2006.04333.x, JNC4333 [pii]

Sian J, Dexter DT, Lees AJ, Daniel S, Agid Y, Javoy-Agid F, Jenner P, Marsden CD (1994) Alterations in glutathione levels in Parkinson's disease and other neurodegenerative disorders affecting basal ganglia. Ann Neurol 36(3):348–355. doi:10.1002/ana.410360305

Singer TP, Ramsay RR, McKeown K, Trevor A, Castagnoli NE Jr (1988) Mechanism of the neurotoxicity of 1-methyl-4-phenylpyridinium (MPP+), the toxic bioactivation product of 1-methyl-4-phenyl-1,2,3,6-tetrahydropyridine (MPTP). Toxicology 49(1):17–23

Sofi F, Cesari F, Abbate R, Gensini GF, Casini A (2008) Adherence to Mediterranean diet and health status: meta-analysis. BMJ 337:a1344. doi:10.1136/bmj.a1344

Sofic E, Riederer P, Heinsen H, Beckmann H, Reynolds GP, Hebenstreit G, Youdim MB (1988) Increased iron (III) and total iron content in post mortem substantia nigra of parkinsonian brain. J Neural Transm 74(3):199–205

Sontag KH, Heim C, Sontag TA, God R, Reichmann H, Wesemann W, Rausch WD, Riederer P, Bringmann G (1995) Long-term behavioural effects of TaClo (1-trichloromethyl-1,2,3,4-

tetrahydro-beta-carboline) after subchronic treatment in rats. J Neural Transm Suppl 46:283–289

Souza JM, Giasson BI, Chen Q, Lee VM, Ischiropoulos H (2000) Dityrosine cross-linking promotes formation of stable alpha -synuclein polymers. Implication of nitrative and oxidative stress in the pathogenesis of neurodegenerative synucleinopathies. J Biol Chem 275(24):18344–18349. doi:10.1074/jbc.M000206200

Spillantini MG, Crowther RA, Jakes R, Hasegawa M, Goedert M (1998) alpha-Synuclein in filamentous inclusions of Lewy bodies from Parkinson's disease and dementia with lewy bodies. Proc Natl Acad Sci U S A 95(11):6469–6473

Stern MB, Lang A, Poewe W (2012) Toward a redefinition of Parkinson's disease. Mov Disord 27(1):54–60. doi:10.1002/mds.24051

Stohs SJ, Bagchi D (1995) Oxidative mechanisms in the toxicity of metal ions. Free Radic Biol Med 18(2):321–336

Storch A, Hwang YI, Bringmann G, Feineis D, Ott S, Bruckner R, Schwarz J (2006) Cytotoxicity of chloral-derived beta-carbolines is not specific towards neuronal nor dopaminergic cells. J Neural Transm 113(12):1895–1901. doi:10.1007/s00702-006-0495-5

Tanner CM (2003) Is the cause of Parkinson's disease environmental or hereditary? Evidence from twin studies. Adv Neurol 91:133–142

Tanner CM, Goldman SM (1996) Epidemiology of Parkinson's disease. Neurol Clin 14(2): 317–335

Tanner CM, Ottman R, Goldman SM, Ellenberg J, Chan P, Mayeux R, Langston JW (1999) Parkinson disease in twins: an etiologic study. JAMA 281(4):341–346. doi:joc81035 [pii]

Tanner CM, Ross GW, Jewell SA, Hauser RA, Jankovic J, Factor SA, Bressman S, Deligtisch A, Marras C, Lyons KE, Bhudhikanok GS, Roucoux DF, Meng C, Abbott RD, Langston JW (2009) Occupation and risk of parkinsonism: a multicenter case–control study. Arch Neurol 66(9):1106–1113. doi:10.1001/archneurol.2009.195, 66/9/1106 [pii]

Tanner CM, Kamel F, Ross GW, Hoppin JA, Goldman SM, Korell M, Marras C, Bhudhikanok GS, Kasten M, Chade AR, Comyns K, Richards MB, Meng C, Priestley B, Fernandez HH, Cambi F, Umbach DM, Blair A, Sandler DP, Langston JW (2011) Rotenone, paraquat, and Parkinson's disease. Environ Health Perspect 119(6):866–872

Tetrud JW, Langston JW, Irwin I, Snow B (1994) Parkinsonism caused by petroleum waste ingestion. Neurology 44(6):1051–1054

Thygesen LC, Flachs EM, Hanehoj K, Kjuus H, Juel K (2011) Hospital admissions for neurological and renal diseases among dentists and dental assistants occupationally exposed to mercury. Occup Environ Med 68(12):895–901. doi:10.1136/oem.2010.064063, oem.2010.064063 [pii]

Tuchsen F, Jensen AA (2000) Agricultural work and the risk of Parkinson's disease in Denmark, 1981–1993. Scand J Work Environ Health 26(4):359–362

Uitti RJ, Snow BJ, Shinotoh H, Vingerhoets FJ, Hayward M, Hashimoto S, Richmond J, Markey SP, Markey CJ, Calne DB (1994) Parkinsonism induced by solvent abuse. Ann Neurol 35(5):616–619. doi:10.1002/ana.410350516

Uversky VN, Li J, Fink AL (2001a) Metal-triggered structural transformations, aggregation, and fibrillation of human alpha-synuclein. A possible molecular NK between Parkinson's disease and heavy metal exposure. J Biol Chem 276(47):44284–44296. doi:10.1074/jbc.M105343200

Uversky VN, Li J, Fink AL (2001b) Pesticides directly accelerate the rate of alpha-synuclein fibril formation: a possible factor in Parkinson's disease. FEBS Lett 500(3):105–108. doi:S0014-5793(01)02597-2 [pii]

Van Den Eeden SK, Tanner CM, Bernstein AL, Fross RD, Leimpeter A, Bloch DA, Nelson LM (2003) Incidence of Parkinson's disease: variation by age, gender, and race/ethnicity. Am J Epidemiol 157(11):1015–1022

van der Mark M, Brouwer M, Kromhout H, Nijssen P, Huss A, Vermeulen R (2012) Is pesticide use related to Parkinson disease? Some clues to heterogeneity in study results. Environ Health Perspect 120(3):340–347

Warner TT, Schapira AH (2003) Genetic and environmental factors in the cause of Parkinson's disease. Ann Neurol 53(Suppl 3):S16–S23. doi:10.1002/ana.10487; discussion S23–15

Watts PM, Riedl AG, Douek DC, Edwards RJ, Boobis AR, Jenner P, Marsden CD (1998) Co-localization of P450 enzymes in the rat substantia nigra with tyrosine hydroxylase. Neuroscience 86(2):511–519. doi:S0306-4522(97)00649-0 [pii]

Weisskopf MG, Weuve J, Nie H, Saint-Hilaire MH, Sudarsky L, Simon DK, Hersh B, Schwartz J, Wright RO, Hu H (2010a) Association of cumulative lead exposure with Parkinson's disease. Environ Health Perspec 118(11):1609–1613. doi:10.1289/ehp.1002339

Weisskopf MG, Knekt P, O''Reilly EJ, Lyytinen J, Reunanen A, Laden F, Altshul L, Ascherio A (2010b) Persistent organochlorine pesticides in serum and risk of Parkinson disease. Neurology 74(13):1055–1061. doi:10.1212/WNL.0b013e3181d76a93, 74/13/1055 [pii]

Weisskopf MG, Knekt P, O'Reilly EJ, Lyytinen J, Reunanen A, Laden F, Altshul L, Ascherio A (2012) Polychlorinated biphenyls in prospectively collected serum and Parkinson's disease risk. Mov Disord. doi:10.1002/mds.25217

Welfinger-Smith G, Minholz JL, Byrne S, Waghiyi V, Gologergen J, Kava J, Apatiki M, Ungott E, Miller PK, Arnason JG, Carpenter DO (2011) Organochlorine and metal contaminants in traditional foods from St. Lawrence Island, Alaska. J Toxicol Environ Health A 74(18): 1195–1214. doi:10.1080/15287394.2011.590099

Wermuth L, von Weitzel-Mudersbach P, Jeune B (2000) A two-fold difference in the age-adjusted prevalences of Parkinson's disease between the island of Als and the Faroe Islands. Eur J Neurol 7(6):655–660

Wermuth L, Bech S, Petersen MS, Joensen P, Weihe P, Grandjean P (2008) Prevalence and incidence of Parkinson's disease in the Faroe Islands. Acta Neurol Scand 118(2):126–131. doi:10.1111/j.1600-0404.2007.00991.x. doi:ANE991 [pii]

Whitton PS (2007) Inflammation as a causative factor in the aetiology of Parkinson's disease. Br J Pharmacol 150(8):963–976. doi:10.1038/sj.bjp.0707167

WHO (1997) World Health Organization – monographs on the evaluation of carcinogenic risks to humans: dry cleaning, some chlorinated solvents and other industrial chemicals. vol 63. Lyon, France: International Agency for Research on Cancer (IARC)

Willis AW, Evanoff BA, Lian M, Galarza A, Wegrzyn A, Schootman M, Racette BA (2010) Metal emissions and urban incident Parkinson disease: a community health study of Medicare beneficiaries by using geographic information systems. Am J Epidemiol 172(12):1357–1363. doi:10.1093/aje/kwq303

Wirdefeldt K, Gatz M, Schalling M, Pedersen NL (2004) No evidence for heritability of Parkinson disease in Swedish twins. Neurology 63(2):305–311

Wirdefeldt K, Gatz M, Reynolds CA, Prescott CA, Pedersen NL (2011) Heritability of Parkinson disease in Swedish twins: a longitudinal study. Neurobiol Aging 32(10):1923 e1921–e1928. doi:10.1016/j.neurobiolaging.2011.02.017, S0197-4580(11)00046-7 [pii]

Wolff MS, Fischbein A, Thornton J, Rice C, Lilis R, Selikoff IJ (1982) Body burden of polychlorinated biphenyls among persons employed in capacitor manufacturing. Int Arch Occup Environ Health 49(3–4):199–208

Wolozin B, Wang SW, Li NC, Lee A, Lee TA, Kazis LE (2007) Simvastatin is associated with a reduced incidence of dementia and Parkinson's disease. BMC Med 5:20. doi:10.1186/1741-7015-5-20

Wu C, Schaum J (2000) Exposure assessment of trichloroethylene. Environ Health Perspect 108(Suppl 2):359–363. doi:sc271_5_1835 [pii]

Yamin G, Glaser CB, Uversky VN, Fink AL (2003) Certain metals trigger fibrillation of methionine-oxidized alpha-synuclein. J Biol Chem 278(30):27630–27635. doi:10.1074/jbc. M303302200, M303302200 [pii]

Zaheer F, Slevin JT (2011) Trichloroethylene and Parkinson disease. Neurol Clin 29(3):657–665. doi:10.1016/j.ncl.2011.05.001, S0733-8619(11)00035-1 [pii]

Zhang J, Zhang Y, Wang J, Cai P, Luo C, Qian Z, Dai Y, Feng H (2010) Characterizing iron deposition in Parkinson's disease using susceptibility-weighted imaging: an in vivo MR study. Brain Res 1330:124–130. doi:10.1016/j.brainres.2010.03.036, S0006-8993(10)00580-9 [pii]

Chapter 7
Neuroimmune Effects of Developmental TCE Exposure

Sarah J. Blossom

Abstract Exposure to certain chemical, biological or physiological risk factors prior to adulthood can alter developmental processes and may in some instances enhance disease risk. This chapter will concentrate on the known effects of exposure to trichloroethylene (TCE) during gestation, lactation, and/or early life on the brain and immune system and discuss how this persistent environmental pollutant may impede immunologic and neurologic development to promote developmental pathology. Possible neuroimmune mechanisms and therapeutic interventions to circumvent the neurotoxic and adverse neurobehavioral effects of developmental TCE exposure are proposed.

Keywords Trichloroethylene • Neurotoxicity • Immunotoxicity • Oxidative stress • Developmental exposure • Locomotor behavior • $CD4^+$ T cells • Cerebellum • Hippocampus • Neuroimmune • Autoimmune-prone mice

7.1 Neurologic and Immunologic Sensitivity to Environmental Exposures During Developmental Periods

The effects of environmental toxicant exposures occurring during fetal development and early life has become an important research focus based on a fetus/child's unique exposure patterns. There is strong evidence to suggest that humans at early stages of development may be more susceptible to environmental exposures than

S.J. Blossom, PhD
Department of Pediatrics, University of Arkansas for Medical Sciences, College of Medicine, Arkansas Children's Hospital Research Institute, 13 Children's Way, Little Rock, AR 72202, USA
e-mail: blossomsarah@uams.edu

K.M. Gilbert, S.J. Blossom (eds.), *Trichloroethylene: Toxicity and Health Risks*, Molecular and Integrative Toxicology, DOI 10.1007/978-1-4471-6311-4_7,

adults. This differential sensitivity is due, in part, to the fact that key developmental processes (e.g., cellular maturation, differentiation and organ development) occur primarily during gestation and postnatally rather than during adulthood.

Humans develop in various stages spanning throughout gestation and postnatally. Human gestational development includes three general stages; peri-conception (2 weeks post-fertilization), embryogenesis (3–7 weeks post conception), and the fetal growth period (8–38 weeks gestation) (Fetal Growth and Development 2010). Postnatally, the neonatal period extends from birth to 1 month. Infancy begins at 1 month and continues to approximately 2 years of age. Childhood begins at 2 years of age and lasts until adolescence. The onset of the adolescent age is extremely variable but typically begins at around 12–13 years of age and ends with the beginning of adulthood (~18 years of age). Aging or senescence is characterized by changes in immunologic and neurological processes over time including a generalized decline in function and activity (McEwen and Morrison 2013; Wong and Goldstein 2013). Generally speaking, most major organ systems fully develop during embryogenesis. The heart, for example, is fully formed by 8 weeks gestation in humans (Bogin 1999). In contrast, the brain and immune system have extensive developmental growth periods that begin during gestation and continue postnatally well into childhood (Bayer et al. 1993; Dietert 2008). Therefore, this extended period of immunologic and neurologic development may increase the likelihood of negative effects due to toxic environmental exposures.

7.2 Developmental TCE Exposure

Most epidemiological studies of TCE toxicity have focused on adult occupational exposure since it is relatively easy to document and often involves relatively high level exposure. In humans the occupational 8 h exposure limit for TCE is 100 ppm or approximately 80 mg/kg/day (A.T.S.D.R.U.S 1995). Human exposure to TCE can occur at low levels in instances of environmental contamination. Aside from occupational exposure, the most common source of human exposure includes ingestion of contaminated drinking water (A.T.S.D.R.U.S 1995). Although TCE levels in water systems are generally monitored, TCE levels in private wells that comprise 10 % of US drinking water supply are often unknown. In addition, exposure to TCE may be elevated for people living near waste facilities where TCE is released, residents of urban or industrialized areas, or individuals using TCE-containing products.

Although adult exposure to TCE has received the most attention, human contact with TCE can occur at all stages of life. TCE and its metabolites can cross the placenta and reach the developing fetus. The United States Environmental Protection Agency has identified quantification of TCE in breast milk as a high priority need for risk assessment. Due to its lipophilic nature, TCE can accumulate in the breast milk (Pellizzari et al. 1982). It is possible that a nursing infant whose mother is exposed to the occupational exposure limit for TCE could receive greater than 80 %

of the daily limit advisable for lifetime exposure for adults (Fisher et al. 1997). In a recent study conducted in a TCE-contaminated area in Nogales, Arizona, TCE was detected in 35 % of the mothers' breast milk samples with the maximum concentration of 6 ng/ml (Beamer et al. 2012). Because TCE concentration in the breast milk was significantly correlated with the concentration in household water, TCE exposure is also a potential concern for bottle-fed infants who also ingest more water on a bodyweight basis than adults. In addition to infants, TCE exposure has been documented in school-aged children. The School Health Initiative: Environment, Learning, and Disease (SHIELD) study, studied school-age children from two inner-city schools in Minneapolis, MN. Samples obtained from the home as well as personal samples using organic vapor monitors attached to the clothes in the breathing zone of the child to detect TCE vapors reached the level of detection in approximately 7 % of subjects 6–10 years of age (Adgate et al. 2004; Sexton et al. 2005). Together these studies confirm that children are exposed to TCE at multiple levels during development.

In terms of functional consequences, studies of mothers exposed to TCE occupationally or in instances of industrial spills have documented increased adverse birth outcomes including low birth weight and cardiac defects (Forand et al. 2012). Although epidemiologic studies have typically focused on birth outcomes, other health effects not studied as extensively may manifest from maternal or early-life exposure.

7.3 Developmental Neurotoxicity of TCE

One system known to be vulnerable to environmental exposures during developmental periods is the central nervous system (CNS). While outside of the focus of this chapter, the development of the brain and its cellular components is a complex process that that extends across the lifespan. The CNS begins to develop during the early embryonic period and continues well into postnatal life. During the third trimester in humans the hippocampal region of the brain involved in learning and memory undergoes a dramatic increase in size and synaptic plasticity by the end of the second postnatal week (Dumas and Foster 1998; Dumas 2005). In the hippocampus, neuronal migration, cell proliferation, and synapse formation continue postnatally from birth through 3 years of age. The process of myelination that involves the development of cellular insulation around nerve fibers continues well into childhood (Rice and Barone 2000). Neurogenesis continues to occur throughout adulthood, albeit to a lesser degree as compared to early development (Semple et al. 2013). In humans, microglia, which are a group of monocyte-derived cells associated with immune and macrophage-like properties, colonize the brain as early as the mid-late trimester (Harry and Kraft 2012). This event corresponds to vascularization, neuronal migration, and myelination. Postnatally, microglia, as well as neuronal and glial cells, continue to disseminate and mature into all regions of the brain including cerebellum and hippocampus (Ponti et al. 2008). Taken together, the dynamic nature and cellular plasticity of the brain throughout gestational and postnatal development and beyond is well established. This unique feature undoubtedly enhances its susceptibility to environmental influences to toxicants like TCE.

TCE was once used as an anesthetic at doses of around 2,000 ppm. Consequently, significant information is available on the acute neurotoxicity of high-level TCE exposure and its metabolites on the brain. A comprehensive assessment of adult neurotoxicity with occupational exposure to TCE in humans and acute, high-level doses in rodents was reviewed in the National Academy of Sciences document and will not be repeated here (Chiu et al. 2006). As far as human populations exposed to lower levels of TCE, one study reported that environmental TCE exposure through consumption of contaminated drinking water by residents living near the TCE-contaminated Rocky Mountain Arsenal Superfund site was associated with higher mean scores for depression, lower intelligence scores, and impaired memory recall, as compared to individuals who did not ingest contaminated water (Reif et al. 2003). Overall, less is known about chronic and/or lower dose exposures on the developing neurologic system (Laslo-Baker et al. 2004; Till et al. 2001a, b). One study found that subjects who were children at the time of TCE exposure by contaminated well water had enhanced cognitive deficits over subjects exposed as adults (White et al. 1997) More recently, studies have shown that children of mothers working with TCE who were exposed both gestationally and postnatally through lactational exposure had poorer visual acuity, as well as impaired motor coordination and behaviors characterized by inattention and hyperactivity (Laslo-Baker et al. 2004; Till et al. 2001a, b).

Experimental studies of developmental TCE-induced neurotoxicity in rodents have focused on adverse effects in the hippocampal region of the brain. In two reports, selective hippocampal damage was documented in rodents exposed developmentally to ~16–32 mg/kg/day of TCE via the drinking water. Both combined prenatal and neonatal, as well as neonatal-only exposure was associated with a decrease in myelinated fibers in the CA1 region of the hippocampus at weaning age (Isaacson et al. 1990). Other studies have reported significant changes in neuronal plasticity in hippocampal slices in vitro with TCE exposure (Altmann et al. 2002; Ohta et al. 2001). Although the exact nature of TCE's mode of action in the brain is not understood, studies in our lab found that TCE-induced alterations in metabolic pathways important in the control of oxidative stress and cellular methylation represent an important feature of developmental TCE-induced neurotoxicity (Blossom et al. 2008, 2012, 2013).

7.4 TCE and Neurologic Redox Imbalance and Oxidative Stress During Development

The cellular maturational processes that occur in the brain during gestation throughout early life increase the need for cellular oxygen, which can result in enhanced free radical and reactive oxygen species (ROS) production leading to an increased sensitivity to cellular damage and oxidative stress. To compensate for this vulnerability, the brain utilizes mechanisms involving the glutathione system to restore redox balance and combat oxidative stress. The tripeptide glutathione (γ-L-glutamyl-L-cysteinylglycine) derived from the transsulfuration pathway functions as the major intracellular antioxidant against oxidative stress and plays an important role in the detoxification of reactive oxygen species (ROS) in the brain (Biswas et al. 2006; Jain et al. 1991). Additional insults such as pro-oxidant environmental exposures have the potential to enhance an already sensitive redox imbalance by decreasing the active form of glutathione (GSH) and increasing the inactive oxidized disulfide form (GSSG) leaving the cell vulnerable to oxidative damage.

Alterations in glutathione redox potential have been shown to modulate the fate of oligodendrocyte precursor cells and maturing cortical neurons in the fetus (Maffi et al. 2008; McLean et al. 2005). This suggests that altered brain redox status and increased oxidative stress resulting from pro-oxidant environmental exposures, including toxicant exposures, could hinder neural development and promote behavioral pathology. Therefore, maintenance of redox status by restorative glutathione levels in the brain is a critical protective mechanism during developmental periods where the brain is more vulnerable to oxidative stress. The clinical significance of these studies is underscored by the presence of altered redox regulation and oxidative stress biomarkers in patients with neurologic disorders including Parkinson's disease (Mythri et al. 2011) Alzheimer's disease (Butterfield et al. 2006) and autism (James et al. 2004; Sajdel-Sulkowska et al. 2011).

In an effort to determine whether TCE impairs glutathione redox imbalance and promotes oxidative stress during developmental periods, our laboratory conducted studies with the MRL+/+ strain of mice. MRL+/+ mice are "autoimmune-prone" but also develop several behavioral deficits and neuropathological changes with age and are considered to be a model of idiopathic neurological lupus (Sakic 2012; Kapadia et al. 2012; Marcinko et al. 2012). In addition, the MRL+/+ strain has been recently identified as a novel model to study hippocampal neurogenesis. MRL+/+ mice apparently display an enhanced response to pharmacologic agents that target neuroplasticity in the hippocampus over the response observed in non-autoimmune C57BL/6 mice (Balu et al. 2009; Hodes et al. 2010). Therefore, this strain of mice may represent a unique and relevant mouse model to examine the neurological impact of TCE exposure.

In the MRL+/+ mouse model, our lab demonstrated that exposure to TCE in the drinking water from birth (postnatal day 0) through early adulthood (postnatal day 42) caused decreased levels of glutathione and an increase in the reduced glutathione (GSH) to oxidized glutathione (GSSG) ratio in both hippocampus and cerebellum indicating cellular redox imbalance (Blossom et al. 2012, 2013). These metabolic changes were accompanied by alterations in the inter-related transmethylation pathway metabolites in the plasma. Figure 7.1 shows the folate-dependent interrelated methionine transmethylation and transsulfuration pathways involved in redox potential cellular methylation. Arrows in the figure demonstrate the effect of TCE (increased or decreased) on key pathway metabolites in plasma, hippocampus, and cerebellum. Also observed in cerebellum, but not hippocampus, was a global decrease in DNA methylation. This finding may implicate potential epigenetic

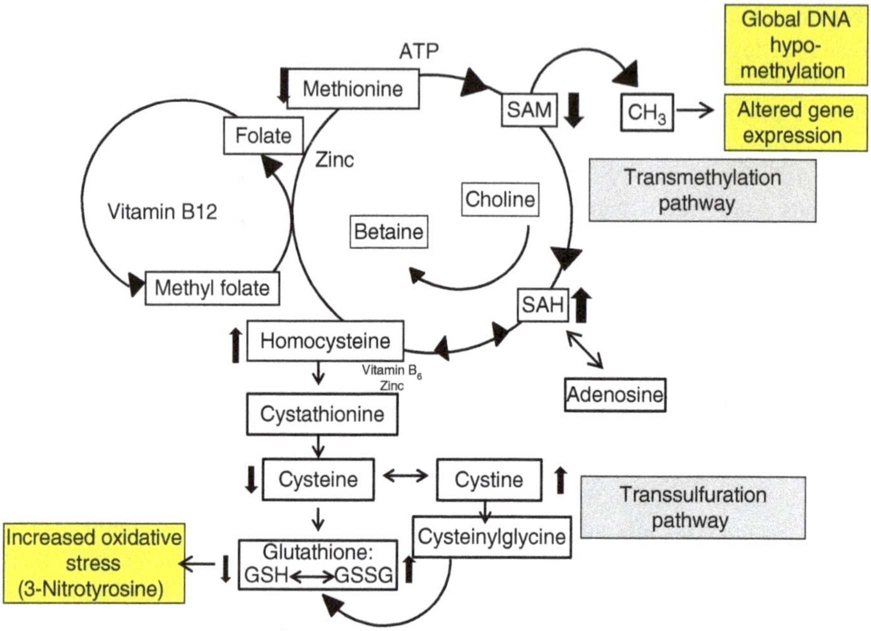

Fig. 7.1 Folate-dependent methionine transmethylation and transsulfuration pathways involved in redox potential and cellular methylation. Block arrows show the impact of postnatal exposure on the metabolite

mechanisms in TCE neurotoxicity. The decreased methionine observed with TCE exposure could indicate a decrease in methyl donors available for cellular methylation events which may have wide-ranging and long-term impacts on behavior.

7.5 Behavioral Changes Associated with Developmental TCE Neurotoxicity

Due to the observed TCE-related effects in cerebellum, a brain region functionally important for coordinating motor activity, including exploratory and social approach behaviors, we examined behavioral parameters using the EthoVision™ video tracking system from Noldus Information Technology (Leesburg, VA). MRL+/+ mice exposed to 28 mg/kg/day postnatally until 6 weeks of age showed significantly increased locomotor activity in the open-field test, as well as increased novelty/exploratory behavior in the novel object/novel mouse testing paradigm (Blossom et al. 2013). Studies by others found that unlike MRL+/+ mice, CD-1 mice exposed to much higher levels of TCE (2,000–8,000 ppm via inhalation) for 6 days in utero did not demonstrate decreased motor activity (Jones et al. 1996). However, TCE levels at this range reach doses that are associated with its anesthetic properties even though the authors did not report a decline in motor function as would be expected.

The discrepancy between the results of these studies and ours could be explained by a number of factors including route of exposure, developmental exposure period, duration of exposure, and strain differences. Thus, the presence of attention deficits and increased hyperactivity with gestational TCE exposure that has been reported in humans points to the relevance of the MRL+/+ model for studying TCE-induced developmental neurotoxicity (Laslo-Baker et al. 2004; Till et al. 2001a, b).

7.6 Effects of Early Postnatal TCE Exposure on Gene Expression in the Brain: Possible Role of Neuroprotective in the Control of Oxidative Stress

From a functional standpoint, redox imbalance, impaired methyl metabolism and epigenetic mechanisms could impact key cellular processes including gene expression in the brain. In particular, epigenetic mechanisms are important for the functional expression of neurotrophic genes (Branchi et al. 2011; Fuchikami et al. 2011; Roth et al. 2011). Changes in the expression of these genes can lead to impaired behavior (Chestnut et al. 2011; Lubin et al. 2008; Numata et al. 2012). Neurotrophic factors including Brain Derived Neurotrophic Factor (BDNF), Nerve Growth Factor (NGF), and Neurotrophin-3 (NT-3) are classically recognized as important mediators of neural growth and plasticity promoting neuronal survival and differentiation (Reichardt 2006). Emerging evidence suggests that neurotrophic factors can maintain control of inflammation in the brain by regulating glutathione redox status (Kapczinski et al. 2008; Sable et al. 2011; Wu et al. 2004). Developmental exposure to the solvent, toluene increased biomarkers of oxidative stress and decreased neurotrophic factors leading to neuroinflammation (Win-Shwe et al. 2010). Similar findings in offspring with mouse models of maternal infection have been demonstrated (Pang et al. 2010). Along this line, antioxidant therapy increased BDNF levels in hippocampus (Xu et al. 2011) and in neurodevelopmental disorders of the CNS, including autism, oxidative stress appear to be linked to the loss of neurotrophic support (Sajdel-Sulkowska et al. 2009, 2011). Thus, normal functioning of the brain appears to involve a positive feedback loop between anti-oxidant processes and neurotrophic expression and function in response to pro-oxidant exposures.

Our lab reported that hippocampal tissue from mice exposed to TCE postnatally expressed lower levels of key neurotrophic factors (e.g., BDNF, NGF, and NT-3) relative to controls confirming the experimental link between impaired redox status, increased oxidative stress with a decrease in neurotrophins observed in our model (Blossom et al. 2012). Based on these intriguing results, we extended our study to include an analysis of gene expression in cerebellum from TCE-treated MRL+/+ mice. We expanded the study to include functionally important gene families that might be impacted by TCE including chemokines/receptors, cytokines/receptors, astrocyte/microglial specific markers, and neurotrophins and receptors. Fluorescence-based quantitative real-time PCR (qRT-PCR) was conducted using methods previously described (Blossom et al. 2012). Gene expression changes in

Table 7.1 qRT-PCR was performed using hippocampus and cerebellum samples from 6 mice per treatment group collected at postnatal day 42

Gene family	Gene name	Hippocampus	Cerebellum
Chemokines	Chemokine (C-C motif) receptor 2 (*CCR2*)	NC	NC
	Chemokine (C-C motif) ligand 2 (*MCP-1*)	NC	NC
	Chemokine (C-C motif) ligand 3 (*MIP1-α*)	NC	NC
	Chemokine (C-C motif) ligand 4 (*MIP1-β*)	NC	2.7[b]
Astrocyte marker	Glial fibrillary acidic protein (*GFAP*)	−1.4[a] −1.7[b]	−2.1[b]
Microglial activation	Allograft inflammatory factor 1 (*Iba-1*)	−1.4[b]	−1.6[b]
Neurotrophins/receptors	Brain derived neurotrophic factor (*BDNF*)	−2.5[b,c]	−2.5[b]
	Neurotrophin 3 (*NTF3*)	−1.8[b,c]	NC
	Nerve growth factor (*NGF*)	−1.9[b,c]	NC
	Neurotrophic tyrosine kinase, receptor type I (*TrkA*)	NC	NC
	Neurotrophic tyrosine kinase receptor, type 2 (*TrkB*)	NC	NC
	Neurotrophic tyrosine kinase receptor, type 2 (*TrkC*)	−2.0[b]	−4.2[a] −3.5[b]

Numbers in the table represent fold change (increase or decrease) difference in gene expression NC indicates no difference in gene expression between TCE and control mice. N compared with a standardized control sample

[a]Statistically different (0.01 mg/ml TCE vs. control)

[b]Statistically different (0.1 mg/ml TCE vs. Control)

[c]Some results have been published previously

both hippocampal and cerebellar tissues from individual mice (n = 6/treatment group) were compared among mice exposed to TCE (0, 2, or 28 mg/kg/day) from postnatal day 1–42. Interestingly, ~50 % of the genes evaluated in the hippocampus were significantly down regulated, as compared with no significant change, with the highest dose of TCE treatment relative to controls (Table 7.1). Expression of glial fibrillary acidic protein (GFAP), a marker associated with astrocyte differentiation was significantly decreased in hippocampal tissue isolated from TCE exposed mice (1.4-fold and 1.7-fold; 2 and 28 mg/kg/day, respectively). It is noteworthy to mention that neurotrophins play an important role in maturation of neurons and glial cells (Abe et al. 2010). Thus, the decrease in neurotrophic factors may represent a plausible explanation for the decreased expression GFAP. Further study to address this question is necessary in order to fully understand developmental neurotoxicity of TCE.

Similarly, in the cerebellum, ~41 % of the genes examined were significantly down regulated in TCE-treated mice (higher dose) relative to control as compared with no change relative to control values. TrkC, the receptor for the neurotrophin,

NGF, was also down regulated in the low TCE exposure groups. Collectively, our data supports an inverse association between increased oxidative stress and altered methyl metabolism with decreased expression of neurotrophic genes and their receptors. A positive correlation between increased oxidative stress and expression of proinflammatory markers would be expected. Studies to explore the proinflammatory cytokines expressed by cultured and activated microglial cells in mice developmentally exposed to TCE are currently underway in our laboratory, and could provide insight and possible mechanisms concerning the role of proinflammatory cytokines in TCE-induced neurotoxicity.

Opposed to all other genes tested, the highest dose of TCE significantly *increased* expression of MIP-1β, but not other important chemokines, relative to controls in the cerebellum. MIP-1β is a chemokine that is expressed in epithelial cells important in regulating traffic of recently activated peripheral T cells across the blood brain barrier (BBB) during inflammation. The functional implication of this finding is not known, but methylmercury exposure has been shown to selectively increase expression of MIP-1β, but not other chemokines, in the cerebellum of mice (Lee et al. 2012). It is possible that TCE and methylmercury alter a common pathway that increases the production of this chemokine in the cerebellum possibly leading to impaired blood brain barrier permeability and enhanced neuroinflammation and/or oxidative stress. This mechanism has not yet been tested in our model, but may represent a plausible mechanism, together with the decrease in neuroprotective factors, leading to effects observed following developmental TCE exposure. Collectively, based on our evidence, many of these neurologic events could represent an effect downstream of TCE's ability to promote immune hyperactivity following developmental exposure as demonstrated by our lab.

7.7 Increased Susceptibility of Developing Immune System to Toxicity

The role of developmental immunotoxicity in the etiology of childhood disease is becoming an important public health concern. The immune system has several well-characterized age-specific developmental stages. The major maturational events occurring during immune system development in humans includes (1) hematopoiesis (gestational week 8–10), (2) stem cell migration and cellular expansion (gestational week 10–16), (3) colonization of the bone marrow and thymus (gestational week 16-birth), (4) maturation to imunocompetence (birth to 1 year), and (5) establishment of immunologic memory (1–13 years) (Dietert 2008).

There is increasing evidence that the developing immune system is more sensitive to toxicant exposure than the adult immune system. More severe effects tend to occur at lower doses and often persist into adult life (Dietert and Piepenbrink 2006). Examples of the more commonly studied developmental suppressive immunotoxicants that induce more severe or persistent immune effects in offspring include the heavy metals (e.g., lead), polycyclic hydrocarbons (e.g., benzo [a] pyrene) and polyhalogenated hydrocarbons (dioxin). A recent review compared early life vs. adult exposure to several immunosuppressive chemicals including lead and tributylin in animal models (Luebke et al. 2006). In all cases, sensitivity was greater if exposure occurred during development. In fact, immune suppression in developmentally exposed offspring often occurred at doses that did not alter adult immune responses.

The immune system's extended period of maturation may leave it especially vulnerable to environmental influences. Thus, in this way, the immune system is similar to the developing brain in terms of vulnerability to environmental insults. Developmental sensitivity to toxicants has also been demonstrated in humans. For example, prenatal exposure to polychlorinated biphenyls decreased the immune response to standard immunizations (Heilmann et al. 2010). Prenatal exposure to polybrominated diphenyl ethers produced a persistent decrease in lymphocyte numbers (Leijs et al. 2009). These studies focused on the ability of toxicants to promote immunologic hyporesponsiveness. Aside from immune suppression, there is increasing evidence that adult onset autoimmune disease can be triggered by pre- and early post-natal toxicant exposure (Colebatch and Edwards 2011; Langer 2010). Children continuously exposed for 3–19 years beginning *in utero* to a water supply contaminated with solvents (including TCE at levels reaching 267 ppb) had altered ratios of T cell subsets and early signs of tissue inflammation (Gist and Burg 1995). Human TCE exposure was associated with a proinflammatory IFN-γ $CD4^+$ T cell response in cord blood isolated from neonates (Lehmann et al. 2002). Thus, unlike the majority of immunotoxicants which tend to suppress the immune system, TCE promotes T cell hyperactivity and proinflammatory responses.

7.8 Immunotoxicity with Developmental TCE Exposure in MRL+/+ Mice

Our lab and others have conducted several studies concerning the immunostimulatory effects of TCE in MRL+/+ mice (Griffin et al. 2000a, b; Khan et al. 1995). Adult female MRL+/+ mice exposed to TCE (0.5 mg/ml) developed autoimmune hepatitis. This pathology was accompanied by expansion of activated ($CD62L^{lo}$) $CD4^+$ T cells that secreted increased levels of the proinflammatory cytokine, IFN-g. Based on the increased sensitivity to toxicants by the developing immune system, our lab used the MRL+/+ mouse model to examine the effects of continuous developmental and early life exposure (gestation through ~6–8 weeks of age) to a substantially lower dose of TCE.

Studying the impact of developmental exposure to different concentrations of TCE is a lengthy and complex process involving multiple breeding pairs. As a first step most likely to demonstrate efficacy, the effects of continuous (gestational throughout adulthood) TCE exposure was examined. This developmental exposure to TCE (126 mg/kg/day) calculated from maternal and direct water consumption increased the production of IFN-γ by $CD4^+$ T cells from the pups as early as

4 weeks of age (Blossom and Doss 2007). TCE exposure also impacted the thymus, the site of T cell development, as early as postnatal day 20, causing an increase in thymus cell numbers as well as an increase in the percentage of mature ($CD24^{lo}$) single-positive $CD4^+$ T cells indicating increased maturational events in the thymus. In a subsequent study, mice continuously exposed to a 5–25-fold lower, more environmentally-relevant dose of TCE showed similar thymus and $CD4^+$ T cell IFN-γ responses in 6 week old mice (Blossom et al. 2008). In addition, TCE enhanced $CD4^+$ T cell TNF-α production in these mice. TNF-α is an inflammatory cytokine secreted by activated T cells and macrophages that plays an important role in many pathological conditions including neurologic disorders. Together these findings suggest that a continuous developmental exposure alters the threshold (decreases the concentration or exposure-time) for TCE-induced T cell hyperactivity.

Other investigators reported that a continuous gestational and early-life exposure to 14,000 ppb TCE in the drinking water of *non-autoimmune* mice induced significantly increased T lymphocyte-mediated delayed-type hypersensitivity (DTH) responses, decreased antibody-mediated responses, and enhanced thymus cellularity in 8 week old mice (Peden-Adams et al. 2006). This group also reported that "life-time" exposure to TCE did not increase the level of anti-dsDNA antibodies in female MRL+/+ mice (Peden-Adams et al. 2008). Their assessment did not start until the mice were 4 months of age, however; a time point at which constitutive production of autoantibodies in untreated MRL+/+ mice can obscure a TCE-induced effect. In addition, since that study was confined to lupus-associated autoantibodies, the effects of lifetime TCE exposure on other types of disease (e.g. autoimmune hepatitis), are unknown.

7.9 Increased Susceptibility of Developing Brain to Neurotoxicity by Peripheral Immune Activation as a Mechanism for TCE's Effects in the Brain

There is emerging evidence that altered neuroimmune mechanisms might play a role in the development of certain neurologic disorders. The brain, once thought to be an immune privileged site, allows small molecules (e.g. cytokines) and lymphocyte trafficking in healthy individuals for immune surveillance during infection or immune responses to a CNS injury (Schwartz et al. 1999). This passage is tightly controlled and regulated by the blood brain barrier (BBB). The BBB provides diffusion restraint in order to control ionic gradients between blood and cerebrospinal fluid (Bito 1969). This restraint is provided by tight junctions located in the BBB interface. The BBB in the embryo, fetus, and newborn is believed to be immature and has been described as poorly formed, "leaky." or even absent (Siegenthaler et al. 2013). Thus a certain level of "cross talk" between the brain and the peripheral immune system occurs during both developmental periods and during adulthood.

During development, an emerging role for peripheral T cells in regulating normal neuronal differentiation and synaptic plasticity has been described (Ziv et al. 2006). In contrast to the positive effect of low level immune interaction in the brain, inflammatory conditions at sites outside of the CNS can lead to neurologic disorders. One of the best characterized peripheral inflammatory insults in this context is maternal and early-life infection. In humans, maternal infection has been linked to autism (Atladottir et al. 2012) attention deficit hyperactivity disorder (ADHD) (Mann and McDermott 2011) and adult-onset schizophrenia in the offspring (Anderson and Maes 2013; Khandaker et al. 2013). Several pieces of evidence in rodent models of linking maternal infection using live virus, viral mimics, the bacterial endotoxin, lipopolysaccharide, and selected inflammatory cytokines with adverse neurologic outcome in the offspring occurring later in life support this human evidence (reviewed in Meyer 2013). Mechanisms for these effects are currently being explored. However, recent evidence suggests that developmental LPS exposure alters neurotrophic factors leading neurobehavioral alterations similar to what is observed in our model (Xu et al. 2013a, b). Whether or not developmental exposure to environmental toxicants, like TCE, that promote immune hyperactivity mediate neurologic effects in a similar manner have not been examined.

7.10 Neuroimmune Impact of TCE and Implications for Neurodevelopmental Disorders

A mechanism involving the pro-inflammatory effect of TCE on the peripheral immune system during developmental periods may be an important consideration in the etiology for some neurologic disorders including autism and ADHD. One specific set of initiating or triggering events in these disorders may involve the immune system. Onore, et.al., indicated in a recent review that sufficient evidence was available to implicate altered immune responses in autism (Onore et al. 2012). There is plenty of supportive evidence of neuroinflammation and oxidative stress in the brains of autistic children involving a marked increase of the inflammatory chemokines together with reduced neurotrophic support (James et al. 2004; Sajdel-Sulkowska et al. 2009, 2011; Ashwood and Wakefield 2006). Thus, many of the characteristics observed in our mouse model of TCE exposure mirror what is observed in autism. One additional compelling link between our model and autism is the association of this disorder with autoimmunity with more than 40 % of autistic children having two or more first-degree family members with an autoimmune disease (Sweeten et al. 2003). The association between autism and parental autoimmunity was recently confirmed in a case-control study (Money et al. 1971). Serological evidence of autoimmunity in the form of anti-brain antibodies have been detected in both mothers of autistic children as well as in the children themselves (Braunschweig et al. 2013; Nordahl et al. 2013; Bauman et al. 2013; Fox et al. 2012). In terms of TCE and neurodevelopmental disorders, one epidemiologic study highlighted the possibility that maternal TCE exposure may be an environmental risk factor for autism (Windham et al. 2006). This study reported increased incidence of autism in children living in areas with the highest quartile (25 %) of TCE in air using EPA HAPS data. Although this linkage needs to be confirmed by a larger more quantitative study, it raises an intriguing possibility that developmental TCE exposure may be a risk factor for the development of autism. We reported and increased exploratory and motor activity in developmentally exposed offspring (Blossom et al. 2013). At this time, studies to address the linkage between TCE exposure and ADHD in humans have not been conducted. The possibility of this association is underscored by reports in the literature showing hyperlocomotor and increased exploratory effects with perinatal alcohol exposure (Brady et al. 2012; Schneider et al. 2011). Both alcohol and TCE share an important metabolite, acetaldehyde. The system that transforms ethanol to acetaldehyde is even more robust in the perinatal rodent, and acetaldehyde itself is capable of enhancing motor activity (March et al. 2013). Thus, the presence of attention deficits and hyperactivity in association with developmental exposure to TCE needs to be studied further.

7.11 Neuroimmune Mechanisms and Future Directions

Collectively our findings demonstrate that developmental exposure to TCE promoted increased maturation of T cells in the thymus, T cell hyperactivity, and increased production of proinflammatory cytokines in association with neurobehavioral alterations. We observed increased locomotor activity and increased novelty/exploratory behavior with TCE exposure. These effects were associated with neural alterations in metabolites in the transsulfuration and transmethylation pathways indicating redox imbalance and altered methylation capacity (Blossom et al. 2008, 2012, 2013; Blossom and Doss 2007).

The neurologic effects of TCE could be a result of a direct effect of TCE and its metabolites in the brain. One potential mechanism may involve the activity of TCE's reactive metabolite trichloroacetaldehyde hydrate (TCAH). TCE is metabolized primarily by the cytochrome P-450 s isoform CYP2E1 to a trichloroethylene oxide intermediate, which spontaneously rearranges to form TCAH. TCAH is a highly reactive aldehyde that has been proposed to spontaneously condense with the biogenic amine tryptamine to produce an alkaloid-type neurotoxin (Bringmann and Hille 1990). Our lab has extensively studied the ability of TCAH to form adducts with T cells and promote their activation in vitro and in vivo (Blossom et al. 2004, 2007). The ability of reactive aldehydes (i.e., from ethanol metabolism) to inhibit methionine synthase activity and subsequently lower glutathione has been documented (Waly et al. 2004, 2011). Decreased methionine synthase activity would therefore result in an accumulation of SAH and inhibition of SAM, and a depletion of GSH similar to what is observed in our model. Therefore it is plausible to hypothesize that TCE, via TCAH, acts in a similar manner.

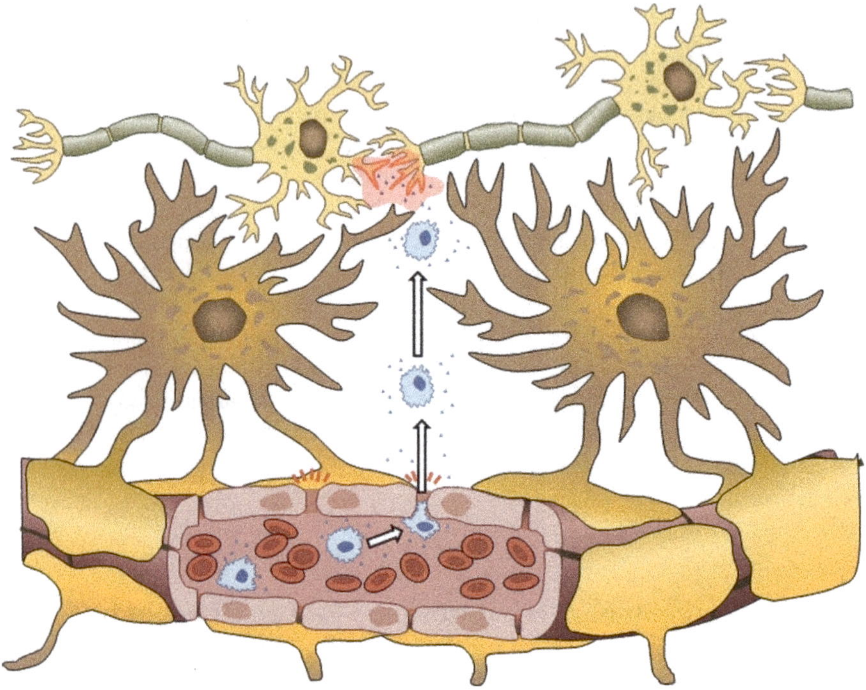

Fig. 7.2 Proposed mechanism. Activated CD4+ T cells from TCE-treated mice may cross a compromised blood brain barrier and promote inflammatory/oxidative stress which could dysregulate neuronal cells or astrocytes leading to adverse behavior. (Illustration courtesy of Mr. Dustyn A. Barnette)

One other attractive hypothesis that will be investigated further is that adverse neurologic and neurobehavioral effects may be secondary to the *early* effects of TCE on CD4+ T cells in early life following developmental exposure. We reported that TCE enhances thymic T cell maturation and $CD4^+$ T cell -oxidant activation (at postnatal day 20–28) in MRL+/+ mice. In contrast, the neurologic effects were only evident 6 weeks of age. It is therefore plausible that activated peripheral $CD4^+$ T cells and/or the cytokines they produce cross the BBB that may already be in a fragile state due to direct effects of TCE or metabolites or possibly by increased cerebellar MIP1β. The cytokines/cells cross the BBB to promote generalized inflammation and decrease the production of neurotrophins which leads to impaired redox status and methylation potential and increased oxidative stress resulting in abnormal behavior. The decrease in neurotrophic factors may also be a consequence of impaired DNA methylation based on our metabolic profile The role of peripheral T cells in adverse neurobehavior could be easily tested in $CD4^+$ T cell depleted mice. This possible scenario is depicted in Fig. 7.2.

Evidence to support our hypothesis is strengthened by emerging evidence that altered neuroimmune mechanisms might play a role in the development of certain neurologic disorders. The brain, once thought to be an immune privileged site, allows small molecules (e.g., cytokines) and $CD4^+$ T cell trafficking in healthy individuals for immune surveillance during infection or immune responses to a CNS injury (Schwartz et al. 1999). This passage is tightly controlled and regulated by the blood brain barrier (BBB). The BBB provides diffusion restraint in order to control ionic gradients between blood and cerebrospinal fluid (Bito 1969). This restraint is provided by tight junctions located in the BBB interface. The BBB in the embryo, fetus, and newborn is immature and has been described as poorly formed, leaky, or even absent (Siegenthaler et al. 2013). Thus a certain level of so-called "cross-talk" between the brain and the peripheral immune system occurs during developmental periods in particular.

Pivotal work has shown that mice deprived of mature $CD4^{+}$ T cells (but not B cells or $CD8^{+}$ T cells) manifested hippocampal-dependent cognitive defects and behavioral abnormalities that were reversed by replenishing T cells (Kipnis et al. 2012; Marin and Kipnis 2013). A later study found that at the interface between the BBB, the epithelial layers of the choroid plexus are populated with $CD4^{+}$ T cell effector memory cells with a T cell receptor repertoire specific to CNS antigens (Baruch and Schwartz 2013; Baruch et al. 2013). This type of immunological control may be lost as a normal part of aging/senescence leading to cognitive decline. As far as development, an emerging role for peripheral $CD4^{+}$ T cells in regulating normal neuronal differentiation and synaptic plasticity has been described (Ziv et al. 2006). Despite these intriguing findings, the interactions between T cells and microglia and/or neurons in the brain and what this may mean in neurodevelopmental disorders where immunological function is abnormal remains a mystery.

In contrast to the positive benefit of low-level immune interaction in the brain, peripheral inflammation has been shown to contribute to the development of neurologic disorders. One of the best characterized peripheral inflammatory insults in this context is infection. In humans, maternal infection has been linked to ASD, ADHD, and adult onset schizophrenia in the offspring (Atladottir et al. 2012; Mann and McDermott 2011; Anderson and Maes 2013). Mechanisms for these effects are currently being explored in animal models (Meyer 2013). In humans, increased peripheral T cells in the brain of Alzheimer's patients have been detected (Liu et al. 2010). Whether developmental exposure to toxicants like TCE that promote $CD4^{+}$ T cell hyperactivity and mediate neurologic and adverse behavioral effects in a similar manner have not been examined.

Additional future experiments to address therapeutic strategies could involve experiments designed to implement a dietary intervention to circumvent the neurologic effects we observe in our model. Methyl-supplemented diets are designed to provide increased amounts of cofactors and methyl donors to support methyl metabolism. The diet will most likely include B12 and folic acid; essential nutrients and cofactors for the production of methyl groups, betaine; a methyl donor to regenerate methionine, choline; an essential nutrient and precursor of betaine, zinc; a cofactor for the mouse DNA methyltransferase and other key enzymes involved in DNA methylation. The diet will provide more methionine which can (via cysteine) increase glutathione production and through effects on SAM and SAH levels affecting DNA methylation (Melnyk et al. 2011; Mosharov et al. 2000; Vitvitsky et al. 2006). Because available data do not indicate that increasing methionine levels will enhance glutathione levels sufficiently (Powell et al. 2010), N-Acetylcysteine (NAC) will be added to the special diet at previously described levels (Filosto et al. 2011; Conaway et al. 1998; Parachikova et al. 2010). NAC is a thiol anti-oxidant form of the amino acid cysteine and is used as a precursor of glutathione. These sets of experiments could potentially lead to novel therapies with real clinical value.

The literature reporting enhanced risk of neurodevelopmental disease after early-life insult to inflammatory insults is still evolving. We have used MRL+/+ mice to model these associations in the context of TCE exposure, and have demonstrated that these mice are sensitive to TCE's neuroimmune effects. Expanding this work to other strains of mice, including knockout mice, and other toxicants that may promote inflammation would truly further our understanding of how toxicant exposure and inflammation increases the risk of neurodevelopmental brain disorders.

References

A.T.S.D.R.U.S. Department of Health and Human Services: Center for Disease Control, Atlanta, Georgia. Agency for Toxic Substances and Disease Registry (1995) Toxicological profile for trichloroethylene. Update Draft for Public Comments. Ref Type: Report.

Abe M, Kimoto H, Eto R et al (2010) Postnatal development of neurons, interneurons and glial cells in the substantia nigra of mice. Cell Mol Neurobiol 30:917–928

Adgate JL, Eberly LE, Stroebel C et al (2004) Personal, indoor, and outdoor VOC exposures in a probability sample of children. J Expo Anal Environ Epidemiol 14(Suppl 1):S4–S13

Altmann L, Welge P, Mensing T et al (2002) Chronic exposure to trichloroethylene affects neuronal plasticity in rat hippocampal slices. Environ Toxicol Pharmacol 12:157–167

Anderson G, Maes M (2013) Schizophrenia: linking prenatal infection to cytokines, the tryptophan catabolite (TRYCAT) pathway, NMDA receptor hypofunction, neurodevelopment and neuroprogression. Prog Neuropsychopharmacol Biol Psychiatry 42:5–19

Ashwood P, Wakefield AJ (2006) Immune activation of peripheral blood and mucosal CD3+ lymphocyte cytokine profiles in children with autism and gastrointestinal symptoms. J Neuroimmunol 173:126–134

Atladottir HO, Henriksen TB, Schendel DE et al (2012) Autism after infection, febrile episodes, and antibiotic use during pregnancy: an exploratory study. Pediatrics 130:e1447–e1454

Balu DT, Hodes GE, Anderson BT et al (2009) Enhanced sensitivity of the MRL/MpJ mouse to the neuroplastic and behavioral effects of chronic antidepressant treatments. Neuropsychopharmacology 34:1764–1773

Baruch K, Schwartz M (2013) CNS-specific T cells shape brain function via the choroid plexus. Brain Behav Immun

Baruch K, Ron-Harel N, Gal H et al (2013) CNS-specific immunity at the choroid plexus shifts toward destructive Th2 inflammation in brain aging. Proc Natl Acad Sci U S A 110:2264–2269

Bauman MD, Iosif AM, Ashwood P et al (2013) Maternal antibodies from mothers of children with autism alter brain growth and social behavior development in the rhesus monkey. Transl Psychiatry 3:e278

Bayer SA, Altman J, Russo RJ et al (1993) Timetables of neurogenesis in the human brain based on experimentally determined patterns in the rat. Neurotoxicology 14:83–144

Beamer PI, Luik CE, Abrell L et al (2012) Correction to concentration of trichloroethylene in breast milk and household water from Nogales. Arizona Environ Sci Technol 46:11483

Biswas S, Chida AS, Rahman I (2006) Redox modifications of protein-thiols: emerging roles in cell signaling. Biochem Pharmacol 71:551–564

Bito LZ (1969) Blood-brain barrier: evidence for active cation transport between blood and the extraceliular fluid of brain. Science 165:81–83
Blossom SJ, Doss JC (2007) Trichloroethylene alters central and peripheral immune function in autoimmune-prone MRL++ mice following continuous developmental and early life exposure. J Immunotoxicol 4:129–141
Blossom SJ, Pumford NR, Gilbert KM (2004) Activation and attenuation of apoptosis of CD4(+) T cells following in vivo exposure to two common environmental toxicants, trichloroacetaldehyde hydrate and trichloroacetic acid. J Autoimmun 23:211–220
Blossom SJ, Doss JC, Gilbert KM (2007) Chronic exposure to a trichloroethylene metabolite in autoimmune-prone MRL+/+ mice promotes immune modulation and alopecia. Toxicol Sci 95:401–411
Blossom SJ, Doss JC, Hennings LJ et al (2008) Developmental exposure to trichloroethylene promotes CD4(+) T cell differentiation and hyperactivity in association with oxidative stress and neurobehavioral deficits in MRL+/+ mice. Toxicol Appl Pharmacol
Blossom SJ, Melnyk S, Cooney CA et al (2012) Postnatal exposure to trichloroethylene alters glutathione redox homeostasis, methylation potential, and neurotrophin expression in the mouse hippocampus. Neurotoxicology 33:1518–1527
Blossom SJ, Cooney CA, Melnyk SB et al (2013) Metabolic changes and DNA hypomethylation in cerebellum are associated with behavioral alterations in mice exposed to trichloroethylene postnatally. Toxicol Appl Pharmacol 269:263–269
Bogin B (ed) (1999) Patterns of human growth, 2nd edn. Cambridge University Press, Cambridge
Brady ML, Allan AM, Caldwell KK (2012) A limited access mouse model of prenatal alcohol exposure that produces long-lasting deficits in hippocampal-dependent learning and memory. Alcohol Clin Exp Res 36:457–466
Branchi I, Karpova NN, D'Andrea I et al (2011) Epigenetic modifications induced by early enrichment are associated with changes in timing of induction of BDNF expression. Neurosci Lett 495:168–172
Braunschweig D, Krakowiak P, Duncanson P et al (2013) Autism-specific maternal autoantibodies recognize critical proteins in developing brain. Transl Psychiatry 3:e277
Bringmann G, Hille A (1990) Endogenous alkaloids in man, VII: 1-trichloromethyl-1,2,3,4-tetrahydro-beta-carboline–a potential chloral-derived indol alkaloid in man. Arch Pharm (Weinheim) 323:567–569
Butterfield DA, Perluigi M, Sultana R (2006) Oxidative stress in Alzheimer's disease brain: new insights from redox proteomics. Eur J Pharmacol 545:39–50
Chestnut BA, Chang Q, Price A et al (2011) Epigenetic regulation of motor neuron cell death through DNA methylation. J Neurosci 31:16619–16636
Chiu WA, Caldwell JC, Keshava N et al (2006) Key scientific issues in the health risk assessment of trichloroethylene. Environ Health Perspect 114:1445–1449
Colebatch AN, Edwards CJ (2011) The influence of early life factors on the risk of developing rheumatoid arthritis. Clin Exp Immunol 163:11–16
Conaway CC, Jiao D, Kelloff GJ et al (1998) Chemopreventive potential of fumaric acid, N-acetylcysteine, N-(4-hydroxyphenyl) retinamide and beta-carotene for tobacco-nitrosamine-induced lung tumors in A/J mice. Cancer Lett 124:85–93
Dietert RR (2008) Developmental immunotoxicology (DIT): windows of vulnerability, immune dysfunction and safety assessment. J Immunotoxicol 5:401–412
Dietert RR, Piepenbrink MS (2006) Perinatal immunotoxicity: why adult exposure assessment fails to predict risk. Environ Health Perspect 114:477–483
Dumas TC (2005) Late postnatal maturation of excitatory synaptic transmission permits adult-like expression of hippocampal-dependent behaviors. Hippocampus 15:562–578
Dumas TC, Foster TC (1998) GABA(b) receptors differentially regulate hippocampal CA1 excitatory synaptic transmission across postnatal development in the rat. Neurosci Lett 248: 138–140
Fetal Growth and Development (2010) Williams obstetrics, vol 23. McGraw-Hill, New York

Filosto S, Castillo S, Danielson A et al (2011) Neutral sphingomyelinase 2: a novel target in cigarette smoke-induced apoptosis and lung injury. Am J Respir Cell Mol Biol 44:350–360
Fisher J, Mahle D, Bankston L et al (1997) Lactational transfer of volatile chemicals in breast milk. Am Ind Hyg Assoc J 58:425–431
Forand SP, Lewis-Michl EL, Gomez MI (2012) Adverse birth outcomes and maternal exposure to trichloroethylene and tetrachloroethylene through soil vapor intrusion in New York State. Environ Health Perspect 120:616–621
Fox E, Amaral D, Van de Water J (2012) Maternal and fetal antibrain antibodies in development and disease. Dev Neurobiol 72:1327–1334
Fuchikami M, Morinobu S, Segawa M et al (2011) DNA methylation profiles of the brain-derived neurotrophic factor (BDNF) gene as a potent diagnostic biomarker in major depression. PLoS One 6:e23881
Gist GL, Burg JR (1995) Trichloroethylene–a review of the literature from a health effects perspective. Toxicol Ind Health 11:253–307
Griffin JM, Blossom SJ, Jackson SK et al (2000a) Trichloroethylene accelerates an autoimmune response by Th1 T cell activation in MRL +/+ mice. Immunopharmacology 46:123–137
Griffin JM, Gilbert KM, Lamps LW et al (2000b) CD4(+) T-cell activation and induction of autoimmune hepatitis following trichloroethylene treatment in MRL+/+ mice. Toxicol Sci 57:345–352
Harry GJ, Kraft AD (2012) Microglia in the developing brain: a potential target with lifetime effects. Neurotoxicology 33:191–206
Heilmann C, Budtz-Jorgensen E, Nielsen F et al (2010) Serum concentrations of antibodies against vaccine toxoids in children exposed perinatally to immunotoxicants. Environ Health Perspect 118:1434–1438
Hodes GE, Hill-Smith TE, Lucki I (2010) Fluoxetine treatment induces dose dependent alterations in depression associated behavior and neural plasticity in female mice. Neurosci Lett 484:12–16
Isaacson LG, Spohler SA, Taylor DH (1990) Trichloroethylene affects learning and decreases myelin in the rat hippocampus. Neurotoxicol Teratol 12:375–381
Jain A, Martensson J, Stole E et al (1991) Glutathione deficiency leads to mitochondrial damage in brain. Proc Natl Acad Sci U S A 88:1913–1917
James SJ, Cutler P, Melnyk S et al (2004) Metabolic biomarkers of increased oxidative stress and impaired methylation capacity in children with autism. Am J Clin Nutr 80:1611–1617
Jones HE, Kunko PM, Robinson SE et al (1996) Developmental consequences of intermittent and continuous prenatal exposure to 1,1,1-trichloroethane in mice. Pharmacol Biochem Behav 55:635–646
Kapadia M, Stanojcic M, Earls AM et al (2012) Altered olfactory function in the MRL model of CNS lupus. Behav Brain Res 234:303–311
Kapczinski F, Frey BN, Andreazza AC et al (2008) Increased oxidative stress as a mechanism for decreased BDNF levels in acute manic episodes. Rev Bras Psiquiatr 30:243–245
Khan MF, Kaphalia BS, Prabhakar BS et al (1995) Trichloroethene-induced autoimmune response in female MRL +/+ mice. Toxicol Appl Pharmacol 134:155–160
Khandaker GM, Zimbron J, Lewis G et al (2013) Prenatal maternal infection, neurodevelopment and adult schizophrenia: a systematic review of population-based studies. Psychol Med 43:239–257
Kipnis J, Gadani S, Derecki NC (2012) Pro-cognitive properties of T cells. Nat Rev Immunol 12:663–669
Langer P (2010) The impacts of organochlorines and other persistent pollutants on thyroid and metabolic health. Front Neuroendocrinol 31:497–518
Laslo-Baker D, Barrera M, Knittel-Keren D et al (2004) Child neurodevelopmental outcome and maternal occupational exposure to solvents. Arch Pediatr Adolesc Med 158:956–961
Lee JY, Hwang GW, Kim MS et al (2012) Methylmercury induces a brain-specific increase in chemokine CCL4 expression in mice. J Toxicol Sci 37:1279–1282
Lehmann I, Thoelke A, Rehwagen M et al (2002) The influence of maternal exposure to volatile organic compounds on the cytokine secretion profile of neonatal T cells. Environ Toxicol 17:203–210

Leijs MM, Koppe JG, Olie K et al (2009) Effects of dioxins, PCBs, and PBDEs on immunology and hematology in adolescents. Environ Sci Technol 43:7946–7951

Liu YJ, Guo DW, Tian L et al (2010) Peripheral T cells derived from Alzheimer's disease patients overexpress CXCR2 contributing to its transendothelial migration, which is microglial TNF-alpha-dependent. Neurobiol Aging 31:175–188

Lubin FD, Roth TL, Sweatt JD (2008) Epigenetic regulation of BDNF gene transcription in the consolidation of fear memory. J Neurosci 28:10576–10586

Luebke RW, Chen DH, Dietert R et al (2006) The comparative immunotoxicity of five selected compounds following developmental or adult exposure. J Toxicol Environ Health B Crit Rev 9:1–26

Maffi SK, Rathinam ML, Cherian PP et al (2008) Glutathione content as a potential mediator of the vulnerability of cultured fetal cortical neurons to ethanol-induced apoptosis. J Neurosci Res 86:1064–1076

Mann JR, McDermott S (2011) Are maternal genitourinary infection and pre-eclampsia associated with ADHD in school-aged children? J Atten Disord 15:667–673

March SM, Cullere ME, Abate P et al (2013) Acetaldehyde reinforcement and motor reactivity in newborns with or without a prenatal history of alcohol exposure. Front Behav Neurosci 7:69

Marcinko K, Parsons T, Lerch JP et al (2012) Effects of prolonged treatment with memantine in the MRL model of CNS lupus. Clin Exp Neuroimmunol 3:116–128

Marin I, Kipnis J (2013) Learning and memory … and the immune system. Learn Mem 20: 601–606

McEwen BS, Morrison JH (2013) The brain on stress: vulnerability and plasticity of the prefrontal cortex over the life course. Neuron 79:16–29

McLean CW, Mirochnitchenko O, Claus CP et al (2005) Overexpression of glutathione peroxidase protects immature murine neurons from oxidative stress. Dev Neurosci 27:169–175

Melnyk S, Fuchs GJ, Schulz E et al (2011) Metabolic imbalance associated with methylation dysregulation and oxidative damage in children with autism. J Autism Dev Disord.

Meyer U (2013) Prenatal Poly(I:C) exposure and other developmental immune activation models in rodent systems. Biol Psychiatry

Money J, Bobrow NA, Clarke FC (1971) Autism and autoimmune disease: a family study. J Autism Child Schizophr 1:146–160

Mosharov E, Cranford MR, Banerjee R (2000) The quantitatively important relationship between homocysteine metabolism and glutathione synthesis by the transsulfuration pathway and its regulation by redox changes. Biochemistry 39:13005–13011

Mythri RB, Harish G, Dubey SK et al (2011) Glutamoyl diester of the dietary polyphenol curcumin offers improved protection against peroxynitrite-mediated nitrosative stress and damage of brain mitochondria in vitro: implications for Parkinson's disease. Mol Cell Biochem 347:135–143

Nordahl CW, Braunschweig D, Iosif AM et al (2013) Maternal autoantibodies are associated with abnormal brain enlargement in a subgroup of children with autism spectrum disorder. Brain Behav Immun 30:61–65

Numata S, Ye T, Hyde TM et al (2012) DNA methylation signatures in development and aging of the human prefrontal cortex. Am J Hum Genet 90:260–272

Ohta M, Saito T, Saito K et al (2001) Effect of trichloroethylene on spatiotemporal pattern of LTP in mouse hippocampal slices. Int J Neurosci 111:257–271

Onore C, Careaga M, Ashwood P (2012) The role of immune dysfunction in the pathophysiology of autism. Brain Behav Immun 26:383–392

Pang Y, Campbell L, Zheng B et al (2010) Lipopolysaccharide-activated microglia induce death of oligodendrocyte progenitor cells and impede their development. Neuroscience 166:464–475

Parachikova A, Green KN, Hendrix C et al (2010) Formulation of a medical food cocktail for Alzheimer's disease: beneficial effects on cognition and neuropathology in a mouse model of the disease. PLoS One 5:e14015

Peden-Adams MM, Eudaly JG, Heesemann LM et al (2006) Developmental immunotoxicity of trichloroethylene (TCE): studies in B6C3F1 mice. J Environ Sci Health A Tox Hazard Subst Environ Eng 41:249–271

Peden-Adams MM, Eudaly JG, Lee AM et al (2008) Lifetime exposure to trichloroethylene (TCE) does not accelerate autoimmune disease in MRL +/+ mice. J Environ Sci Health A Tox Hazard Subst Environ Eng 43:1402–1409

Pellizzari ED, Hartwell TD, Harris BS III et al (1982) Purgeable organic compounds in mother's milk. Bull Environ Contam Toxicol 28:322–328

Ponti G, Peretto P, Bonfanti L (2008) Genesis of neuronal and glial progenitors in the cerebellar cortex of peripuberal and adult rabbits. PLoS One 3:e2366

Powell CL, Bradford BU, Craig CP et al (2010) Mechanism for prevention of alcohol-induced liver injury by dietary methyl donors. Toxicol Sci 115:131–139

Reichardt LF (2006) Neurotrophin-regulated signalling pathways. Philos Trans R Soc Lond B Biol Sci 361:1545–1564

Reif JS, Burch JB, Nuckols JR et al (2003) Neurobehavioral effects of exposure to trichloroethylene through a municipal water supply. Environ Res 93:248–258

Rice D, Barone S Jr (2000) Critical periods of vulnerability for the developing nervous system: evidence from humans and animal models. Environ Health Perspect 108(Suppl 3):511–533

Roth TL, Zoladz PR, Sweatt JD et al (2011) Epigenetic modification of hippocampal Bdnf DNA in adult rats in an animal model of post-traumatic stress disorder. J Psychiatr Res 45:919–926

Sable P, Dangat K, Kale A et al (2011) Altered brain neurotrophins at birth: consequence of imbalance in maternal folic acid and vitamin B metabolism. Neuroscience 190:127–134

Sajdel-Sulkowska EM, Xu M, Koibuchi N (2009) Increase in cerebellar neurotrophin-3 and oxidative stress markers in Autism. Cerebellum

Sajdel-Sulkowska EM, Xu M, McGinnis W et al (2011) Brain region-specific changes in oxidative stress and neurotrophin levels in autism spectrum disorders (ASD). Cerebellum 10:43–48

Sakic B (2012) The MRL, model: an invaluable tool in studies of autoimmunity-brain interactions. Methods Mol Biol 934:277–299

Schneider ML, Moore CF, Adkins MM (2011) The effects of prenatal alcohol exposure on behavior: rodent and primate studies. Neuropsychol Rev 21:186–203

Schwartz M, Cohen I, Lazarov-Spiegler O et al (1999) The remedy may lie in ourselves: prospects for immune cell therapy in central nervous system protection and repair. J Mol Med 77: 713–717

Semple BD, Blomgren K, Gimlin K et al (2013) Brain development in rodents and humans: identifying benchmarks of maturation and vulnerability to injury across species. Prog Neurobiol 106–107:1–16

Sexton K, Adgate JL, Church TR et al (2005) Children's exposure to volatile organic compounds as determined by longitudinal measurements in blood. Environ Health Perspect 113:342–349

Siegenthaler JA, Sohet F, Daneman R (2013) 'Sealing off the CNS': cellular and molecular regulation of blood-brain barriergenesis. Curr Opin Neurobiol.

Sweeten TL, Bowyer SL, Posey DJ et al (2003) Increased prevalence of familial autoimmunity in probands with pervasive developmental disorders. Pediatrics 112:e420

Till C, Westall CA, Rovet JF et al (2001a) Effects of maternal occupational exposure to organic solvents on offspring visual functioning: a prospective controlled study. Teratology 64: 134–141

Till C, Koren G, Rovet JF (2001b) Prenatal exposure to organic solvents and child neurobehavioral performance. Neurotoxicol Teratol 23:235–245

Vitvitsky V, Thomas M, Ghorpade A et al (2006) A functional transsulfuration pathway in the brain links to glutathione homeostasis. J Biol Chem 281:35785–35793

Waly M, Olteanu H, Banerjee R et al (2004) Activation of methionine synthase by insulin-like growth factor-1 and dopamine: a target for neurodevelopmental toxins and thimerosal. Mol Psychiatry 9:358–370

Waly MI, Kharbanda KK, Deth RC (2011) Ethanol lowers glutathione in rat liver and brain and inhibits methionine synthase in a cobalamin-dependent manner. Alcohol Clin Exp Res 35:277–283

White RF, Feldman RG, Eviator II et al (1997) Hazardous waste and neurobehavioral effects: a developmental perspective. Environ Res 73:113–124

Windham GC, Zhang L, Gunier R et al (2006) Autism spectrum disorders in relation to distribution of hazardous air pollutants in the san Francisco bay area. Environ Health Perspect 114: 1438–1444

Win-Shwe TT, Tsukahara S, Yamamoto S et al (2010) Up-regulation of neurotrophin-related gene expression in mouse hippocampus following low-level toluene exposure. Neurotoxicology 31:85–93

Wong C, Goldstein DR (2013) Impact of aging on antigen presentation cell function of dendritic cells. Curr Opin Immunol 25(4):535–541

Wu A, Ying Z, Gomez-Pinilla F (2004) The interplay between oxidative stress and brain-derived neurotrophic factor modulates the outcome of a saturated fat diet on synaptic plasticity and cognition. Eur J Neurosci 19:1699–1707

Xu JX, Yang M, Deng KJ et al (2011) Antioxidant activities of Dracocephalum tanguticum maxim extract and its up-regulation on the expression of neurotrophic factors in a rat model of permanent focal cerebral ischemia. Am J Chin Med 39:65–81

Xu M, Sulkowski ZL, Parekh P et al (2013a) Effects of perinatal lipopolysaccharide (LPS) exposure on the developing rat brain; modeling the effect of maternal infection on the developing human CNS. Cerebellum 12:572–586

Xu M, Sajdel-Sulkowska EM, Iwasaki T et al (2013b) Aberrant cerebellar neurotrophin-3 expression induced by lipopolysaccharide exposure during brain development. Cerebellum 12:316–318

Ziv Y, Ron N, Butovsky O et al (2006) Immune cells contribute to the maintenance of neurogenesis and spatial learning abilities in adulthood. Nat Neurosci 9:268–275

Chapter 8
Environmental Sensitivity to Trichloroethylene (TCE) in the Developing Heart

Ornella I. Selmin, Om Makwana, and Raymond B. Runyan

Abstract Epidemiological reports first suggested a specific teratogenicity of trichloroethylene (TCE) in drinking water but due to other potential contaminants, uncertain exposure levels and selection of controls, these reports were controversial. Early animal studies explored the effects of high doses of TCE on heart development and a drinking water model was established that showed the ability of TCE to produce cardiac malformations. However, these doses were well above typical levels of environmental exposure and the significance of TCE as a cardiac teratogen remained controversial. Using molecular and functional measures of cardiac development, several laboratories began to examine the effects of TCE in animal models at relevant exposure levels. This work demonstrated a non-monotonic dose response where significantly more effects on gene expression and cardiac function were seen just above the current maximum contamination level (5 parts per billion) than at much higher dose levels. An examination of early heart valve formation showed that formation of valve progenitors was impaired. Molecular studies pointed towards changes in expression of several muscle genes

O.I. Selmin, PhD (✉)
Department of Nutritional Sciences, The University of Arizona Cancer Center, The University of Arizona, Tucson, AZ, USA

The University of Arizona Cancer Center, Tucson, AZ, USA

The Sarver Heart Center, The University of Arizona, Tucson, AZ, USA
e-mail: selmin@email.arizona.edu

O. Makwana
Department of Medicine, University of California, San Francisco, CA, USA

Veterans Affairs Medical Center, San Francisco, CA, USA

R.B. Runyan
Department of Cellular and Molecular Medicine, The University of Arizona Cancer Center, Tucson, AZ, USA

The Sarver Heart Center, The University of Arizona, Tucson, AZ, USA

K.M. Gilbert, S.J. Blossom (eds.), *Trichloroethylene: Toxicity and Health Risks*, Molecular and Integrative Toxicology, DOI 10.1007/978-1-4471-6311-4_8,

involved in calcium homeostasis and myocardial contraction. Further studies showed that calcium-mediated contraction in the heart was impaired and that this corresponded to changes in intracellular calcium flux and cardiac output. To explore the non-monotonic dose curve, expression of phase I metabolic enzymes were examined. The early heart was found to be a specific site of cytochrome p450 expression prior to the development of the liver and CYP2C expression was elevated with TCE exposure. Surprisingly, the CYP expression pattern replicated the non-monotonic dose response and did not explain it. This suggests that a metabolite of TCE is likely to be the teratogen. To explore the teratogenicity of TCE, microarray data from exposed chick hearts were analyzed by utilization of interactome analysis. This analysis identified a subset of highly-linked genes that were perturbed by TCE exposure. The most centrally linked gene in the interactome was the transcription factor HNF4a. Though not previously known in the heart, HNF4a expression was confirmed. Its level of expression was unchanged with TCE exposure but its level of phosphorylation was altered. Ongoing studies are investigating the hypothesis that HNF4a is a proximal target of TCE exposure and that misregulation of gene expression by this transcription factor is a significant mediator of congenital heart defects produced by TCE.

Keywords Embryonic heart • Epithelial mesenchymal transition • Gene expression • Cytochrome P450 • Hepatocyte nuclear factor 4 alpha

8.1 Introduction

Recent epidemiological studies reinforced earlier controversial findings indicating that TCE exposure during pregnancy increased the frequency of congenital heart disease (CHD) in newborns. Every year, approximately 1 % of all children are born with heart defects around the world. This number includes 0.6 % of newborns with severe and moderate defects with additional measures of 0.3–6.9 % of newborns with lesser defects as determined by various protocols (Hoffman and Kaplan 2002). It is estimated by the Children's Heart Foundation that inpatient surgery to repair heart defects in the US exceeds $2.2 billion/year. Estimates in TCE-exposed populations show an odds ratio of 2.5–3.0 compared to control groups (The Toxicological Review of TCE; Goldberg et al. 1990; Bove et al. 1995). That only a small proportion of these CHDs can be attributed to specific genetic defects, highlights the apparent importance of environmental factors in embryonic heart maldevelopment. The most convincing data attesting to TCE cardiotoxicity derive from mechanistic studies, conducted in chick embryos, using a wide range of doses during different phases of heart differentiation. In vitro studies unraveled likely mechanisms of action by which TCE may interfere with normal heart development. As the heart is the first functional organ in the embryo, and must continue to develop and remodel throughout embryonic and fetal life, it utilizes a number of specific cellular and molecular processes that are sensitive to TCE. Endothelial-mesenchymal transition is a cellular process that produces the progenitors of the valves and septa while a regulated calcium flux is required to mediate muscle contraction and produce blood flow. As described below, both processes are altered by TCE exposure. Although independent groups have confirmed similar findings in different species, outstanding questions to be addressed remain the bi-modal behavior of TCE dose–response, the role of P450 cytochromes in TCE metabolism in the embryonic heart, and the role of TCE metabolites in determining cardiac toxicity in mammals and avians. In this chapter we review data from epidemiologic, animal and in vitro studies supporting the notion that sensitivity of the embryonic heart to TCE exposure may be responsible for the increased number of congenital heart defects observed in contaminated areas. Lastly, we present a new hypothesis suggesting the role of a nuclear transcription factor in mediating, at least in part, the negative effects of TCE in the developing heart.

8.2 Epidemiological Studies

Very few epidemiological studies analyzed the effects of maternal TCE exposure on the incidence of CHD in children and these show mixed results (The Toxicological Review of TCE). Reports of cardio-teratogenicity first came from human studies (Goldberg et al. 1990) that were later verified by Bove et al. (1995). The spectrum of defects seen in children exposed to up to 270 parts per billion (ppb) TCE in drinking water *in utero* included aortic stenosis, valvular, septal and muscular defects and the odds ratio (OR) for all heart defects was between 3.0 (Goldberg et al. 1990) and 2.5 (Bove et al. 1995). In a retrospective study, looking at ambient air exposure to TCE, Yauck et al. (2004) reported an increase in heart defects among mothers ≥38 years of age living within 1.33 miles of a TCE-emitting site (OR = 3.2; 95 %). More recently, Forand et al. (2012) reported that both birth and cardiac birth defects approximately doubled in populated areas of Endicott, New York that were found to be contaminated with TCE and perchloroethylene (PCE). When the analysis was limited to the conotruncal defects, these, although rare, were nearly five times more prevalent in both the PCE and TCE exposed areas. A few community-based studies have looked at the association between TCE exposure and incidence of birth defects other than cardiac malformations. Lagakos et al. (1986), using a health survey of ~5,000 residents of Walburn (MA), observed an increased rate of eye anomalies (OR = 14.9) and perinatal deaths (OR = 10) associated with exposure to TCE-contaminated water wells. Other studies using the same populations were not conclusive, although a retrospective analysis looking at hospital records reported increased OR for many birth defects including congenital heart diseases (NAS Report 2006).

Studies performed at the US Marine Corps Base at Camp Lejeune, NC, where residents had been exposed to TCE, DCE and PCE for as long as 30 years, found that TCE exposure was associated to smaller male, but not female infants (Agency for Toxic Substances and Disease Registry 1998; Sonnenfeld et al. 2001). Risk was higher for women older than 35 and those with a history of fetal loss. Limitations of these studies included misclassification of the groups (i.e. exposed vs nonexposed, and long term vs short term exposure) lack of information about water consumption, dermal and inhalation exposure, tobacco and alcohol usage. Taken altogether, data collected from available studies provide robust indication that environmentally relevant doses of TCE exposure through drinking water increased the risk of intrauterine growth reduction. These findings are similar to those reported for maternal alcohol consumption (alcohol has a metabolism similar to TCE), and are confirmed by observations gathered from animal studies (Johnson et al. 1998a, b; Fisher et al. 2001; Smith et al. 1992, 1989). In summary, epidemiological studies were complicated by several factors including chemical co-exposure, difficulty in establishing the doses and duration of TCE exposure, small number of cases, aggregation of a wide range of malformations, and the unknown role of genetic variations in determining sensitivity to TCE. However, considered as a whole, these studies indicate a relatively consistent elevation for cardiac defects in the range of 2.5–3.0-fold (Chiu et al. 2013), and other birth defects including eye malformations and low birth weight.

8.3 In Vivo Studies

While the timing varies between species, the developmental processes of heart development are consistent. The embryonic heart develops as a hollow tube consistent of two cellular layers, the inner endothelium and the exterior myocardium, separated by extracellular matrix (Fig. 8.1a–c). The heart tube then loops as part of a developmental process to bring the inflow and outflow portions of the heart to their adult orientation. During the looping process, there is a signal derived from the myocardium that initiates valve formation by endothelial cells lining the atrioventricular canal (Runyan and Markwald 1983). A subset of endothelial cells in this region loses cell-cell adhesions and migrates into the adjacent extracellular matrix, becoming mesenchymal cells. This process, called epithelial-mesenchymal transition (EMT) produces the progenitors of the valves and membranous septa of the mature heart. Factors altering this process of valve and septa formation would affect the size and shape of the mitral and tricuspid valves and/or produce septal defects between the right and left sides of the heart. Valve or septal defects would, in turn, impair normal hemodynamics and produce muscular malformations. Additional developmental processes include muscular protrusions that divide the atria and ventricles into right and left sides. Fusion between the intraventricular muscle tissue and the fused cardiac cushions normally closes the septum between the two ventricles and separates the function of the right and left chambers. The atrial muscular septum is comprised of two layers with offset openings that permit blood to pass from the right to the left side of the heart during fetal life. Changes in blood pressure

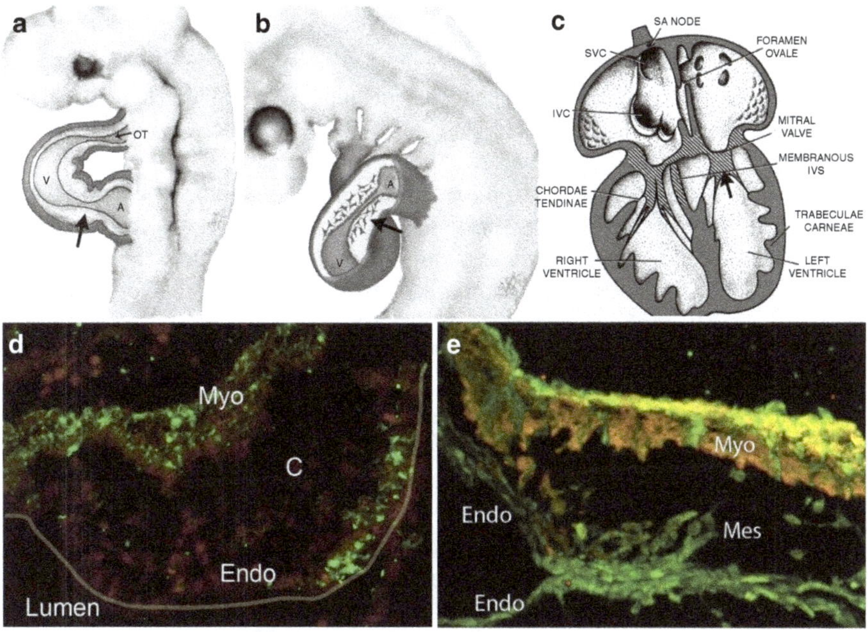

Fig. 8.1 Heart Development and Gene Expression. (**a**) The early embryonic heart is seen at this stage as a single tube E2 (HH14)- chick, E9.25-mouse, and E20- human. The outer layer is the myocardium and the inner layer is endothelium. The intervening space is filled with extracellular matrix and protrudes into the lumen in the canal between the single atrium and the single ventricle (*arrow*) as paired cardiac cushions. (**b**) By HH17 in the chick and E9.5 in the mouse the atrium has continued to rotate and shift towards the head. Endothelial cells lining the AV canal have begun to undergo an epithelial-mesenchymal transition to form the progenitors of the valve fibroblasts within the cushions (*arrow*). (**c**) Later in development, the four-chambered heart has developed from the heart tube. The striped area of the valves and septa in this cross section are derived from the cells that invaded the cushions in (**b**). (**d**) An embryonic chick heart stained for CYP2C after 8 ppb exposure to TCE. A single cardiac cushion is outlined by a *light yellow line*. Staining (*green*) is seen in the myocardium and in a subset of endothelial cells proximal to the atrium. (**e**) A segment of the heart stained for HNF4a (*green*) and cardiac myosin (*orange/red*). HNF4a is seen throughout the embryonic heart. Abbreviations- *A* atrium, *V* ventricle, *OT* outflow tract, *Myo* myocardium, *Endo* endothelium, *Mes* mesenchyme, *C* cushion

at birth cause the two layers to fuse and functionally separate the pulmonary and corporeal blood flows. By the end of the embryonic period of development, the heart is largely in its final form. In the rodent model, the heart starts contracting around 8.5 days, begins looping to bring the inflow and outflow of the heart together around day 9 and the valvular primordial begin to develop around day 9.5. Much of the early morphogenesis of the heart is complete by day 12. In the human, the heartbeat begins about day 18 and looping and early valve formation take place between day 21 and 30. The equivalent period in the chick embryo is between days 2 and 3.5. Studies showing cardio-teratogenic effects of TCE in each of these species appear to consistently target this early organogenic period of heart development.

8.4 Rodents

After the original epidemiological studies in humans, Goldberg and his group (Johnson et al. 1998a, b, 2003; Dawson et al. 1990, 1993) reported on TCE cardiac teratogenicity in rat embryo studies that were controversial largely because of very high exposures (1,100 parts per million, ppm, which approximates the maximum solubility of TCE in water) and atypical dose response curves (Dawson et al. 1993; Johnson et al. 2003). These authors found a correlation between TCE exposure in rat embryos through maternal drinking water and congenital heart defects, but it was statistically significant only for the highest doses and appeared to be protective at the lowest dose tested of 2.5 ppb. Conversely, in an independent study similar high doses of TCE (500 mg/kg) provided by gavage to pregnant rats from day 6 to 15 of pregnancy, produced no defects in embryos (Fisher et al. 2001). The contrasting results were attributed to differences in modality and time of exposure that included maternal drinking water vs. gavage in soybean oil, and exposure starting at gestational day (GD) 0 vs 6. A few studies investigated the effects of TCE metabolites on cardiac malformations in exposed embryos. In humans and other mammals TCE is partially excreted by expiration. The ingested TCE is converted into trichloroethanol (TCEtOH) through an unstable chloral compound by CYP2E1. Conversely, ALDH1 mediates the conversion of the intermediate chloral into trichloroacetic acid (TCA), which is eliminated through the urine, unaltered or after dechlorination in form of di- or mono-chloroacetic acid (DCA and MCA, respectively). TCEEtOH is excreted either as a free compound or as a glucoronide conjugate (Davidson and Beliles 1991). TCE can cross the placenta and CYP2E1 has been detected in rat placenta and embryos.

Smith et al. (1992, 1989) reported that both TCA and DCA administered by gavage to Long-Evans dams between GD 6 and 15, produced significant increase in levocardia at doses of TCA equal or >330 mg/kg-day, and intraventricular septal defects at >800 mg/kg-day. The same types of defects were observed at doses of DCA equal or >900 mg/kg-day for levocardia and equal or >1,400 mg/kg-day for intraventricular septal defects. Epstein et al. (1992) reported that GD between 9 and 12 was particularly sensitive to TCA exposure (one dose of 2,400 mg/kg-day) in Long-Evans rats and produced intraventricular septal defects. The authors explained their findings as a failure of the proliferating intraventricular septal tissue to fuse with the tight tubercle of the atrioventricular (AV) cushion tissue. Subsequent studies from Johnson et al. (1998a, b) found that TCA induced significant increases in cardiac defects at the highest dose (2,730 ppm = 291 mg/kg-day) when administered by maternal drinking water during GD 1–22. Of all the other TCE metabolites tested, only dichloroethylene (DCE) produced cardiac defects in a similar model of exposure at either 0.15 or 110 ppm (= .015 and 10.64 mg/kg-day, respectively), when administered pre- and during pregnancy. No particular type or grouping of cardiac anomaly was observed in these studies.

The same animal model of TCE exposure described by Johnson et al. (1998a), was used to identify proteins whose expression is affected by TCE in embryonic hearts. Using a differential display approach, Collier et al. (2003) reported that moderate doses (100 ppm) of TCE in maternal drinking water altered the expression of genes critical for heart development, including vimentin, a marker of mesenchymal differentiation, and the calcium ATP-ase Serca2a.

8.5 Chicks

Original observations regarding TCE cardiac teratogenicity in avians were made by Loeber et al. (1988) who reported that in ovo exposure of chick embryos led to several cardiac malformations including conotruncal, AV cushion and myocardium abnormalities. Although this study suffered from lack of accurate information concerning time and concentration of TCE used, it provided a solid rationale for using chick embryos as a sensitive model for TCE cardiac sensitivity.

Mishima et al. (2006) provided additional data using a chick whole-embryo culture system in which they were able to determine the TCE concentration in the medium, and pin-point the stage of embryonic development most sensitive to TCE cardiotoxicity. In particular, they observed that when cultures were prepared containing initial concentrations between 25 and 250 ppm of TCE, after equilibration (1 h), about 70 % of the added TCE was lost and none was detected after 24 h. Under these conditions, the authors estimated the chick embryos were exposed to TCE for an average of 6 h and the most sensitive stages were at 2 days when the heart began to loop. After 24 h, more than 80 % of 3 day embryos exposed to 80 ppm TCE were apparently normal. However, a closer examination at the cellular level revealed a significant reduction in the number and altered distribution of mesenchymal cells in the AV cushion.

These results were independently confirmed by Drake et al. (2006a), using low doses of TCE (4 nM = 8 ppb) injected into chick eggs between HH stages 13 and 20 (2–3.5 days), corresponding to the time of development of cardiac cushions. The authors reported a significant reduction in mesenchymal cells and cardiac function, using pulsed-Doppler ultrasound, when embryos were exposed to 8 ppb TCE and similar concentrations of metabolites TCA and trichloroethanol (TCOH). Interestingly, these effects were not observed when embryos at earlier and later stages of cardiac morphogenesis (HH 3+ and HH20) were exposed to TCE (Drake et al. 2006b). In contrast, 4nM TCE or TCA increased myocyte proliferation in these embryos exposed at an early developmental time. The conclusion from these studies is that cardiac sensitivity to TCE in chick is dependent on specific stages of heart development.

8.6 Mechanisms of Action

The notion that TCE can disrupt the early events of cardiac differentiation emerged in an earlier in vitro study (Boyer et al. 2000) looking at the effects of TCE on EMT in chick AV explants by collagen gel assays. Results from Boyer et al. (2000)

indicated that the EMT process was impaired by doses of TCE between 50 and 250 ppm in vitro. As these doses reflected starting levels in plastic dishes with some airspace above each culture, the actual effective dose was likely far lower than reported. In the dose range tested, TCE inhibited the expression of Mox-1 and fibrillin2, which are components of the EMT process during the AV canal formation.

Ou et al. (2003) reported that, in differentiated bovine coronary endothelial cells (BCEC), TCE altered the function of endothelial nitric oxide synthase (eNOS) by inhibiting its interaction with the heat shock protein hsp90, and inducing an increased production of O_2^- in a dose–response fashion. The reduced synthesis of NO in favor of O_2^- may have a negative impact on endothelial cell proliferation and lead to defects in valve and septa development. Consistent with this analysis, Feng et al. (2002) observed that eNOS-KO mice had increased incidence of congenital heart defects. In a microarray analysis looking at changes in gene expression caused by TCE exposure, our group found that molecules involved in regulation of calcium signaling (e.g. CamKII and Ryr2) were affected by TCE exposure as low as 1 and 10 ppb, in mouse embryonal carcinoma cell line P19. These cells were used as a model of undifferentiated cardiac myocytes and provided important clues regarding possible molecular mechanisms mediating TCE action (Selmin et al. 2008). In particular, we observed that many transcripts affected either by TCE or its metabolite, TCA, encoded for calcium-responsive proteins or are dependent on cellular DNA methylation for their function (e.g. rhodopsin-G-proteins).

The importance of molecules involved with regulation of calcium flux during early cardiogenesis was confirmed by subsequent studies looking at the effects of low (10–1,000 ppb) and moderate (10–100 ppm) doses of TCE on vasopressin-induced calcium flux in rat myoblasts, using FURA2 measurements. Using the H9c2 cell line, Caldwell et al. (2008) observed a dose-dependent response in which doses of TCE between 10 and 100 ppb inhibited calcium flux, whereas higher doses (10 ppm TCE) did not alter either calcium flux or expression of proteins involved in calcium regulation (e.g. Serca2a and Ryr2). This bimodal behavior of cell response to low versus high doses of TCE has been confirmed in numerous studies using different systems (Mishima et al. 2006; Drake et al. 2006a) but the reasons have not been elucidated. Makwana et al. (2013) recently proposed a role for p450 cytochrome oxygenases in TCE metabolism as essential to explain this phenomenon, and their findings will be discussed below.

Numerous lines of evidence from research groups examining TCE carcinogenicity in liver and kidney (Tao et al. 1999, 2000; Dow and Green 2000) suggested that TCE might exert its negative action by disrupting the folate/homocysteine pathway and ultimately altering the normal methylation status of proteins and nucleic acids. To test this hypothesis Caldwell et al. (2010) analyzed the transcriptome of mouse embryonic hearts isolated from dams that had been exposed to 10 ppb TCE through maternal drinking water and a diet containing 0, 2, or 8 mg/kg folate. The goal of this study was to identify cellular pathways altered in the developing heart following exposure to low, environmentally significant doses of TCE and to determine whether folic acid supplementation might counteract the effects of TCE. Standard chow for pregnant rats usually contains 5–8 mg of folate per kg of pellet and provides ~200 ug/day of folate. The recommended daily allowance (RDA) for pregnant women is 600 ug/day. A gross examination of the embryos collected from the three folate groups indicated that both high (8 mg/kg) and low (0 mg/kg) folate in the maternal diet lead to similar morphological outcomes. In fact, in both groups we observed an increased rate of resorbed and developmentally delayed embryos, accompanied by a reduced number of normally developed embryos. However, when compared with the TCE exposed groups, we observed an almost perfect inversion of phenotypic outcome, suggesting that (i) any alteration from the optimal level of folate in the pregnant rats may alter fetal development, and (ii) low levels of TCE may facilitate developmental progress through otherwise restrictive developmental checkpoints. In fact, we observed no change in percentage of embryos resorbed or delayed in dams exposed to TCE and low folate supplementation (0 mg/kg) compared with those not exposed to TCE and receiving a normal (2 mg/kg) folate level in their diet (Dow and Green 2000). In addition, results from these studies showed that exposure to the low dose of TCE caused extensive alterations in transcripts encoding proteins involved in transport, ion channel, transcription, differentiation, cytoskeleton, cell cycle, and apoptosis. Exogenous folate did not offset the effects of TCE exposure on normal gene expression, and both high and low levels of folate produced additional significant changes in gene expression. We concluded that a mechanism in which TCE induced a folate deficiency did not explain altered gene expression patterns in the embryonic mouse heart. The data further suggested that use of folate supplementation, in the presence of TCE, might be detrimental and not protective of the developing embryo.

A possible mechanism behind these effects is suggested by recent findings reported by our laboratory (Palbykin et al. 2011). We found that TCE affects DNA methylation of the Serca2 promoter region both in rat myoblasts H9c2 cells, and in murine embryonic hearts exposed to TCE via maternal drinking water. The Serca2 gene encodes for a calcium ATPase that is essential for regulation of the calcium flux in myocyte. In particular, we observed a modest increase in CpG methylation across the Serca2 proximal promoter, accompanied by a dramatic hypermethylation of a CpG dimer adjacent to a SP1 binding site, which had been previously described as involved in Serca2 transcriptional activity (Brady et al. 2003). These changes were paralleled by reduced levels of Serca2a protein both in TCE exposed H9c2 cells and embryonic hearts. Methylation of CpG dinucleotides is catalyzed by DNA methyltransferases and uses S-adenosyl-methionine (SAM) as the methyl-donor group. The reduced amounts of SAM observed in this study were explained by the fact that TCE-induced hypermethylation of the Serca2 promoter, and likely other actively transcribed genes, may deplete the cellular content of SAM, which in turn may lead to delayed embryonic development and possible congenital birth defects, including heart defects (Ifergan and Assaraf 2008). Although these findings reported a cellular depletion of SAM, the methylation status of each promoter is regulated by dynamic and complex processes that determine gene activation or repression during embryonic development. Therefore, in order to better understand the mechanisms of action of TCE, it is necessary to investigate the nature of the interactions between

TCE and crucial transcription factors that mediate gene expression during critical phases of embryonic development.

The hypothesis that TCE exposure may cause a reduction in calcium homeostasis in myocytes and directly impact cardiac morphology, was tested by Makwana et al. (2010) using chick embryos exposed to TCE in ovo. Cardiac myocytes were isolated from E18 embryos exposed to either 8 or 800 ppb TCE at HH stage 13 (E2) by in ovo injection into the yolk. Sarcomeric function was assessed by measuring the rate of contraction after electrical stimulus. A reduced half-width of contraction by the sarcomere was observed in myocytes exposed to 8 ppb, whereas sarcomere length itself was not affected, suggesting the change was due to an alteration in Ca++ handling and not to changes in sarcomere structure. Despite the comparatively early exposure of these myocytes in development, sarcomere function remained altered in these cells 16 days later. The functional loss of sarcomeric contraction was also seen by reduced expression of markers of shear stress, NOS-3 and KLF2 in the heart. These data are consistent with observations that altered Ca++ homeostasis can reduce blood flow and the loss of flow can lead to congenital heart defects (Hogers et al. 1997). Independent observations from Rufer et al. (2010) confirmed these ideas with further experimental data. In their study, chick embryos exposed to 8 ppb TCE in ovo, between HH15 and 17, displayed high embryonic mortality and functional dysmorphologies. In particular, a significant high frequency of ventricular septal defects (VSDs) was observed. These defects occurred in 37.5 % of the exposed embryos, and in none of the controls, exposed to vehicle only. The authors concluded that since cardiac hemodynamics are a major contributor to VSDs, their findings support a mechanism of TCE cardiac teratogenesis based on altered calcium handling and blood flow.

In a recent study, Makwana et al. (2013) explored the possibility that the non-monotonic dose–response curve could reflect issues on cardiac metabolism of TCE. Previously, Lash et al. (2000) had identified specificity of the cytochrome P450 CYP2 family for TCE metabolism in human liver microsomes and in murine systems. Makwana et al. (2013) showed the presence of two members of the CYP2 family in embryonic chick hearts (2H1and 2C45), and more importantly, exposure to 8 ppb TCE, but not 800 ppb, induced the expression of CYP2H1 in particular in myocardial cells and in AV canal endothelial cells most proximal to the atrium. No detectable expression of cytochrome p450s was observed in extracardiac tissues (Fig. 8.1d). These results indicate that the earliest embryonic expression of phase I detoxification enzymes are in the developing heart, which may explain its distinct sensitivity to TCE. Later, cytochrome p450 expression in the heart drops and develops in the newly-formed liver. This roughly coincides with the period of TCE sensitivity observed by Rufer et al. (2010). While the bi-modal response curve remains poorly understood, Makwana et al. (2013) suggest that low doses of TCE are metabolized, causing toxic effects, whereas at higher TCE doses, CYPs are degraded faster than can be synthesized. The question whether in the latter case, TCE metabolites cause any cardiac toxicity in the embryonic heart, is still unanswered.

8.7 Current Studies

Altering the expression of proteins that regulate myocyte calcium flux may be a significant mechanism by which TCE causes heart defects. Consistent with the notion of altered gene regulation is the finding that numerous transcription factors involved in early embryonic cardiac differentiation (e.g. brachyuryT, NFATc1, Hoxa1, Foxa1) were down-regulated in mouse embryonic hearts exposed to TCE (Caldwell et al. 2010). The central question remains: what are the proximal signals induced or inhibited by TCE exposure and how do they regulate the expression of these structural and regulatory proteins? In an attempt to address this question, our group returned to the chick system to explore microarray analysis of embryos treated with 8 ppb TCE. The microarray data were subjected to analysis using a chick interactome database (Konieczka et al. 2009). The chick interactome is a compendium of seven databases derived from biological data across many species. The nodes in the database (Fig. 8.2) are molecules converted to a consistent species nomenclature (avian) and the edges are connections derived from public databases on transcriptional regulation, yeast two hybrid screens, protein interaction immunoprecipitation data and other evidence of biological interaction. Of the approximately

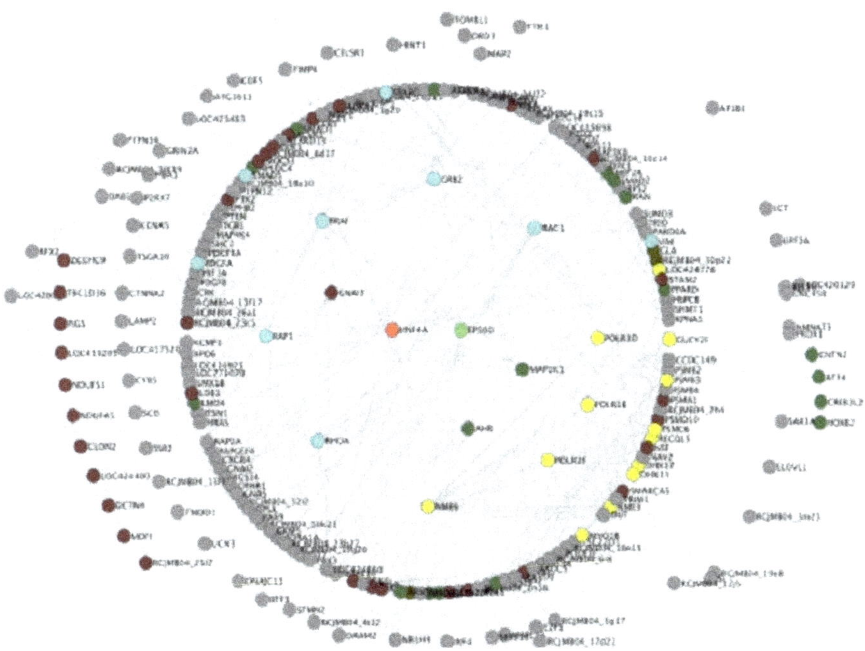

Fig. 8.2 TCE interactome. Circles represent nodes (molecules) and connecting lines are edges showing an identified interaction in one or more of the core databases. The data are displayed in a circular pattern by Cytoscape and the most highly linked nodes were moved to the center of the circle. HNF4a is shown in *red* and directly linked nodes are shown in *maroon*

Table 8.1 Mostly highly linked nodes in interactome analysis of TCE

Name	1′ Neighbors	2′ Neighbors	1′ Sig DN	1′ Sig UP
HNF4A	159	866	35	55
TRAF6	116	778	12	43
IKBKE	100	621	11	28
14-3-3z	89	670	9	25
14-3-3 g	80	555	14	20
EIF1B	63	431	5	16
PRKAB1	56	518	8	13
MAP3K3	54	646	6	17
HSPA8	54	540	4	16

HNF4a shows 159 directly linked (1′ Neighbors) nodes in the TCE dataset and indirect linkage to a further 866 nodes (2′ Neighbors). The combined 1′ and 2′ linkage represents approximately 76 % of the total 1,345 nodes found in the TCE interactome. Of the 159 directly linked nodes, 35 were significantly down-regulated and 55 were up-regulated by 8 ppb TCE. Thus HNF4a is the most highly linked node and is central to the greatest number of regulated targets in the dataset. Most of the regulated primary targets of HNF4a were also found in the mouse dataset. Cytoscape analysis is heuristic and begins randomly in the dataset. Linkage was determined after 5 reiterations. After 20 repeats, EP300 moved to 12th in rank and is not shown in table despite a retained involvement

4,000 genes significantly altered in the chick arrays by TCE exposure, 1,345 genes were found in the interactome dataset that linked to each other. This subset was identified as the TCE interactome. As shown in Table 8.1, we ranked, within the TCE interactome, the set of genes with the highest levels of interconnections in the dataset. The most highly linked molecule was the hepatocyte nuclear factor 4 alpha, HNF4a. In fact, 76 % of the 1,345 molecules included in the TCE interactome are directly or indirectly linked to HNF4a through 1 partner. The transcription factor HNF4a (bright red) and the transcriptional co-activator EP300 (bright green) are shown as highly linked genes. Interestingly, although the HNF4a transcript, itself, is not altered by TCE, in both mouse and chick, the expression of the coactivator EP300, which has acetyl transferase activity, is reduced by TCE, along with other cofactors including HDAC11, HNF1b, and Ppargc1a and 1b (Caldwell et al. 2010). This is consistent with a hypothesis that HNF4a protein is a proximal target of a TCE or its metabolite and that alteration of HNF4a activity would perturb expression of molecules normally regulated by HNF4a.

We used PCR and Western analysis to investigate the expression of HNF4a in mouse heart (embryonic and maternal) and in rat cardiomyoblasts, H9c2 cells. The results illustrated in Fig. 8.3 indicate that HNF4a protein is expressed in both embryonic and maternal heart, and that TCE exposure in H9c2 cells inhibits HNF4a phosphorylation but does not affect its protein level. We had previously shown that Serca2a expression is altered by TCE. Examination of the promoter region of Serca2a showed a candidate HNF4a binding site. After an acute low dose exposure of TCE on H9c2 rat cardiomyocytes at two intervals a CHIP analysis was carried

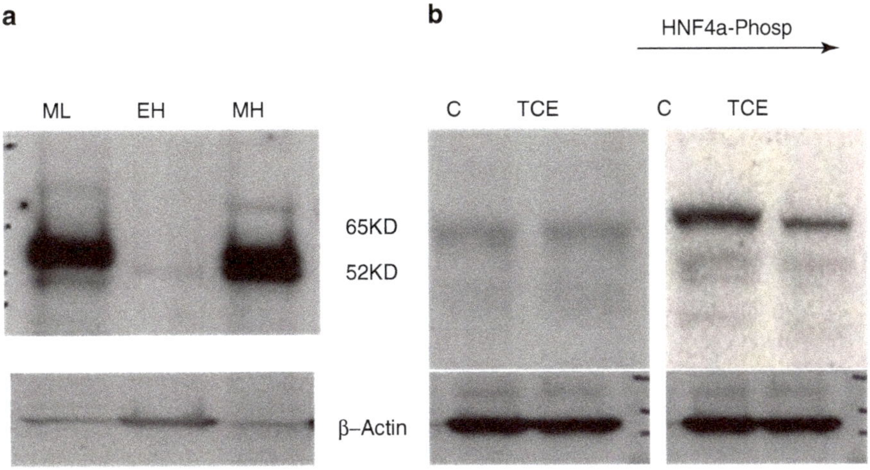

Fig. 8.3 Expression of HNF4a. (**a**) Tissue homogenate from mouse maternal liver (*ML*), Day 18 embryonic heart (*EH*), and maternal heart (*MH*), was analyzed by Western blot using anti HNF4a antibodies which recognized in all tissues a major band of ~52 kD, and a minor band of ~65 kD band in maternal liver and heart only. Lower levels of HNF4a are present in EH, as expected. (**b**) Equal amounts of H9c2 cell lysates were analyzed by Western blot using the same HNF4a antibody in panel **a** (*left panel* in **b**) or antibodies against the phosphorylated form of HNF4a (*right panel* in **b**). H9c2 cells exposed to 10 ppb TCE for 24 h (TCE) show reduced levels of phosphorylated HNF4a compared to control, not exposed cells

out to verify an increased HNF4a expression after TCE exposure. H9C2 cells were fixed, the genomic DNA was sheared and an immunoprecipitation was performed with anti-HNF4a antibodies. The eluted DNA was then measured by quantitative real time PCR to measure changes in HNF4a bound Serca2 promoter region compared to untreated control cells. In Fig. 8.4, we show that 10 ppb TCE exposure after 30 min and 1 h results in an increase in HNF4a transcription factor binding to this region. These data confirm that HNF4a is found in cardiomyocytes and confirm an alteration of activity consistent with a proximal role for HNF4a in mediating some of the effects of TCE on cardiomyocytes. We followed this up by collecting stage embryonic chick heart material from the early, looped heart through a stage nearing completed septation into the 4-chambered heart. These stages are all before a functional liver has developed in the embryo. Quantitative PCR showed a variable pattern of expression during development but a loss of signal about the time that the liver begins to form (data not shown). Immunostaining with the HNF4a antibody (stage 18 hearts) showed that endothelia lining the heart are positive for HNF4a and that there is a slight loss during the transition into valvular mesenchyme in the region where the mitral and tricuspid valves will form. There is also expression in the myocardial cell layer that is heaviest in the outermost (epicardial) layer of muscle and in clusters of myocardial cells that stain poorly for myosin (Fig. 8.1e). These cells that are positive for HNF4a staining likely represent undifferentiated myoblasts. Thus, HNF4a is found in the heart early in development and could be the target of TCE exposure.

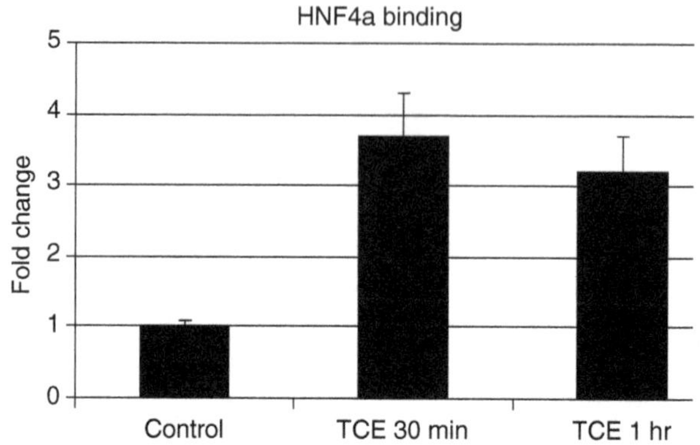

Fig. 8.4 ChIP analysis of HNF4a binding to the Serca2 promoter. H9c2 cells were exposed to 10 ppb TCE for 30 min or 1 h before chromatin was isolated for immunoprecipitation using HNF4a specific antibodies, and primers specific for Serca2 proximal promoter region flanking a HNF4a putative binding site for real time PCR analysis. The graph represents the average of three PCR runs in which each sample was run in quadruplicate. HNF4a binding was normalized by subtracting non specific IgG binding from each sample and value were expressed as average fold change compared with control, non exposed cells. The bars represent standard error

8.8 Conclusions

Earlier studies were controversial, in large part, due to a poorly understood non-monotonic dose curve. In the last 10 years findings from chick and murine models of TCE exposure unequivocally support the embryonic heart sensitivity to low, environmentally relevant doses of TCE. In addition, the embryonic heart is particularly susceptible to TCE toxicity during the phase of transition between a looped heart divided in two compartments separated by the AV cushion and a 4-chambered functional heart. The most current hypothesis corroborated by several independent groups is that TCE affects the ability of the cardiac tissue to regulate calcium flux, thus altering myocyte contraction and maintenance of normal blood flow. Altered hemodynamics likely lead or contribute to development of the variety of congenital heart malformations observed in humans and animal models. Although the molecular mechanisms are still unclear, our current studies suggest a nuclear transcription factor, HNF4alpha, as a most proximal target of TCE action. Future studies will focus on elucidating the mechanisms by which TCE may disrupt transcriptional regulation of HNF4a target genes involved in heart differentiation and metabolism, including CYP2 members found in the embryonic heart (Table 8.2). While the bi-modal or non-monotonic behavior of TCE dose–response is not unique and has been observed for other compounds, including serotonin, opiates, alcohol, and formaldehyde (Calabrese 2001; Calabrese and Baldwin 2003; Gaylor and Aylward 2004; Sari and Zhou 2003), the mechanism underlying this effect is still unclear. However, the finding that low doses of TCE can be metabolized in the embryonic chick heart by members of the CYP2 family, highlight the possibility that cardiac toxicity from TCE and other teratogens may be influenced by a transitory localized metabolizing ability.

Table 8.2 TCE effects on cytochrome P450 expression

Gene	Effect	Tissue/cell	Compound	Mouse homolog
Cyp1b1	UP	Mouse e. h.	10ppbTCE	
Cyp39a1	UP	Mouse e.h.	10ppbTCE	
Cyp26a1	UP	Mouse e.h.	10ppbTCE	
Cyp7b1	UP	Mouse e. h.	10ppbTCE	
Cyp2c66	UP	Mouse e.h.	10ppbTCE	
Cyp2d11	DOWN	Mouse e.h.	10ppbTCE	
Cyp24a1	DOWN	Mouse e.h.	10ppbTCE	
Cyp4f13	DOWN	Mouse e.h.	10ppbTCE	
Cyp4f16	DOWN	Mouse e.h.	10ppbTCE	
Cyp2c37	DOWN	Mouse e.h.	10ppbTCE	
Cyp2g1	DOWN	Mouse e.h.	10ppbTCE	
Cyp2a12	UP	P19 cells	1ppmTCE	
Cyp2r1	UP	P19 cells	1ppmTCA	
Cyp2h1	UP	Chick e. h.	8ppbTCE	Cyp2e1
Cyp1a4	UP	Chick e.h.	8ppbTCE	Cyp1a1
Cyp2c45	UP	Chick e.h.	8ppbTCE	Cyp2c6

Data were extracted from Caldwell et al. (2010), and Makwana et al. (2010)

Level of expression of Cyp transcripts in mouse and chick embryonic heart (e.h.) were determined by microarray or PCR analyses

References

Agency for Toxic Substances and Disease Registry (1998) Volatile organic compounds in drinking water and adverse pregnancy outcomes: U.S. Marine Corps Camp Lejeune, North Carolina. Atlanta: US Department of Health and Human Services

Bove FJ, Fulcomer MC, Klotz JB, Esmart J, Dufficy EM, Savrin JE (1995) Public drinking water contamination and birth outcomes. Am J Epidemiol 141(9):850–862

Boyer AS, Finch WT, Runyan RB (2000) Trichloroethylene inhibits development of embryonic heart valve precursors in vitro. Toxicol Sci 53(1):109–117

Brady M, Koban MU, Dellow KA, Yacoub M, Boheler KR, Fuller SJ (2003) Sp1 and Sp3 transcription factors are required for trans-activation of the human SERCA2 promoter in cardiomyocytes. Cardiovasc Res 60(2):347–354

Calabrese EJ (2001) The future of hormesis: where do we go from here? Crit Rev Toxicol 31(4–5): 637–648

Calabrese EJ, Baldwin LA (2003) Ethanol and hormesis. Crit Rev Toxicol 33(3–4):407–424. Review

Caldwell PT, Thorne PA, Johnson PD, Boitano S, Runyan RB, Selmin O (2008) Trichloroethylene disrupts cardiac gene expression and calcium homeostasis in rat myocytes. Toxicol Sci 104(1):135–143

Caldwell PT, Manziello A, Howard J, Palbykin B, Runyan RB, Selmin O (2010) Gene expression profiling in the fetal cardiac tissue after folate and low-dose trichloroethylene exposure. Birth Defects Res A Clin Mol Teratol 88(2):111–127

Chiu WA, Jinot J, Scott CS, Makris SL, Cooper GS, Dzubow RC, Bale AS, Evans MV, Guyton KZ, Keshava N, Lipscomb JC, Barone S Jr, Fox JF, Gwinn MR, Schaum J, Caldwell JC (2013) Human health effects of trichloroethylene: key findings and scientific issues. Environ Health Perspect 121:303–311

Collier JM, Selmin O, Johnson PD, Runyan RB (2003) Trichloroethylene effects on gene expression during cardiac development. Birth Defects Res A Clin Mol Teratol 67(7):488–495

Davidson IW, Beliles RP (1991) Consideration of the target organ trichloroethylene in terms of metabolite toxicity and pharmacokinetics. Drug Metab Rev 23(5–6):493–599. Review. PubMed PMID: 1802654

Dawson BV, Johnson PD, Goldberg SJ, Ulreich JB (1990) Cardiac teratogenesis of trichloroethylene and dichloroethylene in a mammalian model. J Am Coll Cardiol 16(5):1304–1309

Dawson BV, Johnson PD, Goldberg SJ, Ulreich JB (1993) Cardiac teratogenesis of halogenated hydrocarbon-contaminated drinking water. J Am Coll Cardiol 21(6):1466–1472

Dow JL, Green T (2000) Trichloroethylene induced vitamin B(12) and folate deficiency leads to increased formic acid excretion in the rat. Toxicology 146(2–3):123–136

Drake VJ, Koprowski SL, Lough J, Hu N, Smith SM (2006a) Trichloroethylene exposure during cardiac valvuloseptal morphogenesis alters cushion formation and cardiac hemodynamics in the avian embryo. Environ Health Perspect 114(6):842–847

Drake VJ, Koprowski SL, Hu N, Smith SM, Lough J (2006b) Cardiogenic effects of trichloroethylene and trichloroacetic acid following exposure during heart specification of avian development. Toxicol Sci 94(1):153–162

Epstein DL, Nolen GA, Randall JL, Christ SA, Read EJ, Stober JA, Smith MK (1992) Cardiopathic effects of dichloroacetate in the fetal Long-Evans rat. Teratology 46(3):225–235

Feng Q, Song W, Lu X, Hamilton JA, Lei M, Peng T, Yee SP (2002) Development of heart failure and congenital septal defects in mice lacking endothelial nitric oxide synthase. Circulation 106(7):873–879

Fisher JW, Channel SR, Eggers JS, Johnson PD, MacMahon KL, Goodyear CD, Sudberry GL, Warren DA, Latendresse JR, Graeter LJ (2001) Trichloroethylene, trichloroacetic acid, and dichloroacetic acid: do they affect fetal rat heart development? Int J Toxicol 20(5):257–267

Forand SP, Lewis-Michl EL, Gomez MI (2012) Adverse birth outcomes and maternal exposure to trichloroethylene and tetrachloroethylene through soil vapor intrusion in New York State. Environ Health Perspect 120(4):616–621. Epub 2011 Dec 5

Gaylor DW, Aylward LL (2004) An evaluation of benchmark dose methodology for non-cancer continuous-data health effects in animals due to exposures to dioxin (TCDD). Regul Toxicol Pharmacol 40(1):9–17

Goldberg SJ, Lebowitz MD, Graver EJ, Hicks S (1990) An association of human congenital cardiac malformations and drinking water contaminants. J Am Coll Cardiol 16(1):155–164

Hoffman JL, Kaplan S (2002) The incidence of congenital heart disease. J Am Coll Cardiol 39(12):1890–1900

Hogers B, DeRuiter MC, Gittenberger-de Groot AC, Poelmann RE (1997) Unilateral vitelline vein ligation alters intracardiac blood flow patterns and morphogenesis in the chick embryo. Circ Res 80(4):473–481

Ifergan I, Assaraf YG (2008) Molecular mechanisms of adaptation to folate deficiency. Vitam Horm 79:99–143

Johnson PD, Dawson BV, Goldberg SJ (1998a) Cardiac teratogenicity of trichloroethylene metabolites. J Am Coll Cardiol 32(2):540–545

Johnson PD, Dawson BV, Goldberg SJ (1998b) A review: trichloroethylene metabolites: potential cardiac teratogens. Environ Health Perspect 106(Suppl 4):995–999. Review

Johnson PD, Goldberg SJ, Mays MZ, Dawson BV (2003) Threshold of trichloroethylene contamination in maternal drinking waters affecting fetal heart development in the rat. Environ Health Perspect 111(3):289–292

Konieczka JH, Drew K, Pine A, Belasco K, Davey S, Yatskievych TA, Bonneau R, Antin PB (2009) BioNetBuilder2.0: bringing systems biology to chicken and other model organisms. BMC Genomics 10(Suppl 2):S6

Lagakos SW, Wessen BJ, Zelen M (1986) An analysis of contaminated well water and health effects in Woburn, Massachusetts. J Am Stat Assoc 81:583–596

Lash LH, Fisher JW, Lipscomb JC, Parker JC (2000) Metabolism of trichloroethylene. Environ Health Perspect 108(Suppl 2):177–200. Review

Loeber CP, Hendrix MJ, Diez De Pinos S, Goldberg SJ (1988) Trichloroethylene: a cardiac teratogen in developing chick embryos. Pediatr Res 24(6):740–744

Makwana O, King NM, Ahles L, Selmin O, Granzier HL, Runyan RB (2010) Exposure to low-dose trichloroethylene alters shear stress gene expression and function in the developing chick heart. Cardiovasc Toxicol 10(2):100–107

Makwana O, Ahles L, Lencinas A, Selmin OI, Runyan RB (2013) Low-dose trichloroethylene alters cytochrome P450-2C subfamily expression in the developing chick heart. Cardiovasc Toxicol 13(1):77–84

Mishima N, Hoffman S, Hill EG, Krug EL (2006) Chick embryos exposed to trichloroethylene in an ex ovo culture model show selective defects in early endocardial cushion tissue formation. Birth Defects Res A Clin Mol Teratol 76(7):517–527

NAS Report (2006) Toxicological Review of Trichloroethylene (TCE). http://cfpub.epa.gov/ncea

Ou J, Ou Z, McCarver DG, Hines RN, Oldham KT, Ackerman AW, Pritchard KA Jr (2003) Trichloroethylene decreases heat shock protein 90 interactions with endothelial nitric oxide synthase: implications for endothelial cell proliferation. Toxicol Sci 73(1):90–97

Palbykin B, Borg J, Caldwell PT, Rowles J, Papoutsis AJ, Romagnolo DF, Selmin OI (2011) Trichloroethylene induces methylation of the Serca2 promoter in H9c2 cells and embryonic heart. Cardiovasc Toxicol 11(3):204–214

Rufer ES, Hacker TA, Flentke GR, Drake VJ, Brody MJ, Lough J, Smith SM (2010) Altered cardiac function and ventricular septal defect in avian embryos exposed to low-dose trichloroethylene. Toxicol Sci 113(2):444–452

Runyan RB, Markwald RR (1983) Invasion of mesenchyme into three-dimensional collagen gels: a regional and temporal analysis of interaction in embryonic heart tissue. Dev Biol 95(1):108–114

Sari Y, Zhou FC (2003) Serotonin and its transporter on proliferation of fetal heart cells. Int J Dev Neurosci 21(8):417–424

Selmin OI, Thorne PA, Caldwell PT, Taylor MR (2008) Trichloroethylene and trichloroacetic acid regulate calcium signaling pathways in murine embryonal carcinoma cells p19. Cardiovasc Toxicol 8(2):47–56

Smith MK, Randall JL, Read EJ, Stober JA (1989) Teratogenic activity of trichloroacetic acid in the rat. Teratology 40(5):445–451

Smith MK, Randall JL, Read EJ, Stober JA (1992) Developmental toxicity of dichloroacetate in the rat. Teratology 46(3):217–223

Sonnenfeld N, Hertz-Picciotto I, Kaye WE (2001) Tetrachloroethylene in drinking water and birth outcomes at the US Marine Corps Base at Camp Lejeune, North Carolina. Am J Epidemiol 154(10):902–908

Tao L, Ge R, Xie M, Kramer PM, Pereira MA (1999) Effect of trichloroethylene on DNA methylation and expression of early-intermediate protooncogenes in the liver of B6C3F1 mice. J Biochem Mol Toxicol 13(5):231–237

Tao L, Yang S, Xie M, Kramer PM, Pereira MA (2000) Effect of trichloroethylene and its metabolites, dichloroacetic acid and trichloroacetic acid, on the methylation and expression of c-Jun and c-Myc protooncogenes in mouse liver: prevention by methionine. Toxicol Sci 54(2):399–407

The Toxicological Review of TCE, Chapter 4, CAS-No79-01-6, EPA/635/r-09/011F at www.epa.gov/iris

Yauck JS, Malloy ME, Blair K, Simpson PM, McCarver DG (2004) Proximity of residence to trichloroethylene-emitting sites and increased risk of offspring congenital heart defects among older women. Birth Defects Res A Clin Mol Teratol 70(10):808–814

Chapter 9
Trichloroethylene and Cancer

Daniel Wartenberg and Kathleen M. Gilbert

Abstract This chapter describes the process by which trichloroethylene (TCE) has now been characterized as "reasonably anticipated to be a human carcinogen." It also summarizes the animal studies and human epidemiological results associated with TCE exposure and kidney cancer that have led to that conclusion. The contribution of TCE metabolism to kidney cancer etiology is discussed, as is some speculation concerning the mechanism of action.

Keywords Risk assessment • Kidney • Nephrotoxicity

9.1 Introduction

In September 2011 the United States Environmental Protection Agency (USEPA) finalized its Toxicological Review for Trichloroethylene (TCE) as part of its Integrated Risk Information System (IRIS http://www.epa.gov/iris/toxreviews/0199tr/Chapter6_0199tr.pdf). IRIS was developed in 1985 to provide the agency's best science-based judgment concerning health effects for individual substances to be used as a basis for regulation and to characterize the health risks of

D. Wartenberg, PhD (✉)
Department of Environmental and Occupational Medicine,
Rutgers Robert Wood Johnson Medical School,
170 Frelinghuysen Road, Piscataway, NJ 08854, USA
e-mail: dew@eohsi.rutgers.edu

K.M. Gilbert, PhD
Department of Microbiology and Immunology,
University of Arkansas for Medical Sciences,
Arkansas Children's Hospital Research Institute,
13 Children's Way, Little Rock, AR 72202, USA
e-mail: gilbertkathleenm@uams.edu

K.M. Gilbert, S.J. Blossom (eds.), *Trichloroethylene: Toxicity and Health Risks*,
Molecular and Integrative Toxicology, DOI 10.1007/978-1-4471-6311-4_9,

human exposure. The full review of TCE was protracted and contentious, pitting one federal agency against another. In addition to an extensive review and assessment of non-cancer effects, both acute and chronic, USEPA also concluded that TCE was a cause of human kidney cancer, paving the way for regulating the chemical as a human carcinogen. This outcome represents a significant event in the public health history of TCE. This chapter presents background and context for this judgment.

9.2 Background

TCE is a chlorinated ethylene, a group of six closely related chemicals all built on a common chemical framework, or backbone, which consists of two carbon atoms connected by a double-bond. This leaves room for four more atoms, two on each carbon. When these spots are occupied only by hydrogen atoms, we have the parent hydrocarbon, ethylene. As we successively replace each hydrogen with a chlorine atom we generate in turn, vinyl chloride (VC or monochloroethylene), dichloroethylene (DCE, three different forms), trichloroethylene (TCE) and tetrachloroethylene (PCE). Vinyl chloride (VC) has long been characterized as a confirmed human carcinogen. PCE is characterized by the IRIS database as "likely to be carcinogenic in humans by all routes of exposure." The data for various forms of DCE were insufficient to characterize its carcinogenicity. Thus at least half of the chlorinated ethylenes have now been characterized as human carcinogens or likely to be human carcinogens.

TCE was first discovered in 1864 and patented in 1906 by Imperial Industries (ICI) in Great Britain (Waters et al. 1977). In its heyday, ICI was the largest manufacturing company in the British Empire, and commonly regarded as a "bellwether" of the British economy. TCE dissolves in water and itself dissolves a variety of organic materials. Its first industrial uses were to extract oils from vegetable crops like soy, palm and coconut. It was later used to decaffeinate coffee, as a surgical anesthetic and in the dry cleaning industry, although its role there was almost entirely supplanted by its higher chlorinated cousin, PCE. TCE's main use, however, has been as a degreasing agent, mainly to remove dirt and oils in small machined parts. It was used in facilities of all sizes, including both large factories and small machine shops and garages. Once the degreasing solvent became dirty it was often carelessly disposed of in pits or surfaces. As a result it TCE became one of the most prevalent groundwater contaminants in the US and elsewhere. In spite of the growing body of evidence documenting the health hazards associated with exposure to TCE, the recognition of the potential impacts of the reported health risks were idiosyncratic and slow in garnering acceptance.

The consequences of TCE exposure encompass a wide variety of venues, such as workplace exposures such as those experienced by workers at degreasing plants, those making commercial goods, (e.g., workers at the View Master Facility in Beaverton Oregon) (Environmental and Occupational Epidemiology Oregon Department of Human Services 2004), and those in military service or living at military bases (e.g., soldiers and other personnel at the Marine Corps Base Camp Le Jejune, North Carolina) (Board on Environmental Studies and Toxicology. National Research Council 2013). Unsuspecting home owners also were exposed to TCE via contaminated drinking water, or through a process called vapor intrusion, where TCE in subsurface soils and groundwater evaporates and makes its way into living spaces.

A notorious example of this contamination and suspected health effects was documented in the book, *A Civil Action* by Jonathan Haar (ref. Haar 1996) and a 1998 major motion picture of the same name starring John Travolta (http://www.imdb.com/title/tt0120633/). It chronicled events in Woburn Massachusetts, where the incidence of childhood lymphocytic leukemia between 1964 and 1986 was four-times greater than expected (Cutler et al. 1986). The Woburn cluster appeared just as concern about the carcinogenicity of TCE was beginning to ramp up. TCE was originally viewed as a major industrial breakthrough because of its effectiveness in a range of industrial activities at a variety of workplaces, but in the half century prior to the Woburn cluster, reports and concerns about health effects associated with exposure to TCE continued to be raised and expanded, particularly with respect to cancer. Beginning in 1940 some US state health agencies, such as New York State and the State of California raised concerns about the health issues, although overall actions and interventions were relatively slow to be implemented.

The first major break regarding cancer effects came in 1975 when the US National Cancer Institute (NCI) published a "memorandum of alert" in the *Chemical Engineering News* which stated that "**preliminary tests on mice implicated TCE as the cause of hepatocellular carcinoma with some metastases**" (National Cancer Institute 1975). This unusual alert stirred up considerable controversy among TCE producers, users and federal regulatory agencies (Seltzer 1975). The NCI alert was backed up with final results consisting of more numbers and experimental details published in 1976 (National Toxicology Program 1976). Surprised by the report on what many had been thought to be a relatively benign compound, the Manufacturing Chemists Association (later known as the Chemical Manufacturers' Association and currently called the American Chemistry Council) initiated and conducted a series of inhalation studies to assess the carcinogenic potential of TCE (Bell et al. 1978). But now there was enough evidence to insure that spent or discarded TCE was recognized as a hazardous waste that needed to be managed and should also be subject to regulation under the 1974 Safe Drinking Water Act, where it is currently listed with a maximum contaminant limit of 5 μg/L.

The next milestone in the story of TCE and cancer was reached in 1995, when the International Agency for Research on Cancer (IARC) conducted a full review of the available data on exposures to TCE and cancer incidence, noting that TCE had already been shown to be associated with liver and kidney cancer in experimental animal studies (International Agency for Research on Cancer 1995). IARC is a part of the World Health Organization. Its purpose is "to identify the causes of cancer so that preventive measures may be adopted and the burden of disease and associated suffering reduced." Based on their review of more than 80 published papers and letters that reported on the investigation of the cancer epidemiology of people and animals exposed to TCE, an IARC Working Group of independent international experts concluded that "(***TCE***) **was *probably carcinogenic to humans* (*and formally declared TCE a Group 2A carcinogen*)**" based on limited evidence in humans for the carcinogenicity of TCE and sufficient evidence in experimental animals.

With continued use of TCE occupationally into the 1990s, and in light of reports of adverse health effects among those exposed in the community, public concern became an increasingly serious issue. In response, USEPA initiated a "State of the Science" review to evaluate the possible health impact of exposure to TCE. USEPA solicited scientific perspectives from a range of groups and individuals to provide a broad summary of the post-IARC (1995) epidemiology literature on potential cancer risks, and non-cancer endpoints from various types of TCE exposures. As summarized by Scott and Cogliano (2000) this outreach effort culminated in 2000 with the publication of 16 state-of-the-science (SOS) papers in *Environmental Health Perspectives* (Supplement 2) under the combined sponsorship of the US EPA, the US Air Force, the US Department of Energy, the National Institute of Environmental Health Sciences and the Halogenated Solvents Industry Alliance. The contributing teams focused primarily on human studies, rather than animal or mechanistic studies, identifying more than 80 published papers that evaluated the possible associations of exposures to TCE and any of the cancers under consideration. Study designs included more than 20 reports on worker cohorts, 40 case-control studies, more than a dozen community-based studies, and several commentaries and reviews on the possible association of exposure to TCE and cancer (Wartenberg et al. 2000).

The complexity of the results from the review, involving different exposures, different cancers, and different study designs from the initial IARC Report and the SOS reports commissioned by the US EPA was considerable. One group of researchers included in their report a statistical approach for summarizing this type of data using a technique known as meta-analysis. To simplify, they averaged the results of individual studies of the same type of exposure and same type of cancer, weighting the results by the number of subjects and combining them into a single measure of the strength of the association. This enabled even those with limited statistical experience to make comparisons across results of the TCE- cancer assessments.

The US EPA used the SOS papers to develop its 2001 draft "Trichloroethylene Health Risk Assessment: Synthesis and Characterization" (US 2001). The 2001 US EPA draft report on TCE laid the groundwork for new regulations that would limit human exposure to the chemical but also triggered a dispute between the US EPA and the Department of Defense, the Department of Energy and NASA. TCE had been used in large quantities by the US Armed Forces and NASA to de-grease rocket and airplane engines. These agencies also constituted some of the main polluters. DOD alone faces the daunting task of cleaning up thousands of military bases and other installation across the country with TCE-contaminated soil, water or storage containers (US Government Accountability Office 2007).

In part to stave off costly remediation, DOD, DOE and NASA (with USEPA participating) contracted the National Research Council, a component of the National Academy of Sciences, to produce yet another independent review of the TCE issue. This resulted in the comprehensive 2006 report entitled "Assessing the Human Health Risk of Trichloroethylene; Key Scientific Issues" (Committee on Human Health Risks of Trichloroethylene, N. R. C 2006). The report found the evidence for carcinogenic risk to be even stronger in the few years since the 2001 draft: "**The committee found that the evidence on carcinogenic risk and other health hazards from exposure to trichloroethylene has strengthened since 2001. Hundreds of waste sites in the United States are contaminated with trichloroethylene and it is well documented that individuals in many communities are exposed to the chemical, with associated health risks. Thus, the committee recommends that federal agencies finalize their risk assessment with currently available data so that risk management decisions can be made expeditiously**."

It is not just USEPA that has come to this conclusion. In 2012 IARC upgraded TCE carcinogenicity to Class 1 based on sufficient evidence in both humans and animals (International Agency for Research on Cancer 2013). TCE has been reclassified as a category 2 carcinogen under the European Union Dangerous Substances Directive. The U.S. Department of Health and Human Services National Toxicology Program has TCE on the list of toxicants "reasonably anticipated to be human carcinogens." In addition to regulatory agencies, the American Conference of Governmental Industrial Hygienists (ACGIH) have recently reclassified TCE to category A2: suspected human carcinogen.

The long chapter that began with the 2001 Draft Report was finally brought to a close with the USEPA's exhaustive report Toxicological Review of Trichloroethylene; In Support of Summary Information on the Integrated Risk Information System (IRIS) (US 2011):

> The available epidemiologic studies provide convincing evidence of a causal association between TCE exposure and cancer. The strongest epidemiologic evidence consists of reported increased risks of kidney cancer, with more limited evidence for NHL and liver cancer, in several well-designed cohort and case-control studies.

The basis for the causal judgment about TCE and kidney cancer is described below. In addition to the agency reports there are several recent and comprehensive reviews that describe in some detail the various human and animal studies used in the TCE carcinogenicity designations (Chiu et al. 2013; Purdue 2013; Karami et al. 2012). We briefly summarize the earlier work, and will confine more in-depth discussion to newly published studies.

9.3 TCE and Kidney Cancer

9.3.1 The Role of Metabolism

It is generally believed that TCE needs to be metabolized in order to elicit toxicity in the kidney or other tissues. TCE is metabolized in humans and experimental animal species by both oxidation and glutathione (GSH)-conjugation pathways. Both produce several toxic metabolites (Chiu et al. 2006; Lash et al. 2000). TCE oxidative metabolism by CYP450s, predominantly CYP2E1, yields chloral and chloral hydrate which are in turn metabolized to trichloroethanol (TCOH), trichloroacetic acid (TCA), and dichloroacetic acid (DCA). The glutathione conjugation pathways produces metabolites dichlorovinyl glutathione and dichlorovinyl cysteine (DCVC). The complex assortment of TCE metabolites generated can be transported across multiple tissues, making it difficult to attribute a particular effect to a specific metabolite (Caldwell and Keshava 2006). However, TCE liver toxicity is generally associated with the oxidative pathway (Buben and O'Flaherty 1985; Bull 2000), whereas kidney toxicity is more often correlated with metabolites resulting from GSH conjugation (Lash et al. 2000).

In numerous studies, DCVC has been shown to induce acute kidney toxicity in rats and mice. Mice receiving a single dose of 1 mg/kg DCVC exhibited karyolytic proximal tubular cells in the outer stripe of the outer medulla, and moderate desquamation of the tubular epithelium (Eyre et al. 1995). Although there is not enough in vivo data to assess the relatively sensitivity of different species, it is apparent that multiple species experience DCVC-induced nephrotoxicity (Krejci et al. 1991; Wolfgang et al. 1990; Jaffe et al. 1984; Terracini and Parker 1965).

Only a few studies have examined chronic rather than acute exposure to DCVC. DCVC given in drinking water to rats at a concentration of 0.01 % for 12 weeks (approximately 10 mg/kg-day), produced consistent and time-dependent pathological and histological changes in the kidney (Terracini and Parker 1965). These included tubular necrosis and dilation, and tubular cells exhibiting karyomegaly. Importantly, the histological changes and their location in subchronic and chronic experiments with DCVC are quite similar to those reported in chronic studies of TCE, particularly the prominence of karyomegaly and cytomegaly in the pars recta section of the kidney. Although DCVC appears to induce both acute and chronic nephrotoxicity, it is still not clear whether sufficient DCVC is formed from TCE exposure to account for TCE nephrotoxicity.

In summary, it appears that DCVC and related GSH conjugation metabolites are the active agents of TCE-induced nephrotoxicity. A role for oxidative metabolites from TCE cannot be ruled out, as it is known that substantial TCOH and TCA are formed from TCE exposure, and that TCOH exposure leads to toxicity in the renal tubules. However, TCOH-induced nephrotoxicity does not generate the range of effects observed after TCE exposure, while those of DCVC-induced nephrotoxicity do. Also, TCOH exposure alone does not induce the same pathology as TCE or DCVC. TCA has also been demonstrated to induce peroxisomal proliferation in the kidney, but this has not been associated with kidney cancer (Goldsworthy and Popp 1987). Therefore, although TCOH and TCA may contribute to TCE-induced nephrotoxicity, their contribution is likely to be small compared to that of DCVC. However, the precise metabolic yield of these DCVC following TCE exposure remains uncertain.

9.3.2 Animal Studies

There is evidence that TCE can cause kidney cancer in rodents. Especially noteworthy was the finding of TCE-induced kidney tumors in multiple strains of male rats exposed by gavage (National Toxicology Program 1990). The admittedly low increases in incidence were still considered biologically significant in view of the very low historical incidence of renal tumors in control rats. Others have also noted a low incidence of renal tubule carcinoma in male rats chronically exposed to TCE (Lock and Reed 2006). In inhalation studies TCE was not found to increase kidney tumor incidence in mice or hamsters (Henschler et al. 1980), but did appear to increase renal adenocarcinomas in male rats (4/130) at the high dose (600 ppm) after 2 years of exposure (Maltoni et al. 1988). Thus, TCE has been shown to promote neoplastic lesions in the kidney of rats (mainly in males, with less evidence in females), treated via inhalation and gavage. Although the TCE-induced increase in incidence was low, because of the rarity of these tumors in controls and the repeatability of this result the finding was judged biologically significant.

9.3.3 Human Epidemiology

Given the clear evidence of kidney toxicity and the carcinogenic potential of TCE in animals, it is a natural question to ask if humans exposed to TCE are similarly affected. The available evidence is entirely consistent with a TCE cancer risk in humans. TCE is used in a variety of workplaces, many of them difficult to study epidemiologically because of concomitant exposure to other toxins.

With this in mind, the U.S. EPA reviewed multiple human epidemiologic studies on TCE and cancer (US 2011; Chiu et al. 2013; Scott and Jinot 2011), each evaluated for specific characteristics of epidemiologic design and analysis in order to evaluate whether chance, bias, or confounding might have skewed the study's results. The epidemiologic evidence for TCE-induced kidney cancer was described according to key concepts in a recent summary by Chiu et al. (2013). These concepts include consistency and strength of observed association, specificity, exposure-response relationship, and biological plausibility and coherence. Once stratified by these primary components the epidemiological database for TCE supported a causal association between TCE exposure and kidney cancer in humans. Kidney cancer risk from TCE exposure has been studied related to TCE exposure in cohort, case-control, and geographical studies. These studies have examined TCE in mixed exposures as well as alone. Elevated risks are observed in many of the cohort and case-control studies examining kidney cancer incidence in occupations with historical use of TCE (Moore et al. 2010; Bruning et al. 2003; Dosemeci et al. 1999; Charbotel et al. 2006; Zhao et al. 2005).

Especially convincing was the consistency of increased relative risk (RR) estimates for kidney cancer across the 15 independent epidemiologic studies of different designs and populations from different countries that met the criteria for inclusion in a meta-analysis (Chiu et al. 2013). As suggested by speakers at the 2009 Society for Risk Analysis and followed up by publications from the IARC and the Federation of American Societies for Experimental Biology (FASEB) similar sets of objective study inclusion criteria have been developed (Conrad and Becker

2011). Using updated criteria to strengthen the meta-analysis process, the U.S. EPA conducted new analyses of the epidemiologic data on TCE (US 2011; Scott and Jinot 2011). In addition, the meta-analysis fit the data to both fixed-effect and random-effects models, evaluated statistical heterogeneity across the studies, performed sensitivity analyses, and conducted tests for potential publication bias (which may occur if positive studies are more likely to be published).

The revised meta-analysis by the US EPA provided strong support for a causal association between TCE exposure and kidney cancer. The summary meta-relative-risk (RRm) estimate for kidney cancer was modest: 1.27 [95 % confidence interval (CI): 1.13, 1.43] with a higher RRm for the highest exposure groups (1.58, 95 % CI: 1.28, 1.96). A meta-analysis of TCE-exposed workers by Kelsh et al. similarly showed a positive association across various study groups with an RRm of 1.42 (95 % CI = 1.17–1.77) (Kelsh et al. 2010). However, the possibility of unmeasured potential confounding and lack of quantitative exposure assessment were raised as cautionary notes. A detailed examination by the U.S. EPA of potential confounding by lifestyle factors or other occupational exposures concluded that confounding was not a likely explanation for the observed excesses. A very recent meta-analysis of occupational TCE exposure and kidney cancer reviewed studies published from 1950 to 2011 (Karami et al. 2013). They were stratified by assessment of occupational exposure to TCE specifically, and exposure to any chlorinated solvent. The results revealed that significant and stronger estimates of TCE carcinogenicity were observed in studies that evaluated TCE exposure specifically, while estimates were lower in studies that assessed exposure to a more broad-based category of chlorinated solvents.

A recent study of kidney toxicity in Chinese factory workers exposed to TCE at levels below the current US OSHA permissible exposure limit showed that kidney injury molecule-1 and Pi-glutathione S transferase alpha were elevated among the exposed subjects as compared to unexposed controls (Vermeulen et al. 2012). This finding suggested that even at relatively low occupational exposure levels, TCE induced measurable kidney toxicity. It may also provide biomarkers of early TCE nephrotoxicity that can be used for early detection and reverse this process before it transitions to neoplasia.

9.4 Potential Mechanisms by Which TCE Induces Kidney Cancer

Based primarily on similarities found in studies conducted in animals or *in vitro* (Fig. 9.1), several mechanisms have been proposed for TCE-induced kidney carcinogenicity. These include mutagenicity, cytotoxicity and regenerative proliferation, peroxisome proliferation, α2μ-related nephropathy, and formic acid-related nephropathy. Although cytotoxicity was considered as an alternative mechanism there are inadequate data to suggest it is sufficient to induce kidney tumors. Similarly, potential mechanisms of action relating to peroxisomal proliferation,

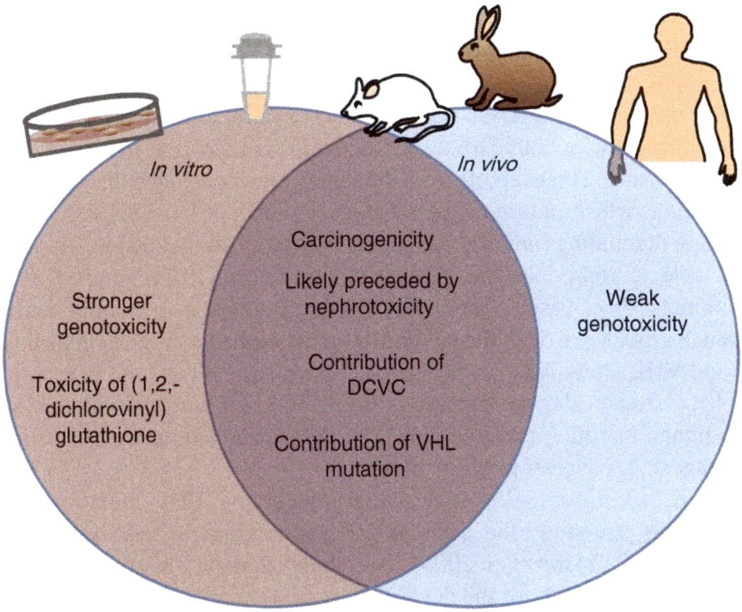

Fig. 9.1 Similarities between data obtained from in vitro and in vivo studies of TCE-induced kidney cancer

α2μ-globulin nephropathy and formic acid-related nephrotoxicity were also deemed unlikely due to limited evidence and/or insufficient experimental support. Although it may not be the only mechanism by which TCE and its metabolites trigger and promote neoplasia, existing evidence supports the conclusion that mutagenesis mediated by the TCE GSH-conjugation metabolites (predominantly DCVC) can induce kidney cancer. This conclusion is supported by evidence of kidney-specific genotoxicity following in vivo exposure to TCE or DCVC. Also consistent with this conclusion, Moore et al. found a statistically significant association between TCE exposure and renal carcinoma risk among TCE-exposed persons with an active GSTT1 (glutathione-*S*-transferase theta-1) enzyme [odds ratio (OR = 1.88; 95 % CI: 1.06, 3.33]) but not among subjects with two deleted alleles for GSTT1 (OR = 0.93; 95 % CI: 0.35, 2.44) (Moore et al. 2010). Although cytotoxicity caused by DCVC may not be sufficient to cause renal carcinogenesis, it may contribute to it by increasing the survival or expansion of mutated cells via regenerative proliferation. A genetic signature for functional effects of an environmental exposure would make the case for a causal association very compelling. In the case of TCE and kidney cancer this approach has focused on the Von Hippel-Lindau (VHL) protein and gene.

Von Hippel-Lindau Disease is a rare autosomal dominant genetic condition that predisposes individuals to a variety of benign and malignant tumors, among them kidney cancers. The mutated gene is called the VHL tumor suppressor gene. Since VHL mutations and loss of heterozygosity have been identified in the majority of renal cancers VHL protein inactivation via germ line sequence alterations is considered a biomarker of early renal carcinogenesis (Gnarra et al. 1994). Homozygous inactivation of the *VHL* gene is linked to the occurrence of renal clear cell carcinoma, the renal carcinoma preferentially induced by trichloroethylene. Bruning et al. (1997) and Brauch et al. (2004) have reported that increase of VHL missense mutations, including a hot spot mutation at nucleotide 454, were correlated with TCE exposure. Three reports from the same group concluded that TCE increases VHL mutations which in turn triggers the development of renal cell carcinomas.

Although the findings are of great interest, a similar study in a French population was not able to reproduce the VHL mutation spectra (Charbotel et al. 2007). Different methods of tissue fixation and DNA extraction may explain some of the discrepancies and leave open the possible association between TCE-induced kidney cancer and VHL alterations. So far the discordant results have not been explained. None of the studies showed mutations in all TCE-exposed individuals, or in all kidney tumors, but other possible means of *VHL* inactivation, and other targets of TCE mutagenesis have yet to be examined.

Although little information is available concerning VHL mutations in TCE-treated animals one study did examine VHL alterations in rats exposed to TCE metabolite DCVC (Mally et al. 2006). This study used the Eker rat model (*Tsc-2*±) which is at increased risk for the development of spontaneous renal cell carcinoma carcinogenesis (Everitt et al. 1995). Another group showed pathway activation in Eker rats similar to that seen in humans with *VHL* mutations leading to Renal Cell Carcinoma (RCC), suggesting that *Tsc-2* inactivation is analogous to inactivation of *VHL* in human RCC (Liu et al. 2003). However, in Mally et al. (2006), male rats carrying the Eker mutation were exposed to TCE (0, 100, 250, 500, or 1,000 mg/kg body weight by gavage, 5 days/week) for 13 weeks. No increase in pre-neoplastic lesions or tumor incidence was found in Eker rat kidneys compared to controls. In addition, no *VHL* gene mutations were found. However, once again it is possible that DCVC inactivates *VHL* by some other method or that *VHL* alterations are caused by other TCE metabolites.

9.5 Summary and Future Challenges

Animal studies have showed that TCE exposure by both gavage and inhalation exposure caused renal toxicity in the form of cytomegaly and karyomegaly of the renal tubules in male rats. Thus kidney cells and the kidney are a target organ for TCE toxicity. Further studies with TCE metabolites have demonstrated a potential role for DCVC, and perhaps TCOH, and TCA in TCE-induced nephrotoxicity. Of these, DCVC induces the renal effects that are most like TCE.

Kidney cancer risk from TCE exposure has been studied in cohort, case-control, and ecological studies. Elevated risks are observed in many of the cohort and case-control studies examining kidney cancer incidence in professions involving occupational exposure to TCE. Greater susceptibility to TCE exposure and kidney cancer is observed among subjects with a functionally active GSTT polymorphism.

The finding of a mutation in the *VHL* gene is potentially supportive, although it would be useful if this finding were replicated in other settings. In terms of mechanism of action it seems most likely that mutagenicity increases the rate of mutation in response to TCE, while regenerative proliferation may enhance the survival or clonal expansion of the mutated cells.

Challenges for the future include a better assessment of the extent to which *S*-(1,2-dichlorovinyl)-L-cysteine and *N*-acetyl-*S*-(1,2-dichlorovinyl)-L-cysteine sulfoxides are formed in human tissues (liver and kidney) following exposure to TCE. The enzymes involved in this process, and their interindividual variability need to be included in this assessment. The contribution of VHL gene mutations to TCE-induced renal carcinogenesis needs more study. This includes validation in other populations and geographic areas.

Identification of additional risk factors including chemical co-exposures that modify the effects of TCE on kidney cancer development needs to be pursued. Aside from mutagenicity, the effects of TCE on epigenetic alterations in oncogenes or other genes that may regulate renal cancinogenesis need to examined. More epidemiological studies with more accurate TCE exposure indices would be helpful. Occupational exposure to TCE is still common in many countries, and the slow pace of remediation means that environmental exposure to TCE is expected to continue for the foreseeable future. This adds urgency to the need for future studies.

These are indeed stiff challenges, but they are challenges related to the details of a broad picture whose outlines are now easily discernible: TCE is a cause of human cancer, specifically kidney cancer. Enough details are now visible to give confidence in this judgment. A number of other cancers have also shown a relationship to TCE exposure, some stronger than others. They include non-Hodgkin's Lymphoma and other hematopoietic cancers, cancer of the liver and biliary tract, breast cancer, bladder cancer and lung cancer. These may be next chapters in the history of TCE-related cancer.

References

Bell ZG, Olson KJ, Benya TJ (1978) Final report of audit findings of the Manufacturing Chemists Association: Administered trichloroethylene chronic inhalation study at Industrial Bio-test Laboratories, Inc., Decatur, IL., Unpublished

Board on Envrionmental Studies and Toxicology. National Research Council (2013) Contaminated water supplies at Camp Lejeune. In: Assessing potential health effects. The National Academies Press, Washington, DC

Brauch H, Weirich G, Klein B, Rabstein S, Bolt HM, Bruning T (2004) VHL mutations in renal cell cancer: does occupational exposure to trichloroethylene make a difference? Toxicol Lett 151:301–310

Bruning T, Weirich G, Hornauer MA, Hofler H, Brauch H (1997) Renal cell carcinomas in trichloroethene (TRI) exposed persons are associated with somatic mutations in the von Hippel-Lindau (VHL) tumour suppressor gene. Arch Toxicol 71:332–335

Bruning T, Pesch B, Wiesenhutter B, Rabstein S, Lammert M, Baumuller A, Bolt HM (2003) Renal cell cancer risk and occupational exposure to trichloroethylene: results of a consecutive case-control study in Arnsberg, Germany. Am J Ind Med 43:274–285

Buben JA, O'Flaherty EJ (1985) Delineation of the role of metabolism in the hepatotoxicity of trichloroethylene and perchloroethylene: a dose-effect study. Toxicol Appl Pharmacol 78:105–122

Bull RJ (2000) Mode of action of liver tumor induction by trichloroethylene and its metabolites, trichloroacetate and dichloroacetate. Environ Health Perspect 108(Suppl 2):241–259

Caldwell JC, Keshava N (2006) Key issues in the modes of action and effects of trichloroethylene metabolites for liver and kidney tumorigenesis. Environ Health Perspect 114(9):1457–1463

Charbotel B, Fevotte J, Hours M, Martin JL, Bergeret A (2006) Case-control study on renal cell cancer and occupational exposure to trichloroethylene. Part II: Epidemiological aspects. Ann Occup Hyg 50:777–787

Charbotel B, Gad S, Caiola D, Beroud C, Fevotte J, Bergeret A, Ferlicot S, Richard S (2007) Trichloroethylene exposure and somatic mutations of the VHL gene in patients with Renal Cell Carcinoma. J Occup Med Toxicol 2:13

Chiu WA, Okino MS, Lipscomb JC, Evans MV (2006) Issues in the pharmacokinetics of trichloroethylene and its metabolites. Environ Health Perspect 114:1450–1456

Chiu WA, Jinot J, Scott CS, Makris SL, Cooper GS, Dzubow RC, Bale AS, Evans MV, Guyton KZ, Keshava N, Lipscomb JC, Barone S Jr, Fox JF, Gwinn MR, Schaum J, Caldwell JC (2013) Human health effects of trichloroethylene: key findings and scientific issues. Environ Health Perspect 121:303–311

Committee on Human Health Risks of Trichloroethylne, N. R. C (2006) Assessing the human health risks of trichloroethylene: key scientific issues. The National Academies Press, Washington, DC

Conrad JW Jr, Becker RA (2011) Enhancing credibility of chemical safety studies: emerging consensus on key assessment criteria. Environ Health Perspect 119:757–764

Cutler JJ, Parker GS, Rosen S, Prenney B, Healey R, Caldwell GG (1986) Childhood leukemia in Woburn, Massachusetts. Public Health Rep 101:201–205

Dosemeci M, Cocco P, Chow WH (1999) Gender differences in risk of renal cell carcinoma and occupational exposures to chlorinated aliphatic hydrocarbons. Am J Ind Med 36:54–59

Environmental and Occupational Epidemiology Oregon Department of Human Services (2004) Feasibility investigation of worker exposure to trichloroethylene at the View-Master factory in Beaverton, Oregon. ATSDR

Everitt JI, Goldsworthy TL, Wolf DC, Walker CL (1995) Hereditary renal cell carcinoma in the Eker rat: a unique animal model for the study of cancer susceptibility. Toxicol Lett 82–83:621–625

Eyre RJ, Stevens DK, Parker JC, Bull RJ (1995) Renal activation of trichloroethene and S-(1,2-dichlorovinyl)-L-cysteine and cell proliferative responses in the kidneys of F344 rats and B6C3F1 mice. J Toxicol Environ Health 46:465–481

Gnarra JR, Tory K, Weng Y, Schmidt L, Wei MH, Li H, Latif F, Liu S, Chen F, Duh FM et al (1994) Mutations of the VHL tumour suppressor gene in renal carcinoma. Nat Genet 7:85–90

Goldsworthy TL, Popp JA (1987) Chlorinated hydrocarbon-induced peroxisomal enzyme activity in relation to species and organ carcinogenicity. Toxicol Appl Pharmacol 88:225–233

Haar J (1996) A civil action. Vintage, New York

Henschler D, Romen W, Elsasser HM, Reichert D, Eder E, Radwan Z (1980) Carcinogenicity study of trichloroethylene by longterm inhalation in three animal species. Arch Toxicol 43:237–248

International Agency for Research on Cancer (1995) Dry cleaning, some chlorinated solvents and other industrial chemicals, vol 63. World Health Organization, Geneva

IARC Monographs on the Evaluation of Carcinogenic Risks to Humans (2013) Trichloroethylene, tetrachloroethylene and some other chlorinated agents, vol 106, Lyon, France

Jaffe DR, Gandolfi AJ, Nagle RB (1984) Chronic toxicity of S-(trans-1,2-dichlorovinyl)-L-cysteine in mice. J Appl Toxicol 4:315–319

Karami S, Lan Q, Rothman N, Stewart PA, Lee KM, Vermeulen R, Moore LE (2012) Occupational trichloroethylene exposure and kidney cancer risk: a meta-analysis. Occup Environ Med 69:858–867

Karami S, Bassig B, Stewart PA, Lee KM, Rothman N, Moore LE, Lan Q (2013) Occupational trichloroethylene exposure and risk of lymphatic and haematopoietic cancers: a meta-analysis. Occup Environ Med 70:591–599

Kelsh MA, Alexander DD, Mink PJ, Mandel JH (2010) Occupational trichloroethylene exposure and kidney cancer: a meta-analysis. Epidemiology 21:95–102

Krejci ME, Ridgewell RE, Koechel DA (1991) Acute effects of the D-isomer of S-(1,2-dichlorovinyl) cysteine on renal function and ultrastructure in the pentobarbital-anesthetized dog: site-specific toxicity involving the S1 and S2 cells of the proximal tubule. Toxicology 69:151–164

Lash LH, Fisher JW, Lipscomb JC, Parker JC (2000) Metabolism of trichloroethylene. Environ Health Perspect 108:177–200

Liu MY, Poellinger L, Walker CL (2003) Up-regulation of hypoxia-inducible factor 2alpha in renal cell carcinoma associated with loss of Tsc-2 tumor suppressor gene. Cancer Res 63:2675–2680

Lock EA, Reed CJ (2006) Trichloroethylene: mechanisms of renal toxicity and renal cancer and relevance to risk assessment. Toxicol Sci 91:313–331

Mally A, Walker CL, Everitt JI, Dekant W, Vamvakas S (2006) Analysis of renal cell transformation following exposure to trichloroethene in vivo and its metabolite S-(dichlorovinyl)-L-cysteine in vitro. Toxicology 224:108–118

Maltoni C, Lefemine G, Cotti G, Perino G (1988) Long-term carcinogenicity bioassays on trichloroethylene administered by inhalation to Sprague–Dawley rats and Swiss and B6C3F1 mice. Ann N Y Acad Sci 534:316–342

Moore LE, Boffetta P, Karami S, Brennan P, Stewart PS, Hung R, Zaridze D, Matveev V, Janout V, Kollarova H, Bencko V, Navratilova M, Szeszenia-Dabrowska N, Mates D, Gromiec J, Holcatova I, Merino M, Chanock S, Chow WH, Rothman N (2010) Occupational trichloroethylene exposure and renal carcinoma risk: evidence of genetic susceptibility by reductive metabolism gene variants. Cancer Res 70:6527–6536

National Cancer Institute (1975) Trichloroethylene is possible carcinogen. Chem Eng News 53:6–12

National Toxicology Program (1976) Carcinogenesis bioassay of trichloroethylene. Natl Cancer Inst Carcinog Tech Rep Ser 2:1–215

National Toxicology Program (1990) NTP Carcinogenesis studies of trichloroethylene (without epichlorohydrin) (CAS No. 79-01-6) in F344/N Rats and B6C3F1 Mice (Gavage Studies). Natl Toxicol Program Tech Rep Ser 243:1–174

Purdue MP (2013) Trichloroethylene and cancer. J Natl Cancer Inst 105:844–846

Scott CS, Cogliano VJ (2000) Trichloroethylene health risks–state of the science. Environ Health Perspect 108(Suppl 2):159–160

Scott CS, Jinot J (2011) Trichloroethylene and cancer: systematic and quantitative review of epidemiologic evidence for identifying hazards. Int J Environ Res Public Health 8:4238–4272

Seltzer RJ (1975) Reactions grow to trichloroethylene alert. Chem Eng News 53:41–43

Terracini B, Parker VH (1965) A pathological study on the toxicity of s-dichlorovinyl-l-cysteine. Food Cosmet Toxicol 3:67–74

US EPA (2001) Health Assessment Document for Trichloroethylene Synthesis and Characterization (External Review Draft). US Environmental Protection Agency, Office of Research and Development, National Center for Environmental Assessment, Washington Office, Washington, DC, EPA/600/P-01/002A

US EPA (2011) Toxicological review of trichloroethylene. In support of summary information on the Integrated Risk Information System (IRIS). National Center for Environmental Assessment, US Environmental Protection Agency, Washington, DC

US Government Accountability Office (2007) Department of defense activities related to trichloroethylene, perchlorate, and other emerging contaminants. GAO-07-1042T, Washington, DC

Vermeulen R, Zhang L, Spierenburg A, Tang X, Bonventre JV, Reiss B, Shen M, Smith MT, Qiu C, Ge Y, Ji Z, Xiong J, He J, Hao Z, Liu S, Xie Y, Yue F, Guo W, Purdue M, Beane Freeman LE, Sabbisetti V, Li L, Huang H, Rothman N, Lan Q (2012) Elevated urinary levels of kidney injury molecule-1 among Chinese factory workers exposed to trichloroethylene. Carcinogenesis 33:1538–1541

Wartenberg D, Reyner D, Scott CS (2000) Trichloroethylene and cancer: epidemiologic evidence. Environ Health Perspect 108(Suppl 2):161–176

Waters EM, Gerstner HB, Huff JE (1977) Trichloroethylene. I. An overview. J Toxicol Environ Health 2:671–707

Wolfgang GH, Gandolfi AJ, Nagle RB, Brendel K, Stevens JL (1990) Assessment of S-(1,2-dichlorovinyl)-L-cysteine induced toxic events in rabbit renal cortical slices. Biochemical and histological evaluation of uptake, covalent binding, and toxicity. Chem Biol Interact 75:153–170

Zhao Y, Krishnadasan A, Kennedy N, Morgenstern H, Ritz B (2005) Estimated effects of solvents and mineral oils on cancer incidence and mortality in a cohort of aerospace workers. Am J Ind Med 48:249–258

Chapter 10
Epigenetic Alterations due to Trichloroethylene

Craig A. Cooney

Abstract Trichloroethylene (TCE) is a volatile, water soluble, chlorinated hydrocarbon used as an intermediate in chemical synthesis. Wider use in the past and inappropriate disposal has resulted in large amounts of TCE in soil and water pollution including in hundreds of Superfund hazardous waste sites. Most TCE in human exposure comes through inhalation and drinking water where the main sources are occupational as well as contaminated ground water and soil.

As heritable modifications of DNA and chromatin, epigenetic changes can occur near the time of toxic exposure and remain for years, eventually contributing to overt disease such as cancer or autoimmunity. TCE could affect epigenetics through effects on metabolism, mitochondrial function, cellular signaling and formation of protein adducts. In this chapter, we mainly consider the epigenetic modifications of DNA and histone methylation and histone acetylation.

TCE can be toxic to many different organ systems in humans and animal models. Epigenetic effects have been demonstrated in animal models of TCE induced cancer, autoimmunity, neuropathy and congenital heart defects. TCE causes DNA hypomethylation in rodent models of liver cancer and interventions that restore methylation can also prevent the cancer. We showed that TCE exposure in a mouse model of autoimmune hepatitis causes increased expression of endogenous retrovirus-like sequences, changed expression of DNA methyltransferases and global DNA hypomethylation in CD4+ cells. Findings in this mouse model are discussed in light of the long-established activation of endogenous retrovirus expression in autoimmune diseases. We also studied the effects of TCE on behavior, gene

C.A. Cooney
Research and Development, Central Arkansas Veterans Healthcare System (CAVHS), John L. McClellan Memorial Veterans Hospital, 4300 West 7th Street, Little Rock, AR 72205, USA
e-mail: cooneycraiga@gmail.com

K.M. Gilbert, S.J. Blossom (eds.), *Trichloroethylene: Toxicity and Health Risks*, Molecular and Integrative Toxicology, DOI 10.1007/978-1-4471-6311-4_10,

expression, metabolism and epigenetics in the plasma and brains of mice. TCE caused a more oxidized cellular environment, compromised methyl metabolism and lower DNA methylation.

Parallel analyses in multiple tissues and the development of biomarkers of TCE exposure are just some of the approaches that will help us understand the long-term health risks of TCE. This should also assist the development of effective interventions to reverse the epigenetic effects of TCE exposure with the goal of preventing diseases such as cancer and autoimmunity.

Keywords Trichloroethylene • Epigenetics • Methylation • Acetylation • Cancer • Autoimmunity • Heart defects • Neuropathy • S-adenosylmethionine • Acetyl coenzyme A • Endogenous retrovirus • Ethanol • Bisphenol A

Abbreviations

5MC	5-methylcytosine
Ac	Acetyl group
AcCoA	Acetyl-coenzyme A
AIH	Autoimmune hepatitis
ATSDR	Agency for Toxic Substances and Disease Registry
B6C3F1	C57B6 strain x C3H strain F1 generation mice
BHMT	Betaine-homocysteine methyltransferase
BPA	Bisphenol A
BPS	Bisphenol S
CH3	Methyl group
DCA	Dichloroacetate
DNMT	DNA methyltransferase
EPA	Environmental Protection Agency (US)
ERV	Endogenous retrovirus
FAS	Fetal alcohol syndrome
GSH	Glutathione (reduced)
GSSG	Glutathione (oxidized)
H4K12	Histone H4 lysine 12
HAT	Histone acetyltransferase
HCY	Homocysteine
HDAC	Histone deacetylase
HERV-K	Human endogenous retrovirus virus K
IAP	Intracisternal A particle
MPTP	N-methyl-4-phenyl-1,2,3,6-tetrahydropyridine
MuERV	Murine endogenous retrovirus
NTP	National Toxicology Program
PD	Parkinson's Disease

RT-PCR	Real time PCR
SAH	*S*-adenosylhomocysteine
SAM	*S*-adenosylmethionine
TaClo	1-trichloromethyl-1,2,3,4-tetrahydro-β-carboline
TCA	Trichloroacetate
TCE	Trichloroethylene

Trichloroethylene (TCE) is a volatile, water soluble, chlorinated hydrocarbon now used primarily as an intermediate chemical for the production of refrigerants. In the past, TCE was used for a wide variety of additional purposes including industrial degreasing, anesthesia, dry cleaning and food processing (including coffee decaffeination) (Doherty, Chap. 1). Despite its declining use, TCE is still widely used as a metal degreaser and in some other applications. The US Environmental Protection Agency (EPA) estimated that 2.7 million pounds of TCE were disposed of, or released, in the United States in 2011 (http://iaspub.epa.gov/triexplorer/tri_release.chemical). According to the Agency for Toxic Substances and Disease Registry (ATSDR), large amounts of TCE remain in soil and water pollution including in over 800 National Priorities List (Superfund) hazardous waste sites (ATSDR 1997 and 2013).

Aside from occupational exposures, most current human exposure comes from contaminated ground water used for drinking, from inhaling TCE while using contaminated water (e.g. showering) or from contaminated indoor air caused by soil vapor intrusion (Forand et al. 2012). Many United States military personnel and their families have been exposed to TCE through drinking water at Camp Lejeune, North Carolina and elsewhere (ATSDR 1997 and 2013, http://www.atsdr.cdc.gov/sites/lejeune/tce_pce.html).

Remediation of TCE-contaminated soil and ground water is costly and time consuming. Thus, existing pollution and the continued production and use of TCE-containing consumer products means that human exposure to this chemical will continue, at least at low levels, for the foreseeable future. This makes understanding the long-term health effects of chronic TCE exposure particularly important. As described in other chapters in this book, TCE has many adverse health effects in humans. These include cancer and probably also neuropathy, heart defects and autoimmunity. Some of these effects are most obvious in animal models where the effects of controlled, intentional exposure can be quantified (Chiu et al. 2013). The effects of toxic compounds on long-term health can include epigenetic changes which occur near the time of toxic exposure but remain for years and contribute to the later overt presentation of disease (Poirier 2002; Waalkes et al. 2004; Cooney 2007; Cooney and Gilbert 2012; Ray and Richardson 2012; Blusztajn and Mellott 2013). Toxicant exposures just before and during gestation are of particular concern since the embryo and fetus are especially sensitive to epigenetic alterations (Wolff et al. 1998; Waalkes et al. 2004; Cropley et al. 2006; Cooney 2009; Davison et al. 2009; Downing et al. 2011).

10.1 Epigenetics

Epigenetics consists of heritable chromatin modifications that affect gene expression and help guide the development and health of plants and animals throughout their lifecycles. A broad range of factors affect epigenetics. These include exposure to certain toxicants such as TCE. However, there are few data on the effects of TCE on DNA methylation and fewer yet on some other epigenetic modifications such as histone methylation and histone acetylation. The available data will be discussed later in this chapter. Epigenetics has been extensively reviewed (Cooney 2007, 2010; Cooney and Gilbert 2012; Dawson and Kouzarides 2012), so only a few general points will be discussed here. Epigenetics has been much studied in cancer and the knowledge and many of the approaches from that work can be applied to toxicology research.

For vertebrates, CG dinucleotides (called CpGs) are the principal targets of DNA methyltransferases (DNMTs) which use the methyl group donor S-adenosylmethionine (SAM) to methylate DNA at the five position of cytosines to form 5-methylcytosine (5MC, Ooi et al. 2009). This reaction also yields the metabolite S-adenosylhomocysteine (SAH) which is often recycled back to SAM by methyl metabolism. Because the CpG sequence is a palindrome, methylation patterns on parental DNA strands can be copied onto daughter strands during cell division by DNMT1. This process is sometimes called maintenance methylation. DNA is also sometimes *de novo* methylated by DNMT1 and by the dedicated *de novo* DNMTs, DNMT3a and DNMT3b. These heritable DNA methylation patterns can propagate long-term changes in gene expression in generations of daughter cells and sometimes in generations of animals (Cropley et al. 2006; Cooney 2007; Champagne and Curley 2009; Li et al. 2011). 5MC near transcription start sites and other nearby regulatory regions tends to silence gene expression (Weaver et al. 2005; Ooi et al. 2009). This works in part by preventing transcription factor access and attracting methylated DNA binding proteins. Silence is maintained by protein complexes that also modify histones to reinforce transcriptional silence in some cases or "poise" a region for activation in other cases (Dawson and Kouzarides 2012). DNA demethylation can occur when 5MC is removed by base excision repair and/or oxidation of the methyl group. A series of increasingly oxidized products of 5MC, namely 5-hydroxymethylcytosine, 5-formylcytosine and 5-carboxycytosine are all found in mammalian DNA (Seisenberger et al. 2013). Demethylation is especially prominent post-fertilization and in primordial germ cells, both times when DNA methylation patterns are extensively rewritten (Hackett et al. 2013; Seisenberger et al. 2013).

Some of the major DNA binding proteins of chromatin, the histones, are also enzymatically methylated using SAM (Kooistra and Helin 2012; Dawson and Kouzarides 2012). Whether histone methylation promotes or silences gene activity depends on the specific methylation site(s). Most histone methyltransferases, demethylases and histone binding proteins are site and methylation specific. There are greater varieties and specificities of histone methyltransferases, histone

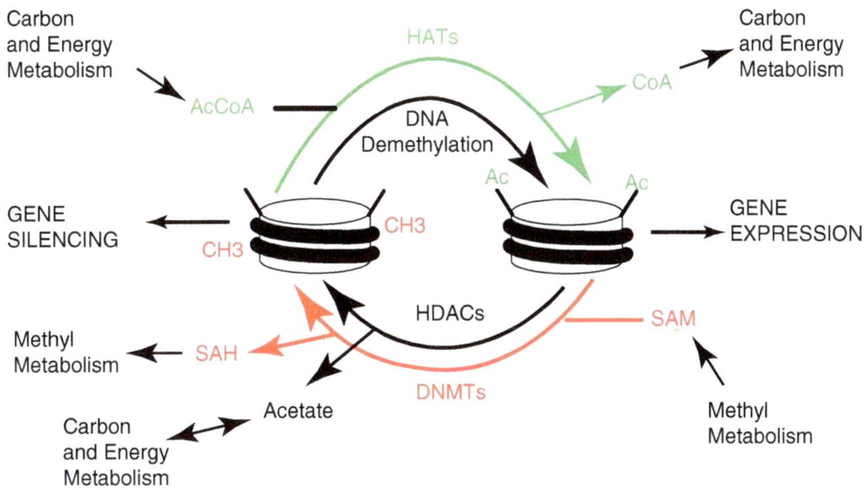

Fig. 10.1 Histone acetylation and DNA methylation: two major epigenetic factors affecting gene expression. Histones in chromatin (nucleosomes) are shown as short cylinders with histone H3 and H4 tails projecting up. Acetyl groups from acetylCoA are used by histone acetyltransferases (*HATs*) to acetylate these histone tails and promote gene expression. This process is reversed by HDACs that remove the acetyl groups. DNA in chromatin (nucleosomes) is shown looping around the histones. Methyl groups from SAM are used to by DNMTs to methylate DNA and silence gene expression. This process is reversed by DNA demethylation processes that result in unmethylated cytosine (Adapted from Cooney (2010))

demethylases and methylated histone binding domains than there are for corresponding enzymes and binding domains for histone acetylation or methylation of DNA.

Gene activity can also be regulated by histone acetylation which depends on histone acetyltransferases (HATs), histone deacetylases (HDACs) and the acetyl donor acetyl-coenzyme A (AcCoA, Cooney 2010). Gene activity is nearly always promoted by histone acetylation. Figure 10.1 shows the dependence of DNA methylation on methyl metabolism and of histone acetylation on carbon and energy metabolism and illustrates how DNA methylation and histone acetylation have largely opposite roles in regulating gene expression.

Epigenetic alterations are primarily mediated by covalent modifying enzymes that add or remove methyl groups, acetyl groups and other groups from chromatin in combination with enzymes and other proteins that recognize these modified sites and recruit protein complexes that then facilitate transcription or silencing. The known epigenetic modifications of DNA are 5MC and its methyl oxidized forms, as described above. In contrast, there are many histone modifications in addition to methylation and acetylation (Dawson and Kouzarides 2012) and some RNAs have prominent roles in epigenetics (Kurokawa et al. 2009; Collins et al. 2010; Lee 2012). Although histone methylation has high specificity and is often thought to lead in epigenetic processes, gene expression is often linked to DNA methylation and/or histone acetylation, and gene expression can often be changed by metabolites or drugs targeting DNA methylation or histone acetylation (Weaver et al. 2005; Eilertsen et al. 2008; Champagne and Curley 2009; Peleg et al. 2010; Cooney 2010).

Many genes will have DNA and histone modifications consistent with their transcriptional activities; however some "bivalent" genes have modifications or complexes with silencing and activating features at the same time (Dawson and Kouzarides 2012). These "bivalent" genes are poised to become transcriptionally active. As discussed in the rest of this article, most epigenetics work to date on TCE has been done on DNA methylation and histone methylation and acetylation.

10.2 Effects of TCE on Metabolism and Other Mechanisms for Epigenetic Change

The likelihood that TCE impacts DNA methylation is bolstered by its previously identified effects on pathways that regulate epigenetic alterations. For example, TCE has been shown to affect methyl metabolism which in turn regulates SAM levels and is important for epigenetics (Gilbert et al. 2009; Blossom et al. 2012, 2013).

TCE affects antioxidant defenses where it depletes reduced glutathione (GSH) and cysteine and causes other changes all consistent with TCE producing oxidative stress (Blossom et al. 2012, 2013, discussed below in Sect. 10.6). Oxidative stress can compromise methyl metabolism when homocysteine (and indirectly methionine via SAM and SAH) is drawn through the transsulfuration pathway to make cysteine for glutathione synthesis and antioxidant defenses. Examples of this have been described in liver (Mosharov et al. 2000), brain (Vitvitsky et al. 2006) and plasma (Melnyk et al. 2012). Compromise of methyl metabolism occurs because production of cysteine for glutathione synthesis depletes levels of SAM and its precursors methionine, homocysteine (HCY) and SAH. This leaves less SAM available for cellular methylation reactions such as DNA and histone methylation.

TCE also affects mitochondrial function including oxidative phosphorylation (Lash et al. 2001; Gash et al. 2008; Sauerbeck et al. 2011) which would be expected to affect overall energy metabolism. Energy metabolism and mitochondrial function could also have effects on epigenetics by, for example, affecting the levels of AcCoA available for histone and transcription factor acetylation and gene activity (Cooney 2008, 2010; Wallace et al. 2010). Zeisel (2013) discusses numerous connections between energy metabolism and methyl metabolism and potential effects on epigenetics. Thus, there are multiple pathways by which TCE could impact the pathways that affect epigenetics. Significant effects of TCE on a variety of normal, endogenous metabolites measured in plasma, serum and urine (Blossom et al. 2012; Fang et al. 2013b) suggest that TCE's effects are systemic and likely affect multiple organs.

When evaluating the effects of TCE on epigenetic pathways, TCE metabolites must also be considered (Lash et al. 2000; Kim et al. 2009; Chiu et al. 2013, Gilbert, Chap. 2). TCE metabolism is complex and has multiple steps including oxidation (by cytochrome P450 enzymes and alcohol dehydrogenase), glucuronidation (by UDP-glucuronosyltransferases) and glutathione conjugation (by glutathione S-transferases) (Lash et al. 2000). Major metabolites include trichloroacetate (TCA), trichloroethanol and chloral hydrate (Lash et al. 2000). However metabolism and therefore metabolite levels differ with routes of exposure, tissue, species, sex and other factors (Abbas and Fisher 1997; Lash et al. 2000; Clewell et al. 2000; Kim et al. 2009). Just a few TCE metabolites are discussed in this chapter because they relate to diseases caused by TCE (such as cancer and Parkinson's disease).

For example, the TCE metabolite dichloroacetate (DCA) has important effects on energy metabolism, mitochondrial function and apoptosis (Bonnet et al. 2007) and tends to activate mitochondria in cancer cells. Because of this, DCA is studied as an anticancer agent (Sun et al. 2010; Strum et al. 2013). Because DCA activates mitochondria and opens the pathway to burn AcCoA (Bonnet et al. 2007) it may well affect epigenetics (especially histone acetylation, Cooney 2008). On the other hand, when given to healthy mice, DCA is a liver carcinogen (Bull et al. 1990; Herren-Freund et al. 1987; Pereira et al. 1997, as discussed below in the Sect. 10.3). Thus the actions of DCA deserve attention so that we can understand its toxic effects as well as its therapeutic potential.

There are many ways that toxicants can interfere with epigenetics. These include alterations in metabolism, cell signaling, mitochondrial function, enzyme activities, and endogenous retrovirus (ERV) activity (Cooney and Gilbert 2012). TCE could affect epigenetics in these ways and through changes in the expression of genes involved in epigenetic control and direct modification of proteins causing changes in their enzymatic or other activities.

The effects of toxicants such as TCE on the various pathways that regulate epigenetics are just beginning to be identified. Once obtained, this important mechanistic information will help predict long-term effects of toxicant exposure, which may lead to the development of interventions. Many clues concerning possible TCE-induced epigenetic alterations come from cancer studies which will be discussed next.

10.3 Cancer

The EPA considers TCE a known carcinogen (USEPA 2011). Numerous studies indicate that occupational exposure to TCE causes cancer including kidney cancer (Jollow et al. 2009; Karami et al. 2012; Hansen et al. 2013, Wartenberg, Chap. 9). Studies, including those of the National Toxicology Program (NTP), show that TCE causes hepatocarcinoma in mice but not in rats (NTP 1988 and 1990; Bull 2000). For example, in a 2 year study of B6C3F1 mice given 1,000 mg/kg TCE by corn oil gavage, hepatocellular carcinoma rates for males were 8/48 in controls and 31/50 in TCE dosed ($P<0.001$) and for females 2/48 in controls and 13/49 in TCE dosed ($P<0.005$) (NTP 1990). Cytomegaly of the kidney (toxic nephrosis) was seen in nearly all TCE dosed B6C3F1 mice and Fischer 344 rats (both sexes) but in none of the controls. Some kidney cancer was seen Fischer 344 rats but its incidence was too low to prove carcinogenicity of TCE (NTP 1990). Some other studies report TCE as a cause of cancer in rats (NTP 2011).

TCE's identification as carcinogenic but not mutagenic is important since in general, all cancers studied in humans and rodents have extensive epigenetic changes (Cooney 2008; Fernandez et al. 2012). Cancers typically also have extensive metabolic, genetic and chromosomal rearrangements (Bonnet et al. 2007; Cooney 2010; Wallace et al. 2010; Vogelstein et al. 2013). Few data are available on the epigenetics of human cancers that likely arose from TCE exposure (Brauch et al. 1999; Banks et al. 2006). However, promising approaches are being developed to address this. For example, Ellsworth et al. (2012) used mutational profiling to find significant differences between brain cancers in human subjects exposed to chlorinated solvents and subjects judged to have sporadic brain cancers.

However, our greater interest is in early epigenetic changes due to TCE that may later lead to cancer. Identifying these changes gives us the opportunity to reverse them early on and prevent the cancer. Mice have been used for such studies because, as discussed above, chronic TCE exposure causes hepatocarcinoma (NTP 1990). To address early events that may lead to cancer, mouse models have been used to look at acute effects of TCE. Several genes that promote cell proliferation, often called oncogenes, tend to be highly expressed in cancers and, importantly, in precancerous lesions. In two of these genes, *c-jun* and *c-myc*, TCE in acute high-doses (1,000 mg/kg/day), given to female B6C3F1 mice for 5 days, decreased DNA methylation of *c-jun* and *c-myc* promoters in the liver. This correlated with increased *c-jun* and *c-myc* gene expression in the liver (Tao et al. 2000a). When mice were given methionine (to increase their SAM levels, Wang et al. 2001) just after TCE dosing, they did not show this *c-jun* and *c-myc* hypomethylation. Similar results were seen when giving mice either DCA or TCA, both metabolites of TCE. In related (but not identical) study designs, TCE and TCA increased DNMT activity in liver when female B6C3F1 mice were first treated with the tumor initiator *N*-methyl-*N*-nitrosourea (Tao et al. 2000b). Overall, this indicates that in the female B6C3F1 mouse model of hepatocarcinoma, short-term effects can be reversed with methionine. Because many human exposures to TCE are in the past, it would be practical to know what treatments would reverse TCE's effects in this mouse model in the weeks and months after TCE exposure.

TCE metabolites DCA and TCA are also liver carcinogens in mice (Bull et al. 1990; Herren-Freund et al. 1987; Pereira et al. 1997). Methionine supplementation prevented most DCA induced cancer and hypomethylation (Pereira et al. 2004). This pattern of oncogene hypomethylation, oncogene overexpression, compromised methyl metabolism and increased DNMT enzyme activity (although sometimes lower expression of *Dnmt* genes) have been shown for liver cancer in rodent models treated with toxicants or methyl deficient diets (Wainfan and Poirier 1992, Pascale et al. 2002; Phillips et al. 2009; Pogribny et al. 2012; Frau et al. 2013). Often early effects predisposing to cancer can be prevented or reversed by methyl donors, folate or similar treatments which often, but not always, reduce cancer incidence (Tao et al. 2000a; Pascale et al. 2002; Sie et al. 2011; Gonda et al. 2012; Fang et al. 2013a; Frau et al. 2013).

One of several hallmarks of cancers is extensive genome rearrangement (Vogelstein et al. 2013) which may be promoted by several factors including transcriptional activation and transposition of ERVs and other interspersed DNA repeats (Romanish et al. 2010). This activation of ERVs probably results from loss of epigenetic silencing normally found on most ERVs and interspersed DNA repeats (Cherkasova et al. 2011). However in most cases it is unknown whether this deregulation occurs prior to cancer, or during cancer development and progression. Of course, the timing of deregulation will help determine whether ERV activation causes some cancers or is mainly involved in later processes such as progression or metastasis (Downey et al. 2012). Although data is not available on ERV activation in TCE induced cancers, when investigating the causes of autoimmunity, we found ERV activation in the T-cells of TCE-treated mice (Gilbert et al. 2012). The activation of ERVs by TCE in tissues prone to TCE-induced cancer has yet to be reported.

Starting from these data, interventions (methionine, folate and other model epigenetic effectors) should be designed and tested in animal models to prevent long-term adverse effects of TCE, including cancer, and to quantify likely effectors and markers (*c-jun*, *c-myc*, *Dnmt1* and *Dnmt3a&b* gene expression, ERV expression and others) to provide guidance for testing similar interventions in people exposed to TCE.

10.4 Immune Disease

Some forms of immune disease including those associated with autoimmunity have been studied for their epigenetic alterations (reviewed by Cooney and Gilbert 2012). In particular, lupus, whether idiopathic or drug-induced, has been well studied in this regard. In pioneering work, Bruce Richardson and colleagues have shown a range of epigenetic effects from global DNA hypomethylation to gene specific DNA hypomethylation in lupus (Gorelik and Richardson 2010). Further, they reproduced similar effects in mouse models of lupus (Quddus et al. 1993; Yung et al. 1996). Some recent studies have used massively parallel surveys to look for DNA methylation changes in humans with autoimmune disease (Fernandez et al. 2012). These show that, unlike cancer and aging which cause mainly gene-specific hypermethylation, autoimmunity causes a preponderance of gene-specific *hypo*methylation.

TCE exposure has been associated with autoimmune disease in both humans and mouse models (Gilbert, Chap. 2). However, the mechanism by which TCE causes immune disease and other long-term health effects has not been determined and only a few studies have addressed the metabolic and epigenetic effects with TCE induced autoimmunity.

Gilbert et al. (2009) studied autoimmune hepatitis (AIH) in female MRL+/+ mice exposed to TCE (0.5 mg/ml in drinking water). AIH in this mouse model is observable at 26 weeks of TCE exposure and resembles idiopathic AIH in humans. Gilbert et al. measured liver gene expression, endogenous metabolite levels, oxidized proteins, liver microsomal protein specific antibodies and histopathology. Gene array results showed that of 200 genes whose expression was significantly altered by TCE, 85 % of these showed increases in gene expression. Genes whose expression was increased included several for the metabolism and detoxification of TCE including alcohol dehydrogenases, cytochrome oxidases and glutathione *S*-transferases. At least one of each of these enzyme types was confirmed by real-time PCR (RT-PCR). They also found that betaine-homocysteine methyltransferase (BHMT) gene expression was increased in the array data but this did not reach significance with RT-PCR data. Interestingly however, metabolite analysis showed that SAH levels were significantly decreased and *N,N*-dimethylglycine levels were significantly increased, both of which would be an expected outcome of increased BHMT activity. They found no evidence of increased oxidative stress in the livers of TCE treated animals. Overall these results suggest that chronic TCE may improve methyl metabolism in the liver.

Using this same mouse model of AIH, we recently compared control mice with mice treated with TCE (Gilbert et al. 2012). Among other endpoints, we studied expression of *Dnmt* genes and the murine endogenous retrovirus (MuERV) and the related, ERV-like intracisternal A particle (IAP) repeats. ERVs are often overexpressed in autoimmunity (Perl et al. 2010) and their expression is controlled by epigenetics including 5MC (Walsh et al. 1998; Gaudet et al. 2004; Kato et al. 2007). We compared splenic CD4+ T cells from control mice with those from mice that we treated for 12 weeks with 0.5 mg/ml TCE in drinking water. After stimulating both groups of cells for 24 h we observed IAP expression that was over 8-fold higher and MuERV expression that was over 2.5-fold higher in TCE treated mice ($p<0.05$). *Dnmt1* expression was significantly higher and *Dnmt3a* expression several fold lower with TCE treatment. In some tissues and some developmental stages of mouse, DNMT1 and DNMTs 3a and 3b are needed for IAP methylation (Gaudet et al. 2004; Kato et al. 2007). This indicates that changes in the epigenetic machinery and in the expression of ERVs may be important in murine AIH. ERV expression may contribute to the disease process in autoimmunity and specific mechanisms have been proposed for this (Perl et al. 2010). However, of these numerous possible mechanisms, definitive experiments have not been done to identify one or more specific mechanisms as causal. It is possible that the decline in Dnmt3a expression contributes to the activation of IAPs following TCE dosing, however additional research is needed to understand how TCE affects IAP and MuERV expression.

To look at epigenetic alterations directly, we compared total DNA methylation in splenic CD4+ T cells from control mice with those from TCE treated mice (in this case after 17 weeks of TCE exposure). After stimulation, total DNA methylation was lower in the cells from TCE treated mice. In other words, TCE caused global hypomethylation in mouse CD4+ T cells (Gilbert et al. 2012). This occurred after treatment with either 0.01 or 0.1 mg/ml TCE-containing drinking water. Consistent with hypomethylation, we observed a nearly 3-fold increase ($p<0.05$) in IAP expression in these same cells for mice treated at the 0.1 mg/ml TCE level.

Compared to earlier studies of female mice with possibly improved liver methyl metabolism on TCE exposue (Gilbert et al. 2009), our recent studies (Blossom et al. 2012 and 2013) of metabolite levels in plasma and brain of male mice showed impaired methyl metabolism in response to chronic TCE. These more recent studies were done in male MRL+/+ mice so that we cannot make a comparison of TCE's effects on multiple organs (liver, brain, plasma). Future studies are needed to study metabolism, gene expression and epigenetics in the liver, brain, T-cells, plasma, urine etc. from MRL+/+ mice of the same TCE exposures, same time points and same sex.

These various effects show that epigenetic alterations occur in TCE-treated mice, however the mechanisms are unclear. Effects on methyl metabolism could explain some or all of these effects as could effects on the expression of Dnmts. Further work is needed to determine how TCE causes these metabolic, gene expression and epigenetic changes and which changes come first. A better understanding of these effects may contribute to deciphering the mechanisms for other TCE induced diseases (in general, methyl metabolism affects cancer, heart defects and neuropathies (Tao et al. 2000a; Pascale et al. 2002; Hobbs et al. 2005a, b). Mechanisms will then allow us to design interventions that may reverse or ameliorate the long-term effects of TCE as has been done for some other toxic exposures (Tao et al. 2000a; Pascale et al. 2002; Downing et al. 2011, Otero et al. 2012, Bekdash et al. 2013).

10.5 Heart Defects

TCE exposure during gestation causes heart defects in the offspring of rodents (Caldwell et al. 2010; Palbykin et al. 2011) and birds (Loeber et al. 1988; Rufer et al. 2010; Makwana et al. 2010) and probably in humans (Yauck et al. 2004; Forand et al. 2012, Selmin, Chap. 8). Yauck et al. (2004) compared infants with congenital heart defects with infants without congenital heart defects for whether or not their mothers lived near TCE-emitting sites. They also compared other factors including maternal age, alcohol use, chronic hypertension, and preexisting diabetes. After adjusting for other factors, they found that, in older women (>37 years of age), the proximity of residence to TCE-emitting sites was associated with a three-fold increased risk of offspring congenital heart defects. In a more recent study, Forand et al. (2012) also found offspring cardiac defects more prevalent when mothers were exposed by soil vapor intrusion of TCE which contaminated their indoor air.

In rat studies, TCE reduced expression of the cardiac gene *Serca2a* in association with hypermethylation of its promoter in embryonic heart after maternal exposure to low concentrations (10 ppb) of TCE in drinking water (Palbykin et al. 2011). In this same study, SAM concentrations were lower in embryos whose mothers were exposed to TCE. In an earlier study by this same group, mouse embryonic heart gene expression was surveyed with DNA microarrays to show broad effects of TCE on cardiac gene expression (Caldwell et al. 2010). They further showed that maternal folate supplementation had its own pattern of altered gene expression and did not reverse the broad effects on embryonic gene expression caused by maternal TCE exposure. Thus while methyl metabolism is changed by TCE, specific gene methylation is not necessarily altered in the same direction as metabolism.

Although much TCE metabolism occurs in the livers of adults, in early bird embryos the liver and brain are not yet developed and the heart must metabolize TCE directly (Makwana et al. 2013). This may begin to explain the adverse cardiac effects of TCE in early development (Selmin, Chap. 8).

Several studies show that maternal diets affecting methyl metabolism such as broadly methyl supplemented diets (Wolff et al. 1998) or folate supplemented diets may positively affect health outcomes for offspring who are exposed to toxicants or have specific genetic defects (Downing et al. 2011; Cho et al. 2012; Billington et al. 2013). Folate supplementation alone did not reverse TCEs effects on gene expression in mouse embryos (Caldwell et al. 2010). Additional interventions combined with massively parallel assays (e.g. next generation sequencing for transcription, DNA methylation etc.) may reveal effective interventions. Maternal interventions to ameliorate TCE's effects on the developing heart have yet to be developed.

10.6 Neurological Effects

Acute TCE exposure, usually as occupational inhalation, can cause intoxication including dizziness, confusion, headaches, numbness, loss of consciousness, and in unusual circumstances, even death (ATSDR 1997). In addition, there can be many longer term neurological effects including memory loss and trigeminal nerve neuropathy (ATSDR 1997). Neurological effects of TCE are reviewed in two chapters of this book on neurotoxicity, by Bale (Chap. ___) and Goldman (Chap. 6) and here I select just a few examples to discuss and to emphasize the important role, in general, of epigenetics in memory, behavior, dementia and neurological function.

Epigenetics and especially DNA methylation and histone acetylation have key roles in animal behavior (Weaver et al. 2005; Champagne and Curley 2009), memory (Miller and Sweatt 2007; Sweatt 2012; Feng et al. 2010) and in dementia (Peleg et al. 2010; Pavlopoulos et al. 2013, reviewed by Cooney 2010). In rats, maternal behavior toward pups in the first postnatal week has lifelong effects on pup behavior (Weaver et al. 2005; Champagne and Curley 2009). Because female pups, once grown, will show different nursing behavior toward their pups, these behavioral effects might be passed on to multiple generations (reviewed by Cooney 2007). These behavioral effects can be attributed at least in part to changes in DNA methylation and histone acetylation in the hippocampal glucocorticoid receptor promoter and can be modified by treatments that target these pathways (Weaver et al. 2005; Champagne and Curley 2009).

Miller and Sweatt (2007) showed that epigenetics and especially DNA methylation are essential for normal memory formation. They showed that hippocampal RNA levels for the DNMTs 3a and 3b were increased with fear conditioning in rats and that DNMT inhibitors prevented memory formation (Miller and Sweatt 2007; Sweatt 2012). Subsequent studies knocking out either or both Dnmt1 and Dnmt3a just in forebrain excitatory neurons showed that a single knock out allowed memory formation but the knockout of both Dnmts interfered with synaptic plasticity, learning and memory (Feng et al. 2010).

To study epigenetics and memory in aged animals, Peleg et al. (2010) tested hippocampus-dependent associative learning in mice at various ages up to 16 months of age. In 3 month old mice, learning upregulated histone H4 lysine 12 (H4K12) acetylation and changed expression of over 2,000 genes whereas in 16 month old mice changes in H4K12 acetylation were insignificant and only 6 genes were differentially expressed. Injection of HDAC inhibitors into the hippocampus of 16-month old mice increased H4K12 acetylation and improved learning. This study shows that reversible histone acetylation changes are important parts of age-related memory loss. Recently, RbAp48, a natural histone deacetylase inhibitor protein, has been shown to help regulate both histone acetylation and memory (Pavlopoulos et al. 2013). In young mice, RbAp48 in the dentate gyrus of the hippocampus helped maintain histone acetylation and normal memory. In aged mice, RbAp48 levels are low and correspond to lower histone acetylation and poor memory performance. Experimental manipulations to lower RbAp48 in young mice adversely affected their memories and manipulations to increase RbAp48 in old mice improved their memories.

TCE and some of its metabolites cause dopaminergic neurodegeneration and may be a cause of Parkinson's Disease (PD, Gash et al. 2008; Liu et al. 2010; Sauerbeck et al. 2012, Goldman, Chap. 6). The TCE metabolite 1-trichloromethyl-1,2,3,4-tetrahydro-β-carboline (TaClo) inhibits mitochondrial complex 1 (Janetzky et al. 1995) diminishing energy production. TaClo is a structural analog of the dopaminergic neurotoxin N-methyl-4-phenyl-1,2,3,6-tetrahydropyridine (MPTP) (Bringmann et al. 1995; Akundi et al. 2004). MPTP is an established cause of PD in humans and in animal models. This work provides a plausible mechanism by which TCE may cause Parkinson's disease in people and should be sufficient to warrant avoiding TCE and related chlorinated solvents. Additional research is needed to establish a clear cause and effect.

Alpha-synuclein, a main component of Lewy bodies, is overexpressed in PD patients. A range of epigenetic effects have been found affecting the alpha-synuclein gene and other genes *in vitro* and in PD patients including alpha-synuclein gene hypomethylation in PD patients (reviewed by Coppedè 2012). Metabolic changes in PD patients could affect epigenetics as serum HCY is above normal and methylation potential (SAM/SAH ratio) varies with higher SAM/SAH correlating with better cognitive function (Obeid et al. 2009).

These recent studies indicate that factors affecting epigenetics could have important effects on behavior and memory. We studied several effects of TCE on behavior, gene expression, metabolism and epigenetics in the hippocampi and cerebella of male mice (Blossom et al. 2012, 2013). In assays of their hippocampi as well as separate assays of cerebella, TCE treated male mice had decreased GSH, decreased ratio of reduced glutathione/oxidized glutathione (GSH/GSSG), decreased cysteine/cystine and increased cystine (the oxidized form of cysteine) and increased 3-nitrotyrosine (Blossom et al. 2012, 2013). These measures indicate that TCE treated mice have a more oxidized environment in their hippocampi and cerebella.

In blood plasma, TCE-treated male mice had higher HCY and SAH levels and lower methionine and SAM levels and lower SAM/SAH ratios (Blossom et al. 2012).

All of these measures suggest lower methylation capacity. These TCE-treated mice also had lower plasma SAM + SAH and HCY + methionine levels suggesting that more HCY is being converted to cysteine, presumably in response to oxidative stress. These changes are all in the directions that would be expected to compromise cellular methylation reactions including DNA methylation (Cooney 2006)

In cerebella, TCE treated male mice had lower methionine levels and lower total 5MC than untreated controls (Blossom et al. 2013). Overall this data indicates that TCE produces oxidative stress and lower methylation capacity in these brain regions and possibly in many tissues based on the plasma values we observed. Studies in female mice suggest that chronic TCE may increase methylation capacity in the liver (Gilbert et al. 2009), which is the opposite of the direction we observed in the brains and plasma of male mice and the CD4 + T-cells of female mice. Future experiments will measure the responses of several tissues and plasma in the same mice to ascertain the effects of chronic TCE on methyl and antioxidant metabolism. These measures will provide data on which we can base interventions with the aim of ameliorating the adverse health effects of chronic TCE.

10.7 Endogenous Retroviruses (ERVs)

ERVs are endogenous components of our genomes (Walsh et al. 1998; Bannert and Kurth 2004; Perl et al. 2010) which are inherited though the germline (Mendelian inheritance). They are thought to have arisen over evolutionary time from repeated retroviral infections of germline cells (Bannert and Kurth 2004). While retroviral infections often lead to integration of the proviral genome into host DNA, only infection of the germline leads to Mendelian inheritance to subsequent generations. Examples of ERVs in humans include endogenous retrovirus virus K (HERV-K) and, in mice, IAPs and MuERVs. Most ERVs are largely silent in healthy cells and tend to remain in place in the genome for extended times and, apparently, for many generations (Wolff et al. 1998; Morgan et al. 1999). In certain developmental stages and in some diseases, ERVs can become transcriptionally active and sometimes transpose in the genome (Romanish et al. 2010). Activation of ERVs can result in their transcription and cause interference with expression of nearby "host" genes. Effects on nearby "host" genes can be over expression, deregulated expression, silencing and other forms of dysregulation (Wolff et al. 1998; Rakyan et al. 2003; Druker et al. 2004). Following transcription, the expression of ERV-encoded proteins can lead to ERV transposition and the promotion of other aberrant processes that disrupt the genome (e.g. reverse transcription of "host" RNAs) (Romanish et al. 2010).

ERVs are controlled by epigenetic silencing including 5MC (Walsh et al. 1998; Cooney et al. 2002; Gaudet et al. 2004; Schulz et al. 2006; Reiss et al. 2010; Cherkasova et al. 2011). Many nutritional (Cooney et al. 2002), metabolic and genetic factors (Gaudet et al. 2004) can affect ERV expression.

ERV activity is clearly correlated, and may be causal, in some cancers and some forms of autoimmunity. Increased ERV expression is found in several types of autoimmune diseases in both humans and mice (Balada et al. 2010; Baudino et al. 2010).

We find that TCE activates expression of two ERVs in mice after 12 weeks of exposure and long before the development of overt disease (autoimmune hepatitis) (Gilbert et al. 2012). This was the first report of TCE-induced ERV overexpression and the increase of IAP transcripts we observed was the strongest transcript induction we have observed with TCE. Because of their high copy number (e.g. about 1000 IAP copies in the mouse genome), small increases in the expression of individual ERVs could have large overall effects if tens or hundreds of ERVs per genome increase their expression.

Retroviral expression seems to have a direct role in autoimmune pathology (Perl et al. 2010) but its role in TCE-induced pathology remains to be determined. The role of ERV expression in other TCE-induced diseases such as heart defects, cancer and neuropathies should be investigated.

10.8 Potential for Epigenetic Effects from TCE Coexposure with Other Toxicants

Coexposure with TCE and other toxicants could result in broader or additive epigenetic effects in some cases or a cancellation of effects in others. Here I discuss two compounds, ethanol and bisphenol A (BPA), to which people are routinely exposed and thus significant coexposure with TCE is likely.

The use of ethanol is widespread. Ethanol has clear epigenetic effects on the fetus during pregnancy and these epigenetic effects may explain much of fetal alcohol syndrome (FAS, Ramsay 2010). Several studies in mice show that 5MC patterns on imprinted genes (such as *Igf2* and *H19*) are changed by maternal alcohol consumption (Haycock and Ramsay 2009; Stouder et al. 2011; Downing et al. 2011; Veazey et al. 2013; Resendiz et al. 2013). Culture of embryos *in vitro* with alcohol also shows extensive changes in 5MC (Liu et al. 2009). Other fetal alcohol mouse studies, some including genes for neural development, show changes in 5MC and histone modifications (Veazey et al. 2013). Supplementation of maternal diets with a combination dietary methyl supplement (folic acid, vitamin B12, betaine, choline, methionine and zinc, Downing et al. 2011) or a dietary choline supplement (Otero et al. 2012, Bekdash et al. 2013) ameliorated some epigenetic and other effects of maternal alcohol on rodent fetuses.

Chronic alcohol use has been shown to cause hepatocellular carcinoma in mice (Tsuchishima et al. 2013). This occurred without tumor initiation by another carcinogen. Ethanol also affects detoxification pathways (Lu and Cederbaum 2008) which can change the detoxification of xenobiotics including TCE and some other carcinogens (Nakajima et al. 1988; Klotz and Ammon 1998). However, the interactions of ethanol and TCE with respect to epigenetics have not been reported.

In adult human subjects, DNA from peripheral blood shows methylation differences between alcohol dependent and control American subjects (Zhang et al. 2013a) and Chinese subjects (Zhang et al. 2013b). In Americans, two genes, *GABRB3* and *POMC* were differentially methylated in African-American subjects while several other genes were differentially methylated in European-American subjects.

Bisphenols were originally developed as estrogen agonists but found widespread use as the building blocks of plastics (Vogel 2009). A variety of bisphenols are agonists or antagonists for estrogen receptors and other nuclear receptors (Molina-Molina et al. 2013). Bisphenols and other estrogen disrupting chemicals can have effects at low doses that are not revealed by more traditional studies of high exposures (Vandenberg et al. 2012).

Bisphenol-A (BPA) and its analog bisphenol-S (BPS) are widely used as the main component in some plastics and as a component in numerous other consumer products including thermal paper cash register receipts and the inside lining of metal food cans (Biedermann et al. 2010; Liao and Kannan 2011). In some cases, including food can lining, these uses stretch back to the 1960s. Use in receipts leads to contact with paper currency which is then subsequently handled by many individuals (Liao et al. 2012).

Mouse studies of BPA and epigenetics have yielded varying results. Using the yellow-agouti mouse model, some small studies show changes in epigenetically determined coat color with BPA and the naturally occurring soy estrogen, genistein (Dolinoy et al. 2006, 2007). Yellow-agouti mouse studies with well-controlled coat color quantification and scoring, including a recent large study, find no change in epigenetically determined coat color with soy protein isolate (Badger et al. 2008), genistein or BPA (Rosenfeld et al. 2013). Studies using other models show effects of BPA on epigenetics, including effects on imprinting (Susiarjo et al. 2013) and effects in the brain (Kundakovic et al. 2013). The route of BPA dosage is important because realistic exposure models give much different results than models using artificial exposures (Vandenberg et al. 2013). Coat color studies are best using quantitative methods (Badger et al. 2008; Ounpraseuth et al. 2009, Rosenfeld et al. 2013) or where unique phenotypes are produced by the treatment (Wolff et al. 1998; Cooney et al. 2002). Clearly more research is needed to determine if BPA affects mainly specific tissues at specific life stages or if its effects are more pervasive involving most tissues (including the periphery e.g. skin and hair) and most life stages (e.g. fetal exposure and adult exposure at multiple ages).

Alcohol and bisphenols are just a few of the more common compounds affecting epigenetics which are likely coexposures with TCE. Other compounds include genistein (from soy products), sulforaphane (from broccoli, Watson et al. 2013). Several nutrients such as folates, betaine and methionine will be found in all subjects as they are nutrients and metabolites. However, these nutrients will be found in greatly varying levels in diets (Cooney 2006) which will likely affect subjects' responses to TCE.

10.9 Conclusions

TCE remains a widespread environmental pollutant and human exposure will continue for the foreseeable future. TCE has a wide range of health effects covering multiple major organs. These health effects can occur during gestation or in adults following chronic exposure.

By investigating epigenetic alterations we expect to decipher the early molecular effects that lead to later disease. Genome wide analyses are needed using next generation sequencing and similar broad measures to understand the extent of TCE's effects. Likewise, effects in multiple organs and at multiple life stages require study. It is important to find predictive biomarkers for the disease(s) to which TCE-exposed individuals are most susceptible.

Understanding the metabolic and cellular signaling effects of TCE exposure will also help us understand how epigenetic alterations occur in the first place. Knowing metabolic and cellular signaling effects may allow us to design interventions for those still exposed to TCE. Coexposures with other toxicants, with phytochemicals and with varying nutritional states need to be measured, and then appropriately addressed. Various therapeutic strategies have been developed for cancer and aging (Cooney 2010; Dawson and Kouzarides 2012). Some of these, especially nutrients and well-tolerated drugs, may be good candidates for preclinical studies (e.g. animal models) to reverse TCE effects.

Understanding epigenetics of TCE exposure will help us understand a wide range of health problems associated with TCE. Further, this understanding will help us design interventions to reverse epigenetic changes before disease develops.

Aknowledgements This article is dedicated to my dear friend and colleague George W. Wolff who passed away in late 2012. I thank Kimberly Cooney for designing Fig. 10.1. Supported by NIEHS grant R01 ES021484-01A1 to Kathleen Gilbert and Sarah Blossom and by an Arkansas Biosciences Institute grant to Kathleen Gilbert.

References

Abbas R, Fisher JW (1997) A physiologically based pharmacokinetic model for trichloroethylene and its metabolites, chloral hydrate, trichloroacetate, dichloroacetate, trichloroethanol, and trichloroethanol glucuronide in B6C3F1 mice. Toxicol Appl Pharmacol 147(1):15–30

Akundi RS, Macho A, Munoz E, Lieb K, Bringmann G, Clement HW, Hull M, Fiebich BL (2004) 1-trichloromethyl-1,2,3,4-tetrahydro-beta-carboline-induced apoptosis in the human neuroblastoma cell line SK-N-SH. J Neurochem 91:263–273

ATSDR toxicological profile for trichloroethylene. U.S. Department of Health and Human Services. 1997 and 2013. http://www.atsdr.cdc.gov/ToxProfiles/tce_addendum.pdf

Badger TM, Ronis MJJ, Wolff G, Stanley S, Ferguson M, Shankar K, Jo CH (2008) Soy protein isolate reduces hepatosteatosis in yellow Avy/a mice without altering coat color phenotype. Exp Biol Med 233(10):1242–1254

Balada E, Vilardell-Tarrés M, Ordi-Ros J (2010) Implication of human endogenous retroviruses in the development of autoimmune diseases. Int Rev Immunol 29:351–370

Banks RE, Tirukonda P, Taylor C, Hornigold N, Astuti D, Cohen D, Selby PJ (2006) Genetic and epigenetic analysis of von Hippel-Lindau (VHL) gene alterations and relationship with clinical variables in sporadic renal cancer. Cancer Res 66(4):2000–2011

Bannert N, Kurth R (2004) Retroelements and the human genome: new perspectives on an old relation. Proc Natl Acad Sci U S A 101(Suppl 2):14572–14579

Baudino L, Yoshinobu K, Morito N, Santiago-Raber ML, Izui S (2010) Role of endogenous retroviruses in murine SLE. Autoimmun Rev Vol 10:27–34

Bekdash RA, Zhang C, Sarkar DK (2013) Gestational choline supplementation normalized fetal alcohol-induced alterations in histone modifications, DNA methylation, and proopiomelanocortin (POMC) gene expression in β-endorphin-producing POMC neurons of the hypothalamus. Alcohol Clin Exp Res 37:1133–1142

Biedermann S, Tschudin P, Grob K (2010) Transfer of bisphenol A from thermal printer paper to the skin. Anal Bioanal Chem 398(1):571–576

Billington CJ, Schmidt B, Zhang L, Hodges JS, Georgieff MK, Schotta G, Petryk A (2013) Maternal diet supplementation with methyl donors and increased parity affect the incidence of craniofacial defects in the offspring of twisted gastrulation mutant mice. J Nutr 143(3): 332–339

Blossom SJ, Melnyk S, Cooney CA, Gilbert KM, James SJ (2012) Postnatal exposure to trichloroethylene alters glutathione redox homeostasis, methylation potential, and neurotrophin expression in the mouse hippocampus. Neurotoxicology 33:1518–1527

Blossom SJ, Cooney CA, Melnyk SB, Rau JL, Swearingen CJ, Wessinger WD (2013) Metabolic changes and DNA hypomethylation in cerebellum are associated with behavioral alterations in mice exposed to trichloroethylene postnatally. Toxicol Appl Pharmacol 269:263–269

Blusztajn JK, Mellott TJ (2013) Neuroprotective actions of perinatal choline nutrition. Clin Chem Lab Med 51(3):591–599

Bonnet S, Archer SL, Allalunis-Turner J, Haromy A, Beaulieu C, Thompson R, Michelakis ED (2007) A mitochondria-K + channel axis is suppressed in cancer and its normalization promotes apoptosis and inhibits cancer growth. Cancer Cell 11(1):37–51

Brauch H, Weirich G, Hornauer MA, Störkel S, Wöhl T, Brüning T (1999) Trichloroethylene exposure and specific somatic mutations in patients with renal cell carcinoma. J Natl Cancer Inst 91(10):854–861

Bringmann G, God R, Feineis D, Wesemann W, Riederer P, Rausch WD, Reichmann H, Sontag KH (1995) The TaClo concept: 1-trichloromethyl-1,2,3,4-tetrahydro-beta-carboline (TaClo), a new toxin for dopaminergic neurons. J Neural Transm 46:235–244

Bull RJ (2000) Mode of action of liver tumor induction by trichloroethylene and its metabolites, trichloroacetate and dichloroacetate. Environ Health Perspect 108(Suppl 2):241

Bull RJ, Sanchez IM, Nelson MA, Larson JL, Lansing AJ (1990) Liver tumor induction in B6C3F1 mice by dichloroacetate and trichloroacetate. Toxicology 63(3):341–359

Caldwell PT, Manziello A, Howard J, Palbykin B, Runyan RB, Selmin O (2010) Gene expression profiling in the fetal cardiac tissue after folate and low-dose trichloroethylene exposure. Birth Defects Res A Clin Mol Teratol 88:111–127

Champagne FA, Curley JP (2009) Epigenetic mechanisms mediating the long-term effects of maternal care on development. Neurosci Biobehav Rev 33:593–600

Cherkasova E, Malinzak E, Rao S, Takahashi Y, Senchenko VN, Kudryavtseva AV, Childs RW (2011) Inactivation of the von Hippel–Lindau tumor suppressor leads to selective expression of a human endogenous retrovirus in kidney cancer. Oncogene 30(47):4697–4706

Chiu WA, Jinot J, Scott CS, Makris SL, Cooper GS, Dzubow RC, Caldwell JC (2013) Human health effects of trichloroethylene: key findings and scientific issues. Environ Health Perspect 121:303–311

Cho K, Mabasa L, Bae S, Walters MW, Park CS (2012) Maternal high-methyl diet suppresses mammary carcinogenesis in female rat offspring. Carcinogenesis 33(5):1106–1112

Clewell HJ 3rd, Gentry PR, Covington TR, Gearhart JM (2000) Development of a physiologically based pharmacokinetic model of trichloroethylene and its metabolites for use in risk assessment. Environ Health Perspect 108(Suppl 2):283–305

Collins LJ, Schonfeld B, Chen XS (2011) The epigenetics of non coding RNA. In: Tollefsbol T (ed) Handbook of epigenetics: the new molecular and medical genetics. Academic Press, London, pp 49–61

Cooney CA (2006) Maternal nutrition: nutrients and control of expression. In: Kaput J, Rodriguez RL (eds) Nutrigenomics: concepts and technologies. Wiley, Hoboken, pp 219–254

Cooney CA (2007) Epigenetics – DNA-based mirror of our environment. Dis Markers 23: 121–137

Cooney CA (2008) Cancer and aging: the epigenetic connection. In: Tollefsbol T (ed) Cancer epigenetics. CRC Press, Boca Raton, pp 303–316

Cooney CA (2009) Nutrients, epigenetics, and embryonic development. In: Sang Woon C, Simonetta F (eds) Nutrients and epigenetics. CRC Press, Boca Raton, pp 155–174

Cooney CA (2010) Drugs and supplements that may slow aging of the epigenome. Drug Dis Today Ther Strateg 7:57–64

Cooney CA, Gilbert KM (2012) Toxicology, epigenetics and autoimmunity. In: Sahu SC (ed) Toxicology and epigenetics. John Wiley & Sons, Ltd, Chichester, UK pp 241–260

Cooney CA, Dave AA, Wolff GL (2002) Maternal methyl supplements in mice affect epigenetic variation and DNA methylation of offspring. J Nutr 132:2393S–2400S

Coppedè F (2012) Genetics and epigenetics of Parkinson's disease. Scientific World Journal 2012

Cropley JE, Suter CM, Beckman KB, Martin DI (2006) Germ-line epigenetic modification of the murine Avy allele by nutritional supplementation. Proc Natl Acad Sci 103(46): 17308–17312

Davison JM, Mellott TJ, Kovacheva VP, Blusztajn JK (2009) Gestational choline supply regulates methylation of histone H3, expression of histone methyltransferases G9a (Kmt1c) and Suv39h1 (Kmt1a), and DNA methylation of their genes in rat fetal liver and brain. J Biol Chem 284(4):1982–1989

Dawson MA, Kouzarides T (2012) Cancer epigenetics: from mechanism to therapy. Cell 150(1):12–27

Dolinoy DC, Weidman JR, Waterland RA, Jirtle RL (2006) Maternal genistein alters coat color and protects Avy mouse offspring from obesity by modifying the fetal epigenome. Environ Health Perspect 114(4):567

Dolinoy DC, Huang D, Jirtle RL (2007) Maternal nutrient supplementation counteracts bisphenol A-induced DNA hypomethylation in early development. Proc Natl Acad Sci 104(32): 13056–13061

Downey R, Burke A, Giles FJ, Sullivan F, Wang-Johanning F, Mee B, Glynn SA (2012) Abstract A62: human endogenous retrovirus activation in prostate cancer: association with disease progression. Cancer Res 72(4 Supplement):A62–A62

Downing C, Johnson TE, Larson C, Leakey TI, Siegfried RN, Rafferty TM, Cooney CA (2011) Subtle decreases in DNA methylation and gene expression at the mouse Igf2 locus following prenatal alcohol exposure: effects of a methyl-supplemented diet. Alcohol 45(1):65–71

Druker R, Bruxner TJ, Lehrbach NJ, Whitelaw E (2004) Complex patterns of transcription at the insertion site of a retrotransposon in the mouse. Nucleic Acids Res 32(19):5800–5808

Eilertsen KJ et al (2008) The epigenetics of adult (somatic) stem cells. Crit Rev Eukaryot Gene Expr 18:189–206

Ellsworth EM, Palma JF, Spence WC, Bleicher JM, Smith DM Jr, Finkelstein SD (2012) Mutational profiling of sporadic versus toxin-associated brain cancer formation: initial findings using loss of heterozygosity profiling. Int J Hyg Environ Health 215(3):427–433

Fang JY, Gao QY, Chen HM, Chen Y, Wang ZH, Ge ZZ, Zheng P (2013a) Folic acid prevents the initial occurrence of sporadic colorectal adenoma in Chinese over 50 years of age: a randomized clinical trial. Cancer Prev Res. doi:10.1158/1940-6207.CAPR-13-0013, Epub May 16, 2013

Fang ZZ, Krausz KW, Tanaka N, Li F, Qu A, Idle JR, Gonzalez FJ (2013b) Metabolomics reveals trichloroacetate as a major contributor to trichloroethylene-induced metabolic alterations in mouse urine and serum. Arch Toxicol 87:1975–1987

Feng J, Zhou Y, Campbell S, Le T, Li E, Sweatt JD, Silva A, Fan G (2010) Dnmt1 and Dnmt3a maintain DNA methylation and regulate synaptic function in adult forebrain neurons. Nat Neurosci 13:423–430

Fernandez AF, Assenov Y, Martin-Subero JI, Balint B, Siebert R, Taniguchi H, Esteller M (2012) A DNA methylation fingerprint of 1628 human samples. Genome Res 22(2):407–419

Forand SP, Lewis-Michl EL, Gomez MI (2012) Adverse birth outcomes and maternal exposure to trichloroethylene and tetrachloroethylene through soil vapor intrusion in New York State. Environ Health Perspect 120(4):616

Frau M, Feo F, Pascale RM (2013) Pleiotropic effects of Methionine adenosyltransferases deregulation as determinants of liver cancer progression and prognosis. J Hepatol 59:830–41

Gash DM, Rutland K, Hudson NL, Sullivan PG, Bing G, Cass WA, Prince TS (2008) Trichloroethylene: Parkinsonism and complex 1 mitochondrial neurotoxicity. Annals Neurol 63(2):184–192

Gaudet F, Rideout WM 3rd, Meissner A, Dausman J, Leonhardt H, Jaenisch R (2004) Dnmt1 expression in pre- and postimplantation embryogenesis and the maintenance of IAP silencing. Mol Cell Biol 24(4):1640–1648

Gilbert KM, Przybyla B, Pumford NR, Han T, Fuscoe J, Schnackenberg LK, Blossom SJ (2009) Delineating liver events in trichloroethylene-induced autoimmune hepatitis. Chem Res Toxicol 22(4):626–632

Gilbert KM, Nelson AR, Cooney CA, Reisfeld B, Blossom SJ (2012) Epigenetic alterations may regulate temporary reversal of $CD4^+$ T cell activation caused by trichloroethylene exposure. Toxicol Sci 127:169–178

Gonda TA, Kim YI, Salas MC, Gamble MV, Shibata W, Muthupalani S, Tycko B (2012) Folic acid increases global DNA methylation and reduces inflammation to prevent Helicobacter-associated gastric cancer in mice. Gastroenterology 142(4):824–833

Gorelik G, Richardson B (2010) Key role of ERK pathway signaling in lupus. Autoimmunity 43:17–22

Hackett JA, Sengupta R, Zylicz JJ, Murakami K, Lee C, Down TA, Surani MA (2013) Germline DNA demethylation dynamics and imprint erasure through 5-hydroxymethylcytosine. Science 339(6118):448–452

Hansen J, Sallmén M, Seldén AI, Anttila A, Pukkala E, Andersson K, McLaughlin JK (2013) Risk of cancer among workers exposed to trichloroethylene: analysis of three Nordic cohort studies. J Natl Cancer Inst 105(12):869–877

Haycock PC, Ramsay M (2009) Exposure of mouse embryos to ethanol during preimplantation development: effect on DNA methylation in the h19 imprinting control region. Biol Reprod 81(4):618–627

Herren-Freund SL, Pereira MA, Khoury MD, Olson G (1987) The carcinogenicity of trichloroethylene and its metabolites, trichloroacetic acid and dichloroacetic acid, in mouse liver. Toxicol Appl Pharmacol 90(2):183–189

Hobbs CA, Cleves MA, Melnyk S, Zhao W, James SJ (2005a) Congenital heart defects and abnormal maternal biomarkers of methionine and homocysteine metabolism. Am J Clin Nutr 81(1):147–153

Hobbs CA, Cleves MA, Zhao W, Melnyk S, James SJ (2005b) Congenital heart defects and maternal biomarkers of oxidative stress. Am J Clin Nutr 82(3):598–604

Janetzky B, God R, Bringmann G, Reichmann H (1995) 1-Trichloromethyl-1,2,3,4-tetrahydro-beta-carboline, a new inhibitor of complex I. J Neural Transm 46:265–273

Jollow DJ, Bruckner JV, McMillan DC, Fisher JW, Hoel DG, Mohr LC (2009) Trichloroethylene risk assessment: a review and commentary. Crit Rev Toxicol 39(9):782–797

Karami S, Lan Q, Rothman N, Stewart PA, Lee KM, Vermeulen R, Moore LE (2012) Occupational trichloroethylene exposure and kidney cancer risk: a meta-analysis. Occup Environ Med 69(12):858–867

Kato Y, Kaneda M, Hata K, Kumaki K, Hisano M, Kohara Y, Okano M, Li E, Nozaki M, Sasaki H (2007) Role of the Dnmt3 family in de novo methylation of imprinted and repetitive sequences during male germ cell development in the mouse. Hum Mol Genet 16(19):2272–2280

Kim S, Kim D, Pollack GM, Collins LB, Rusyn I (2009) Pharmacokinetic analysis of trichloroethylene metabolism in male B6C3F1 mice: Formation and disposition of trichloroacetic acid, dichloroacetic acid,< i> S</i>–(1, 2-dichlorovinyl) glutathione and < i> S</i>–(1, 2-dichlorovinyl)-l-cysteine. Toxicol Appl Pharmacol 238(1):90–99

Klotz U, Ammon E (1998) Clinical and toxicological consequences of the inductive potential of ethanol. Eur J Clin Pharmacol 54(1):7–12

Kooistra SM, Helin K (2012) Molecular mechanisms and potential functions of histone demethylases. Nature Rev Mole Cell Biol 13(5):297–311

Kundakovic M, Gudsnuk K, Franks B, Madrid J, Miller RL, Perera FP, Champagne FA (2013) Sex-specific epigenetic disruption and behavioral changes following low-dose in utero bisphenol A exposure. Proc Natl Acad Sci 110(24):9956–9961

Kurokawa R, Rosenfeld MG, Glass CK (2009) Transcriptional regulation through noncoding RNAs and epigenetic modifications. RNA Biol 6(3):233–236

Lash LH, Fisher JW, Lipscomb JC, Parker JC (2000) Metabolism of trichloroethylene. Environ Health Perspect 108(Suppl 2):177–200

Lash LH, Qian W, Putt DA, Hueni SE, Elfarra AA, Krause RJ, Parker JC (2001) Renal and hepatic toxicity of trichloroethylene and its glutathione-derived metabolites in rats and mice: sex-, species-, and tissue-dependent differences. J Pharmacol Exp Ther 297(1):155–164

Lee JT (2012) Epigenetic regulation by long noncoding RNAs. Science 338(6113):1435–1439

Li CC, Cropley JE, Cowley MJ, Preiss T, Martin DI, Suter CM (2011) A sustained dietary change increases epigenetic variation in isogenic mice. PLoS Genet 7(4):e1001380

Liao C, Kannan K (2011) High levels of bisphenol A in paper currencies from several countries, and implications for dermal exposure. Environ Sci Technol 45(16):6761–6768

Liao C, Liu F, Kannan K (2012) Bisphenol s, a new bisphenol analogue, in paper products and currency bills and its association with bisphenol a residues. Environ Sci Technol 46(12):6515–6522

Liu Y, Balaraman Y, Wang G, Nephew KP, Zhou FC (2009) Alcohol exposure alters DNA methylation profiles in mouse embryos at early neurulation. Epigenetics 4(7):500–511

Liu M, Choi DY, Hunter RL, Pandya JD, Cass WA, Sullivan PG, Kim HC, Gash DM, Bing G (2010) Trichloroethylene induces dopaminergic neurodegeneration in Fisher 344 rats. J Neurochem 112:773–783

Loeber CP, Hendrix MJ, Diez De Pinos S, Goldberg SJ (1988) Trichloroethylene: a cardiac teratogen in developing chick embryos. Pediatr Res 24(6):740–744

Lu Y, Cederbaum AI (2008) CYP2E1 and oxidative liver injury by alcohol. Free Radic Biol Med 44(5):723–738

Makwana O, King NM, Ahles L, Selmin O, Granzier HL, Runyan RB (2010) Exposure to low-dose trichloroethylene alters shear stress gene expression and function in the developing chick heart. Cardiovasc Toxicol 10(2):100–107

Makwana O, Ahles L, Lencinas A, Selmin OI, Runyan RB (2013) Low-dose trichloroethylene alters cytochrome P450-2C subfamily expression in the developing chick heart. Cardiovasc Toxicol 13(1):77–84

Melnyk S, Fuchs GJ, Schulz E, Lopez M, Kahler SG, Fussell JJ et al (2012) Metabolic imbalance associated with methylation dysregulation and oxidative damage in children with autism. J Autism Dev Disord 42:367–377

Miller C, Sweatt JD (2007) Covalent modification of DNA regulates memory formation. Neuron 53(6):857–869

Molina-Molina JM, Amaya E, Grimaldi M, Sáenz JM, Real M, Fernández MF, Olea N (2013) In vitro study on the agonistic and antagonistic activities of bisphenol-S and other bisphenol-A congeners and derivatives via nuclear receptors. Toxicol Appl Pharmacol 272:127–136

Morgan HD, Sutherland HG, Martin DI, & Whitelaw E (1999) Epigenetic inheritance at the agouti locus in the mouse. Nature genetics, 23(3):314–318.

Mosharov E, Cranford MR, Banerjee R (2000) The quantitatively important relationship between homocysteine metabolism and glutathione synthesis by the transsulfuration pathway and its regulation by redox changes. Biochemistry 39:13005–13011

Nakajima T, Okino T, Okuyama S, Kaneko T, Yonekura I, Sato A (1988) Ethanol-induced enhancement of trichloroethylene metabolism and hepatotoxicity: difference from the effect of phenobarbital. Toxicol Appl Pharmacol 94(2):227–237

NTP (1988) Toxicology and carcinogenesis studies of trichloroethylene (CAS No. 79-01-6) in four strains of rats (ACI, August, Marshall, Osborne-Mendel) (Gavage Studies). Technical Report Series no. 273. National Toxicology Program, Research Triangle Park, 303 pp

NTP (1990) Carcinogenesis studies of trichloroethylene (Without Epichlorohydrin) (CAS No. 79-01-6) in F344/N rats and B6C3F mice (Gavage Studies). Technical Report Series no. 243. National Toxicology Program, Research Triangle Park, 176 pp

NTP (2011) Trichloroethylene. Report on carcinogens, 12 edn. http://ntp.niehs.nih.gov/ntp/roc/twelfth/profiles/Trichloroethylene.pdf

Obeid R, Schadt A, Dillmann U, Kostopoulos P, Fassbender K, & Herrmann W (2009) Methylation status and neurodegenerative markers in Parkinson disease. Clinical chemistry, 55(10):1852–1860
Ooi SK, O'Donnell AH, Bestor TH (2009) Mammalian cytosine methylation at a glance. J Cell Sci 122(16):2787–2791
Otero NK, Thomas JD, Saski CA, Xia X, Kelly SJ (2012) Choline supplementation and DNA methylation in the hippocampus and prefrontal cortex of rats exposed to alcohol during development. Alcohol Clin Exp Res 36(10):1701–1709
Ounpraseuth S, Rafferty TM, McDonald-Phillips RE, Gammill WM, Siegel ER, Wheeler KL, Cooney CA (2009) A method to quantify mouse coat-color proportions. PloS One 4(4):e5414
Palbykin B, Borg J, Caldwell PT, Rowles J, Papoutsis AJ, Romagnolo DF, Selmin OI (2011) Trichloroethylene induces methylation of the Serca2 promoter in H9c2 cells and embryonic heart. Cardiovasc Toxicol 11(3):204–214
Pascale RM, Simile MM, De Miglio MR, Feo F (2002) Chemoprevention of hepatocarcinogenesis: S-adenosyl-L-methionine. Alcohol 27(3):193–198
Pavlopoulos E, Jones S, Kosmidis S, Close M, Kim C, Kovalerchik O, Small SA, Kandel ER (2013) Molecular mechanism for age-related memory loss: the histone-binding protein RbAp48. Sci Transl Med 5:200ra115
Peleg S, Sananbenesi F, Zovoilis A, Burkhardt S, Bahari-Javan S, Agis-Balboa RC, Fischer A (2010) Altered histone acetylation is associated with age-dependent memory impairment in mice. Science 328(5979):753–756
Pereira MA, Li K, Kramer PM (1997) Promotion by mixtures of dichloroacetic acid and trichloroacetic acid of N-methyl-N-nitrosourea-initiated cancer in the liver of female B6C3F1 mice. Cancer Lett 115(1):15–23
Pereira MA, Wang W, Kramer PM, Tao L (2004) Prevention by methionine of dichloroacetic acid–induced liver cancer and DNA hypomethylation in mice. Toxicol Sci 77(2):243–248
Perl A, Fernandez D, Telarico T, Phillips PE (2010) Endogenous retroviral pathogenesis in lupus. Curr Opinion Rheumatol 22(5):483–492
Phillips JM, Burgoon LD, Goodman JI (2009) Phenobarbital elicits unique, early changes in the expression of hepatic genes that affect critical pathways in tumor-prone B6C3F1 mice. Toxicol Sci 109(2):193–205
Pogribny IP, James SJ, Beland FA (2012) Molecular alterations in hepatocarcinogenesis induced by dietary methyl deficiency. Mol Nutr Food Res 56(1):116–125
Poirier LA (2002) The effects of diet, genetics and chemicals on toxicity and aberrant DNA methylation: an introduction. J Nutr 132:2336S–2339S
Quddus J, Johnson KJ, Gavalchin J, Amento EP, Chrisp CE, Yung RL, Richardson BC (1993) Treating activated CD4+ T cells with either of two distinct DNA methyltransferase inhibitors, 5-azacytidine or procainamide, is sufficient to cause a lupus-like disease in syngeneic mice. J Clin Invest 92:38–53
Rakyan VK, Chong S, Champ ME, Cuthbert PC, Morgan HD, Luu KV, Whitelaw E (2003) Transgenerational inheritance of epigenetic states at the murine AxinFu allele occurs after maternal and paternal transmission. Proc Natl Acad Sci 100(5):2538–2543
Ramsay M (2010) Genetic and epigenetic insights into fetal alcohol spectrum disorders. Genome Med 2(4):27
Ray D, Richardson BC (2012) Toxicoepigenomics in lupus. In: Sahu SC (ed) Toxicology and epigenetics. Wiley-Blackwell, Oxford, pp 261–274
Reiss D, Zhang Y, Rouhi A, Reuter M, Mager DL (2010) Variable DNA methylation of transposable elements: the case study of mouse early transposons. Epigenetics 5:68–79
Resendiz M, Chen Y, Oztürk NC, Zhou FC (2013) Epigenetic medicine and fetal alcohol spectrum disorders. Epigenomics 5(1):73–86
Romanish MT, Cohen CJ, Mager DL (2010) Potential mechanisms of endogenous retroviral-mediated genomic instability in human cancer. Semin Cancer Biol 20(4):246–253. Academic Press

Rosenfeld CS, Sieli PT, Warzak DA, Ellersieck MR, Pennington KA, Roberts RM (2013) Maternal exposure to bisphenol A and genistein has minimal effect on Avy/a offspring coat color but favors birth of agouti over nonagouti mice. Proc Natl Acad Sci 110(2):537–542

Rufer ES, Hacker TA, Flentke GR, Drake VJ, Brody MJ, Lough J, Smith SM (2010) Altered cardiac function and ventricular septal defect in avian embryos exposed to low-dose trichloroethylene. Toxicol Sci 113(2):444–452

Sauerbeck A, Pandya J, Singh I, Bittman K, Readnower R, Bing G, & Sullivan P (2011) Analysis of regional brain mitochondrial bioenergetics and susceptibility to mitochondrial inhibition utilizing a microplate based system. Journal of neuroscience methods, 198(1):36–43.

Sauerbeck A, Hunter R, Bing G, Sullivan PG (2012) Traumatic brain injury and trichloroethylene exposure interact and produce functional, histological, and mitochondrial deficits. Exp Neurol 234(1):85–94

Schulz WA, Steinhoff C, Florl AR (2006) Methylation of endogenous human retroelements in health and disease. In: DNA methylation: development, genetic disease and cancer. Springer, Berlin/Heidelberg, pp 211–250

Seisenberger S, Peat JR, Hore TA, Santos F, Dean W, Reik W (2013) Reprogramming DNA methylation in the mammalian life cycle: building and breaking epigenetic barriers. Philos Trans R Soc B Biol Sci 368:20110330. http://dx.doi.org/10.1098/rstb.2011.0330

Sie KK, Medline A, Van Weel J, Sohn KJ, Choi SW, Croxford R, Kim YI (2011) Effect of maternal and postweaning folic acid supplementation on colorectal cancer risk in the offspring. Gut 60(12):1687–1694

Stouder C, Somm E, Paoloni-Giacobino A (2011) Prenatal exposure to ethanol: a specific effect on the H19 gene in sperm. Reprod Toxicol 31(4):507–512

Strum SB, Adalsteinsson Ö, Black RR, Segal D, Peress NL, Waldenfels J (2013) Case report: sodium dichloroacetate (DCA) inhibition of the "Warburg Effect" in a human cancer patient: complete response in non-Hodgkin's lymphoma after disease progression with rituximab-CHOP. J Bioenerg Biomembr 45(3):307–315

Sun RC, Fadia M, Dahlstrom JE, Parish CR, Board PG, Blackburn AC (2010) Reversal of the glycolytic phenotype by dichloroacetate inhibits metastatic breast cancer cell growth in vitro and in vivo. Breast Cancer Res Treatment 120(1):253–260

Susiarjo M, Sasson I, Mesaros C, Bartolomei MS (2013) Bisphenol A exposure disrupts genomic imprinting in the mouse. PLoS Genet 9(4):e1003401

Sweatt JD (2012) DNA methylation in memory formation. In: Sassone Corsi P, Christen Y (eds) Epigenetics, brain and behavior, Series: research and perspectives in neurosciences. Springer, Heidelberg, pp 81–96

Tao L, Yang S, Xie MI, Kramer PM, Pereira MA (2000a) Effect of trichloroethylene and its metabolites, dichloroacetic acid and trichloroacetic acid, on the methylation and expression of c-Jun and c-Myc protooncogenes in mouse liver: prevention by methionine. Toxicol Sci 54(2): 399–407

Tao L, Yang S, Xie M, Kramer PM, Pereira MA (2000b) Hypomethylation and overexpression of c-jun and c-myc protooncogenes and increased DNA methyltransferase activity in dichloroacetic and trichloroacetic acid-promoted mouse liver tumors. Cancer Lett 158(2):185–193

Tsuchishima M, George J, Shiroeda H, Arisawa T, Takegami T, Tsutsumi M (2013) Chronic ingestion of ethanol induces hepatocellular carcinoma in mice without additional hepatic insult. Dig Dis Sciences 58(7):1923–1933

United States Environmental Protection Agency (2011) EPA 635 (R-09/011F). http://www.epa.gov/iris/toxreviews/0199tr/0199tr.pdf

Vandenberg LN, Colborn T, Hayes TB, Heindel JJ, Jacobs DR, Lee DH, Myers JP (2012) Hormones and endocrine-disrupting chemicals: low-dose effects and nonmonotonic dose responses. Endocr Rev 33(3):378–455

Vandenberg LN, Hunt PA, Myers JP, vom Saal FS (2013) Human exposures to bisphenol A: mismatches between data and assumptions. Rev Environ Health 28(1):37–58

Veazey KJ, Carnahan MN, Muller D, Miranda RC, Golding MC (2013) Alcohol-induced epigenetic alterations to developmentally crucial genes regulating neural stemness and differentiation. Alcohol Clin Exp Res 37:1111–1122

Vitvitsky V, Thomas M, Ghorpade A, Gendelman HE, Banerjee R (2006) A functional transsulfuration pathway in the brain links to glutathione homeostasis. J Biol Chem 281: 35785–35793

Vogel SA (2009) The politics of plastics: the making and unmaking of bisphenol a "safety". Am J Public Health 99(S3):S559–S566

Vogelstein B, Papadopoulos N, Velculescu VE, Zhou S, Diaz LA, Kinzler KW (2013) Cancer genome landscapes. Science 339(6127):1546–1558

Waalkes MP, Ward JM, Diwan BA (2004) Induction of tumors of the liver, lung, ovary and adrenal in adult mice after brief maternal gestational exposure to inorganic arsenic: promotional effects of postnatal phorbol ester exposure on hepatic and pulmonary, but not dermal cancers. Carcinogenesis 25(1):133–141

Wainfan E, Poirier LA (1992) Methyl groups in carcinogenesis: effects on DNA methylation and gene expression. Cancer Res 52(7 Suppl):2071s–2077s

Wallace DC, Fan W, Procaccio V (2010) Mitochondrial energetics and therapeutics. Ann Rev Pathol 5:297

Walsh CP, Chaillet JR, Bestor TH (1998) Transcription of IAP endogenous retroviruses is constrained by cytosine methylation. Nature Genet 20(2):116–117

Wang W, Kramer PM, Yang S, Pereira MA, Tao L (2001) Reversed-phase high-performance liquid chromatography procedure for the simultaneous determination of S–adenosyl-l-methionine and S–adenosyl-l-homocysteine in mouse liver and the effect of methionine on their concentrations. J Chromatogr B Biomed Sci Appl 762(1):59–65

Watson GW, Beaver LM, Williams DE, Dashwood RH, Ho E (2013) Phytochemicals from cruciferous vegetables, epigenetics, and prostate cancer prevention. AAPS J 15(4):951–961

Weaver IC, Champagne FA, Brown SE, Dymov S, Sharma S, Meaney MJ, Szyf M (2005) Reversal of maternal programming of stress responses in adult offspring through methyl supplementation: altering epigenetic marking later in life. J Neurosci 25(47):11045–11054

Wolff GL, Kodell RL, Moore SR, Cooney CA (1998) Maternal epigenetics and methyl supplements affect agouti gene expression in Avy/a mice. FASEB J 12:949–957

Yauck JS, Malloy ME, Blair K, Simpson PM, McCarver DG (2004) Proximity of residence to trichloroethylene-emitting sites and increased risk of offspring congenital heart defects among older women. Birth Defects Res A Clin Mol Teratol 70(10):808–814

Yung R, Powers D, Johnson K, Amento E, Carr D, Laing T, Yang J, Chang S, Hemati N, Richardson B (1996) Mechanisms of drug-induced lupus. II. T cells overexpressing lymphocyte function-associated antigen 1 become autoreactive and cause a lupuslike disease in syngeneic mice. J Clin Invest 97:2866–2871

Zeisel SH (2013) Metabolic crosstalk between choline/1-carbon metabolism and energy homeostasis. Clin Chem Lab Med 51(3):467–475

Zhang H, Herman AI, Kranzler HR, Anton RF, Zhao H, Zheng W, Gelernter J (2013a) Array-based profiling of DNA methylation changes associated with alcohol dependence. Alcohol Clin Exp Res 37(s1):E108–E115

Zhang R, Miao Q, Wang C, Zhao R, Li W, Haile CN, Zhang XY (2013b) Genome-wide DNA methylation analysis in alcohol dependence. Addict Biol 18:392–403

Chapter 11
Mathematical Modeling and Trichloroethylene

Brad Reisfeld and Jaime H. Ivy

Abstract Mathematical modeling has been used extensively to quantify and characterize the disposition, fate, and risk associated with the volatile organic chemical trichloroethylene (TCE). Here, we summarize many of these models that have been developed and applied across the exposure-dose-effect continuum, ranging from pharmacokinetic and pharmacodynamic models to quantitative structure-activity relationships. We conclude by reviewing some future directions in computational modeling that are increasingly used to inform an understanding of the adverse health effects associated with exposure to TCE, and introduce elements of a first-generation systems biology model of TCE-induced autoimmune disease.

Keywords Mathematical modeling • Computational modeling • Exposure • Pharmacokinetics • Pharmacodynamics • QSAR • Systems biology • Omics • Dose response

Acronyms and Abbreviations

ADME	Absorption, distribution, metabolism, and excretion
AIH	Autoimmune hepatitis
ARR	Arrest of mitosis in *Aspergillus nidulans*

B. Reisfeld (✉)
Department of Chemical and Biological Engineering and School of Biomedical Engineering, Colorado State University, 1370 Campus Delivery, Fort Collins, CO 80523-1370, USA
e-mail: brad.reisfeld@colostate.edu

J.H. Ivy
Department of Chemical and Biological Engineering, Colorado State University, 1370 Campus Delivery, Fort Collins, CO 80523-1370, USA
e-mail: jhivy@engr.colostate.edu

K.M. Gilbert, S.J. Blossom (eds.), *Trichloroethylene: Toxicity and Health Risks*, Molecular and Integrative Toxicology, DOI 10.1007/978-1-4471-6311-4_11,

AUC	Area under the curve
BBDR	Biologically-based dose response
BBPD	Biologically-based pharmacodynamic
BDM	Benchmark dose method
BEI	Biological exposure index
CNS	Central nervous system
CPK	Compartmental pharmacokinetic
D37	Measure of lethality in *Aspergillus nidulans*
DCA	Dichloroacetic acid
DCVC	*S*-(1,2-dichlorovinyl)-L-cysteine
DCVG	*S*-(1,2-dichlorovinyl)glutathione
DIFF	Difference between the highest occupied molecular orbital and the lowest unoccupied molecular orbital
DNAPL	Dense nonaqueous phase liquid
EDR	Exposure-dose-response
EPA	US Environmental Protection Agency
HBA	H-bonding acceptor ability
HBD	H-bonding donor ability
IC50	Chemical concentration that inhibits some endpoint in 50 % of the test animals in a given time
IRIS	Integrated Risk Information System
Kow	Octanol-water partition coefficient
LC50	Chemical concentration that kills 50 % of the test animals in a given time
LCA	Life cycle assessment
LEC	Induction of chromosome malsegregation leading to aneuploidy in *Aspergillus nidulans*
LOEC	Lowest observed effect concentration
logP	The log of the ratio of concentration of neutral species in octanol divided the concentration of neutral species in water
MCL	Maximum contaminant level
MOE	Margin of exposure
MR	Molar refractivity
MRL	Minimal risk level
NAPL	Non-aqueous phase liquid
NCPK	Non-compartmental pharmacokinetic
PBPK	Physiologically-based pharmacokinetic
PC	Partition coefficient
PCE	Tetrachloroethylene, perchloroethylene
PD	Pharmacodynamics
PEL	Permissible exposure limit
PK	Pharmacokinetics
QSAR	Quantitative structure activity relationship
RfC	Reference concentration

$t_{1/2}$	Chemical half life
TAI	TCE-induced autoimmunity
TBARS	Thiobarbituric acid reactive substance
TCA	Trichloroacetic acid
TCE	Trichloroethylene
TCOH	Trichloroethanol
TLV	Threshold limit value
VOC	Volatile organic compound

11.1 Introduction

Mathematical/computational models in toxicology have the potential to integrate information and data from a variety of sources to help in the prediction and understanding of the adverse health effects caused by exposure to foreign chemicals (xenobiotics); moreover, mathematical modeling can provide researchers and practitioners with a "virtual lab" in which to explore hypotheses, conduct complex multifactorial studies that would be impractical or impossible using conventional experimental techniques, and rapidly analyze, extrapolate, and evaluate results, all while reducing the need for animal experimentation. Further, these models can be used to investigate the interactions of chemical agents and biological organisms across many scales (e.g., population, individual, cellular, and molecular) and may be used to inform the hazard and risk prioritization of chemicals (Reisfeld and Mayeno 2012a, b).

In this chapter, we review many of the mathematical models and approaches that have been used to analyze and quantify the exposure-dose-effect continuum (National Center for Environmental Assessment 2011) for trichloroethylene (TCE). As shown in Fig. 11.1, these include models for exposure, pharmacokinetics, pharmacodynamics, and quantitative structure activity relationships. We conclude the chapter with a summary of several future directions in modeling and then introduce a potential approach to developing a systems biology model describing TCE-induced autoimmune hepatitis.

Although the modeling of the source, emission, and transport of TCE (Brusseau et al. 2007, 2012; Asher et al. 2007; Chambon et al. 2010, 2011; Johnson et al. 2003; Pederson et al. 2001; Atteia and Höhener 2010; Reynolds and Kueper 2001; Oostrom et al. 1999; McKone 1996; Clement et al. 1998; Cohen and Ryan 1985) are important considerations in an overall chemical risk assessment, those modeling aspects and approaches are not covered here. Moreover, the references cited in the following sections are not intended to be comprehensive, but are representative of the body of work in each modeling category. Many additional references are available in the scientific literature and in the excellent comprehensive risk assessment (Environmental Protection Agency 2001) and Integrated Risk Information System (IRIS) (Environmental Protection Agency 2011) for TCE.

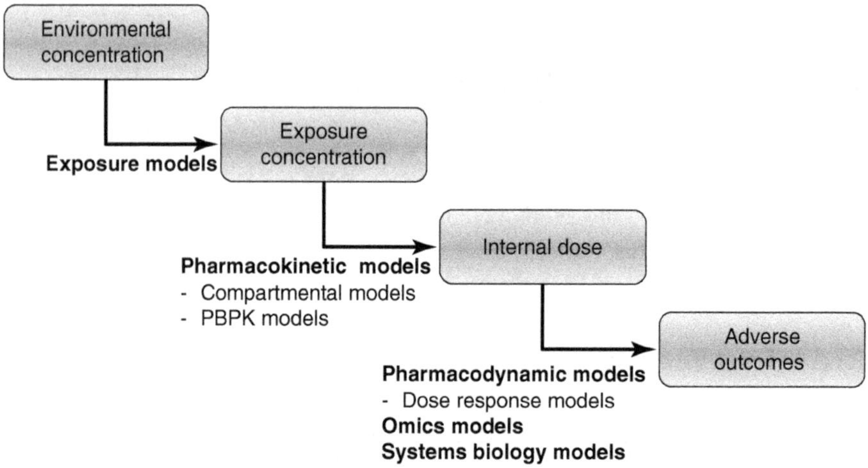

Fig. 11.1 Computational models relevant at each stage along the exposure-dose-effect continuum

11.2 Exposure Models

Exposure is the contact between a contaminant or pollutant and an individual or population through environmental media, such as air, water, soil, dust, and food. The modeling of exposure focuses on the prediction and quantitative description of the spatial and temporal characteristics of this contact. Thus, exposure modeling can be used to help inform our understanding of how the properties of chemical contaminants and the media and pathways in which they move affect pollutant exposure, and can provide quantitative measures of this exposure that can then be used to estimate dose and other metrics useful for risk assessment.

A number of computational models have been developed and utilized to simulate exposure to TCE through various environmental media. In particular, several studies have employed computational models to assess the impact of the transport and release of TCE from source zones above and below the water table into indoor air. Vapor intrusion, in which TCE vapors move from contaminated groundwater and soil into the indoor air of overlying buildings, is an significant route of environmental exposure (Office of Solid Waste and Emergency Response 2012). Yu and coworkers (2009a) used the multi-phase compositional model CompFlow Bio (Yu et al. 2009b) to simulate the transport of TCE into the indoor air of residential dwellings from a dense non-aqueous phase liquid (DNAPL) source zone located below the water table. Of interest in these studies was the role of heterogeneity in the subsurface permeability structure of the aquifer and a determination of the relative importance of variability in factors such as source zone location and pressure drop within the dwelling. Through model simulations, these investigators found that the simulated indoor air concentrations of TCE were extremely sensitive to assumptions made about the aquifer heterogeneity and that pressure fluctuations in the soil gas beneath the foundation slab had a significant effect on these contaminant concentrations. Motivated by the desire to simulate indoor air concentrations of TCE in houses with locations that were offset from the groundwater plume flow, Wang et al. (2012) extended the above analysis of Yu et al. to include a fully three-dimensional geometry. Following a similar set of simulation and sensitivity studies using CompFlow Bio (Yu et al. 2009b), these researchers determined that houses that are laterally offset from the groundwater plume are less affected by vapor intrusion than those located directly above the plume. They also noted that characterizing the site stratigraphy is a first-order priority when attempting to assess the impact of the fate and transport of TCE from an observed source zone to the indoor air.

Although the above studies examined exposure to TCE through multiple environmental media, other studies have focused on exposure to this pollutant through air alone. In particular, to better understand and quantify non-occupational exposure to TCE and other VOCs, Sexton et al. (2007) developed a new modeling approach to estimate the concentrations of these pollutants in five relevant microenvironments: indoors at home, indoors at work/school, indoors in other locations, outdoors in any location, and in transit. Employing hierarchical Bayesian techniques, they predicted that concentrations would be highest in "other" indoor microenvironments, intermediate in the indoor work/school and residential microenvironments, and lowest in the outside and in-transit microenvironments. Based on a series of comparisons with biomonitoring measurements, they found that predicted concentrations of all VOCs examined were in reasonable agreement with experimental median concentrations in the indoor residential microenvironment. They further suggested that since personal monitoring is often impractical in many situations, their modeling approach would be a promising alternative for estimating VOC concentrations in seldom monitored microenvironments.

In the context of general biomonitoring of TCE exposure through the air, Droz and Fernández (1978) used their previously developed pharmacokinetic model (Fernández et al. 1977) to investigate the impact of biomarker selection and sample collection timing on predicted exposure. Based on results of systematic model simulations, they found that for maximum usefulness, sampling and analysis of alveolar air for TCE and trichloroethanol (TCOH) must be carried out at least 6 h after the end of the exposure. In contrast, they noted that the timing for the collection and analysis of urine for trichloroacetic acid (TCA) was unimportant, but also suggested that this biomarker is of limited value for the biological monitoring of exposure because of its lack of sensitivity as an indicator of a single exposure to TCE.

Another area in which the use of mathematical models has provided quantitative information and insights into the risks of TCE is the estimation of exposure to this chemical at or near hazardous waste sites. Using a combination of computational modeling techniques, Maslia et al. (1996) conducted a health assessment for the Gratuity Road site in the town of Groton, Massachusetts, which had been contaminated with TCE and several other environmental pollutants. These researchers first used an environmental transport model to create a spatial and temporal mapping of pollutant concentrations and flow in the region. They next used these contaminant levels to carry out a computational analysis of probable exposures routes and levels. Based on these studies, Maslia and coworkers concluded that (i) predicted groundwater concentrations of TCE in the area typically exceeded the US Environmental Protection Agency (EPA) value for the maximum contaminant level (MCL) for TCE, (ii) despite direct remediation of the waste site, historical contamination can cause nearby populations to experience significant exposure, and (iii) because the predicted exposure to TCE through inhalation during showering was nearly identical to that through ingestion of contaminated domestic water, both of these routes should be considered when conducting exposure analyses of contamination from VOCs such as TCE.

Using a different exposure modeling methodology, Johnston and Gibson (2011) estimated residential indoor air concentrations of TCE and perchloroethylene (PCE) resulting from plumes of groundwater contamination from the former Kelly Air Force Base in San Antonio, Texas. For this study, the authors developed a probabilistic exposure model, based on the Johnson-Ettinger algorithm (Johnson and Ettinger 1991), and compared predicted results with measurements taken in a subset of homes in the affected area. From these comparisons, they noted that the model systemically underestimated high exposures, but that the 95th percentile of the predicted value would be a more useful indicator of the risk. An overall analysis of simulation results and sampling data led these researchers to conclude that homes above the contaminant groundwater plume surrounding the Kelly Air Force Base are still at risk of vapor intrusion and that the probabilistic approach used in their model could better identify priority areas for further sampling than current deterministic approaches.

11.3 Pharmacokinetic Models

Pharmacokinetics is the study of the absorption, distribution, metabolism, and excretion (ADME) of xenobiotics. Pharmacokinetics (PK) is frequently referred to as "what the body does to the chemical". Models of pharmacokinetics are often designed to answer questions like "Given a dose of a chemical, where does it go in the body, what is the time-course blood or tissue concentration of the parent chemical and/or its metabolites, and how quickly is it metabolized and eliminated?". PK models are also used to compute derived quantities, such as the chemical clearance, area under the curve (AUC), and half-life ($t_{1/2}$). In the context of toxicant exposure, pharmacokinetics are often referred to as toxicokinetics. Pharmacokinetic modeling is critical in the field of toxicology because it allows investigators to predict time-dependent quantities that are highly relevant in assessing chemical toxicity: biodistribution, internal dose, and clearance.

There are several types of pharmacokinetic models that have been created and used to predict chemical disposition:

(i) non-compartmental pharmacokinetic (NCPK) models: These models are useful for the estimation of certain PK parameters, such as area under the curve (AUC), and half-life ($t_{1/2}$). NCPK approaches use mathematical and statistical techniques to derived these parameters using a minimal amount of experimental data (typically, chemical levels in the blood or plasma over time). However, since they do not contain mechanistic underpinnings, such models are not useful for any type of extrapolation. Non-compartmental models have been applied extensively for drugs, but their utility in toxicology is limited and hence are not discussed here.

(ii) compartmental pharmacokinetic (CPK) models: By "lumping" major tissues, organs, or regions of the body together, these models treat the body as one or more compartments comprising "apparent volumes of distribution" (Shen 2007) (conventionally, the dose administered divided by the resultant plasma concentration). CPK models typically require data on concentration of the chemical species over time in the blood and use that information to estimate certain kinetic parameters related to transport and elimination within and between the compartments. In general, like NCPK models, CPK models are not useful outside of the datasets for which they have been developed and will not be useful for extrapolations across individuals and doses.

(iii) physiologically-based pharmacokinetic (PBPK) models: In contrast to NCPK and CPK models, physiologically based pharmacokinetic (PBPK) models incorporate the anatomical entities and physiological and biochemical processes of organisms. Because of this, PBPK models can be used to perform inter-species, inter-route, and/or inter-dose extrapolations and to describe concentration-time profiles in individual tissues or organs and in the plasma or blood. While PBPK models can provide a wealth of information, they require extensive data for parameterization and validation, including anatomical, physiological, and biochemical data, as well as experimentally-derived time-course concentration levels in multiple tissues, ideally at varying dose levels and routes of exposure. The need for this amount and detail of information makes PBPK modeling impractical and/or overly burdensome in many situations.

11.3.1 Compartmental Pharmacokinetic Models

To date, there have been few CPK models constructed for the analysis of TCE pharmacokinetics. Nevertheless, one such model was constructed by Kim et al. (2009) to characterize and quantify the pharmacokinetics of TCE metabolites in male B6C3F1 mice exposed to TCE. Specifically, these researchers created a two-compartment model to predict the time course concentrations of TCE, TCA, dichloroacetic acid (DCA), *S*-(1,2-dichlorovinyl)-L-cysteine (DCVC), and *S*-(1,2-dichlorovinyl)glutathione (DCVG) formation, and used data acquired from a novel analytical method to calibrate and validate the model. The authors found that following calibration, model predictions agreed well with the acquired data, and through a mechanistic pathway analysis, suggested that TCE-oxide is the most likely source of the hepatotoxicant DCA. They concluded by noting that the results of their analyses could be used to reassess existing models of TCE and ultimately inform the risk assessment for this important chemical.

11.3.2 Physiologically-Based Pharmacokinetic Models

Aside from its use in predictive tissue dosimetry described above, PBPK models have been used for a large variety of applications (Reisfeld et al. 2007, 2013; Reddy et al. 2005; Lyons et al. 2008; Bois et al. 1996; Bois 2000; Hack et al. 2006; Caldwell et al. 2012), including risk assessment, development of dose metrics, biomarker characterization, regulatory review, chemical prioritization, chemical mixture toxicity assessment, uncertainty and variability analyses, and dose reconstruction.

A large number of PBPK models for TCE have been developed for virtually all of the above applications. Table 11.1 contains an extensive list of many of these models and their principal features. Some of the distinctive classes of PBPK models listed are in the areas of cancer and cancer risk assessment (Clewell et al. 1995, 2000; USAF-EPA TCE PBPK workgroup 2004; Evans et al. 2009; Chiu 2011; Cronin et al. 1995), non-cancer effects and risk assessment (Clewell et al. 1997, 2000; Barton and Clewell 2000; Simmons et al. 2005; Fisher and Allen 1993; Bushnell et al. 2005), development of acute exposure guidelines (Bruckner et al. 2004; Boyes et al. 2005), equation and data harmonization (Hack et al. 2006; USAF-EPA TCE PBPK workgroup 2004), combined pharmacokinetic/pharmacodynamic models (Clewell et al. 1997; Bushnell et al. 2005; Clewell and Andersen 1994; Simon 1997), and population effects (Bois 2000; Hack et al. 2006; Simon 1997; Sohn et al. 2004; Chiu et al. 2009).

One of the most recent and comprehensive models is that of Chiu and coworkers (2009). The modeling framework developed by these investigators comprises a PBPK model for TCE and its major metabolites (see Fig. 11.2) and uses Bayesian inference to account for population variability and experimental and model uncertainty. In developing, calibrating, and validating this model, they used data from mice, rats, and humans, and considered a wider range of physiological, chemical, *in vitro*, and *in vivo* data than any previously published analysis of TCE. Owing to the above features, this PBPK model may represent the most complete, and thoroughly parameterized and validated, PBPK model for TCE to date.

11.4 Pharmacodynamic Models

Pharmacodynamics is the study of the biochemical and physiological effects of xenobiotics and the mechanisms of their actions. Pharmacodynamics (PD) is frequently referred to as "what the chemical does to the body". Models of pharmacodynamics incorporate information about how, and to what extent, the toxicant and/or its metabolites interact with relevant biomolecules or structures (e.g., receptors, enzymes, macromolecules, membranes). These models are often designed to answer questions like "Given a concentration (internal dose) of a chemical contaminant at some site of action, what is the level of the biological response over time and how does this response depend on the internal dose?". For example, a researcher interested in understanding the carcinogenic potential of a new chemical may develop a

Table 11.1 Physiologically-based pharmacokinetic (PBPK) models for trichloroethylene

Year	Study authors	Modeling features	Reference
1987	Andersen, Gargas, Clewell, and Severyn	Simulation of gas uptake studies for TCE and 1,1-dichloroethylene (1,1-DCE) in male Fischer 344 rats using PBPK modeling	Andersen et al. (1987)
1989	Koizumi	Amalgamation of information obtained in rats and man by various routes of exposure to TCE and PCE using PBPK modeling	Koizumi (1989)
1990	Fisher, Whittaker, Taylor, Clewell, and Andersen	Prediction of TCE kinetics in a lactating rat and nursing pup using PBPK modeling	Fisher et al. (1990)
1991	Sato, Endoh, Kaneko, and Johanson	Investigation of the effect of physiological factors on the pharmacokinetic behavior of inhaled TCE	Sato et al. (1991)
1991	Staats, Fisher, and Connolly	Simulation of TCE, methylene chloride, chloroform, and dichloroethane toxicokinetics using a two-compartment description of GI absorption	Staats et al. (1991)
1993	Allen and Fisher	Prediction of TCE and TCA disposition in humans using PBPK modeling	Allen and Fisher (1993)
1993	Fisher and Allen	Simulation of gavage and inhalation bioassays with TCE using PBPK modeling and linkage with plausible dose-metrics for carcinogenesis	Fisher and Allen (1993)
1994	Clewell and Andersen	Overview of several PBPK models, including one for TCE	Clewell and Andersen (1994)
1995	Barton, Creech, Godin, Randall, and Seckel	Simulation of the pharmacokinetics of a mixture of TCE and vinyl chloride in rats using a PBPK model	Barton et al. (1995)
1995	Clewell, Gentry, Gearhart, Allen, and Andersen	Cancer risk estimation for human exposure to TCE using a PBPK model coupled with a linearized multistage model	Clewell et al. (1995)
1995	Cronin, Oswald, Shelley, Fisher, and Flemming	Risk assessment for TCE using a PBPK model coupled with a linearized multistage model to derive human carcinogenic risk extrapolations	Cronin et al. (1995)
1996	El-Masri, Constan, Ramsdell, and Yang	Investigation of an interaction threshold between TCE and 1,1-dichloroethylene in Fischer 344 rats using PBPK modeling	el-Masri et al. (1996a)
1996	El-Masri, Tessari, and Yang	Investigation of mechanism of interaction between TCE and 1,1-dichloroethylene using data from gas uptake experiments and a PBPK model	El-Masri et al. (1996b)

(continued)

Table 11.1 (continued)

Year	Study authors	Modeling features	Reference
1996	Thomas, Bigelow, Keefe, and Yang	Comparison of simulations results with existing biological exposure indices (BEIs) for six industrial solvents (TCE, benzene, chloroform, carbon tetrachloride, methylene chloride, methyl and chloroform) using PBPK and Monte Carlo modeling.	Thomas et al. (1996)
1997	Abbas and Fisher	Simulation of the pharmacokinetics of TCE and its metabolites in the B6C3F1 mouse using a six-compartment PBPK model	Abbas and Fisher (1997)
1997	Bogen and Gold	Prediction of maximum concentration level for cytotoxic end points using PBPK modeling	Bogen and Gold (1997)
1997	Clewell, Gentry, and Gearhart	Non-cancer risk assessment incorporating both mechanistic and delivered dose information using a PBPK model along with the benchmark dose method	Clewell et al. (1997)
1997	Simon	Simulation of occupational exposure to TCE using Monte Carlo population distribution sampling and PBPK modeling	Simon (1997)
1998	Fisher, Mahle, and Abbas	Prediction of blood, urine, and exhaled breath concentrations using PBPK modeling and comparison to data from human volunteers	Fisher et al. (1998)
1998	Lipscomb, Fisher, Confer, and Byczkowski	Extrapolation of TCE pharmacokinetics to humans using in vitro data and a PBPK model	Lipscomb et al. (1998)
1998	Stenner, Merdink, Fisher, and Bull	Investigation of the role of enterohepatic recirculation on the pharmacokinetics of major metabolites of TCE using PBPK modeling	Stenner et al. (1998)
1999	Greenberg, Burton, and Fisher	Prediction of the disposition of inhaled TCE for mice; PBPK model contains submodels for chloral hydrate, free and glucuronide-bound TCOH, TCA, and DCA	Greenberg et al. (1999)
2000	Bois	Estimation of both variability between experimental groups and uncertainty in toxicokinetics using Bayesian analyses of a PBPK model for rodents and humans, including	Bois (2000)
2000	Barton and Clewell	Utilization of a PBPK model within a framework for evaluation of chronic exposure limits for non-cancer effects	Barton and Clewell (2000)
2000	Clewell, Gentry, Covington, and Gearhart	Prediction of the kinetics of TCE, TCOH, and TCA, in the mouse, rat, and human using a PBPK model, for both oral and inhalation exposure; dose metrics provided for cancer risk assessment	Clewell et al. (2000)

Table 11.1 (continued)

Year	Study authors	Modeling features	Reference
2002	Albanese, Banks, Evans, and Potter	Investigation of TCE pharmacokinetics in adipose tissue using three different PBPK models	Albanese et al. (2002)
2002	Dobrev, Andersen, and Yang	Simulation of interaction thresholds for human exposure to mixtures of TCE, PCE, and methyl chloroform using PBPK modeling	Dobrev et al. (2002)
2002	Hissink, Bogaards, Freidig, Commandeur, Vereulen, and van Bladeren	Risk assessment for TCE using *in vitro* metabolic parameters and PBPK modeling	Hissink et al. (2002)
2002	Simmons, Boyes, Bushnell, Raymer, Limsakun, McDonald, Sey, and Evans	Evaluation of neurotoxicity data aided by the development of a PBPK specifically for the Long Evans rat	Simmons et al. (2002)
2003	Keys, Bruckner, Muralidhara, and Fisher	Expansion and extensive tissue dosimetry validation of rodent PBPK models for TCE exposure	Keys et al. (2003)
2004	Bruckner, Keys, and Fisher	Estimation of acute exposure guideline levels based on PBPK model predictions of time course concentrations for TCE in the blood and/or brain of rats and humans	Bruckner et al. (2004)
2004	Clewell and Andersen	Estimate target tissue doses for the three principal animal tumors associated with TCE exposure (liver, lung, and kidney) using PBPK modeling	Clewell and Andersen (2004)
2004	Isaacs, Evans, and Harris	Investigation of the mechanism of metabolic interactions during simultaneous exposures to TCE and chloroform using a PBPK model incorporating mixed enzyme inhibition	Isaacs et al. (2004)
2004	Sohn, McKone, and Blancato	Identification of some of the difficulties in reconstructing population-scale exposures when using Bayesian inference and PBPK models	Sohn et al. (2004)
2004	USAF-EPA TCE PBPK workgroup	Prediction of the kinetics of TCE, TCOH, and TCA, in the mouse, rat, and human, for both oral and inhalation exposure; dose metrics provided for cancer risk assessment	USAF-EPA TCE PBPK workgroup (2004)
2005	Beliveau and Krishnan	Simulation of the pharmacokinetics of inhaled TCE and other VOCs in humans using a spreadsheet-based PBPK model, and the estimation of its parameters based on quantitative structure-property relationships (QSPRs)	Béliveau and Krishnan (2005)

(continued)

Table 11.1 (continued)

Year	Study authors	Modeling features	Reference
2005	Boyes, Evans, Eklund, Janssen, and Simmons	Development of acute exposure guideline level recommendations for various exposure durations and levels of severity using arterial blood concentrations predicted using a PBPK model	Boyes et al. (2005)
2005	Bushnell, Shafer, Bale, Boyes, Simmons, Eklund, and Jackson	Prediction of the neurotoxicity of TCE and other VOCs using an exposure–dose–response (EDR) model comprising a PBPK model linked to a toxicodynamic component	Bushnell et al. (2005)
2005	Simmons, Evans, and Boyes	Determination of dose metrics predictive of the acute neurotoxic effects of TCE using PBPK modeling	Simmons et al. (2005)
2006	Hack, Chiu, Jay Zhao, and Clewell	Population analysis of a harmonized PBPK model for TCE using Bayesian inference	Hack et al. (2006)
2006	Haddad, Tardif, and Tardif	Characterization of the influence of different routes of exposure to volatile organic chemicals present in drinking water using PBPK models for trihalomethanes and TCE	Haddad et al. (2006)
2007	Liao, Tan, and Clewell	Estimation of exposures to volatile organic compounds that correspond to levels measured in fluids and/or tissues using a generic PBPK model coupled with exposure pattern characterization, Monte Carlo analysis, and quantitative structure property relationships	Liao et al. (2007)
2007	Rodriquez, Mahle, Gearhart, Mattie, Lipscomb, Cook, and Barton	Prediction of age-appropriate pharmacokinetics of TCE, PCE, benzene, chloroform, methylene chloride, or methyl ethyl ketone in the rat utilizing physiologically based pharmacokinetic modeling	Rodriguez et al. (2007)
2007	Yokley and Evans	Evaluation and comparison of two alternative PBPK models for TCE based on parameter sensitivity analyses	Yokley and Evans (2007)
2008	Easterling, Evans, and Kenyon	Comparison of SimuSolv and MATLAB for PBPK modeling of TCE	Easterling et al. (2000)
2008	Li, Schultz, Keys, Campbell, and Fisher	Prediction of dichloroacetic acid (DCA) biotransformation and kinetics in humans administered DCA by intravenous infusion and oral ingestion using PBPK modeling	Li et al. (2008)
2009	Chiu, Okino, and Evans	Development of a comprehensive, Bayesian, PBPK model-based analysis of the population toxicokinetics of TCE and its metabolites in mice, rats, and humans, considering a wider range of physiological, chemical, in vitro, and in vivo data than any previously published analysis of TCE	Chiu et al. (2009)

Table 11.1 (continued)

Year	Study authors	Modeling features	Reference
2009	Evans, Chiu, Okino, and Caldwell	Investigation of the role of TCA in the liver in TCE-induced hepatomegaly in mice using PBPK modeling and exposure data for TCE, TCA, and DCA.	Evans et al. (2009)
2010	Chen, Shih, and Wu	Estimation of inhalation exposure to TCE using a PBPK model based on repeated measurements in venous blood along with a hierarchical Bayesian approach	Chen et al. (2010)
2010	Csanády, Göen, Klein, Drexler, and Filser	Development of a two-compartment PBPK model to simulate the disposition of the TCE metabolite TCA based on concentration of inhaled TCE in humans	Csanády et al. (2010)
2011	Chiu	Analysis of the role of TCA in hepatomegaly by PBPK modeling that incorporates non-linear changes in internal TCA dose due to dose-dependent fractional absorption	Chiu (2011)
2011	Price and Krishnan	Prediction of the inhalation toxicokinetics of chemicals in a mixture containing TCE using an integrated QSAR-PBPK modeling approach	Price and Krishnan (2011)
2011	Valcke and Krishnan	Assessment of the impact of exposure route on the human kinetic adjustment factor used in non-cancer risk assessment using a multi-route PBPK model appropriate TCE and several other VOCs	Valcke and Krishnan (2011)
2012	Chen, Shih, and Wu	Reconstruction of exposure to TCE using a physiologically based toxicokinetic model with cumulative amount of metabolite in urine	Chen et al. (2012)
2012	Mumtaz, Ray, Crowell, Keys, Fisher, and Ruiz	Evaluation of minimal risk levels for volatile organic compounds, including TCE, using a generic seven-compartment PBPK model	Mumtaz et al. (2012)

PD model to predict DNA adduct levels as a function of the internal dose of this compound. In the context of toxicant exposure, pharmacodynamics are often referred to as toxicodynamics.

As described below, pharmacodynamic models for TCE have focused on predictions for both cancer and non-cancer endpoints. In addition, a number of studies related to TCE risk assessment have utilized pharmacodynamic models coupled to pharmacokinetic components (Clewell et al. 1997; Bushnell et al. 2005; Clewell and Andersen 1994; Simon 1997); these studies are not described in this section, but were discussed previously (*vide supra*) and/or are listed in the PBPK model table (Table 11.1).

To delineate and quantify the effects of TCE on oxidative stress in the liver, Byczkowski and coworkers (1999) developed a biologically based pharmacodynamic (BBPD) model. Focusing on chemically-induced lipid peroxidation (a process associated with nephrotoxicity (Cojocel et al. 1989), autoimmune diseases

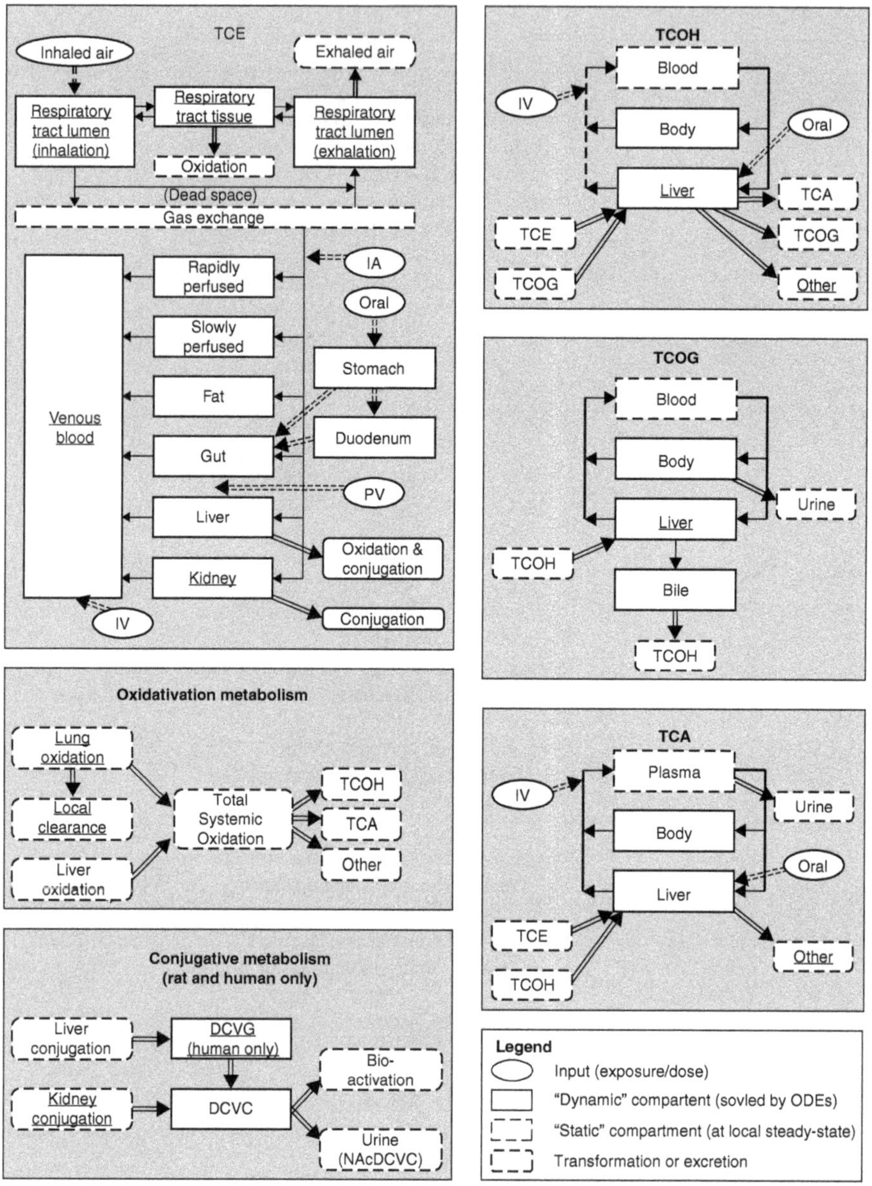

Fig. 11.2 Structural diagram for the comprehensive PBPK model for TCE and its metabolites developed by Chiu et al. (2009) (Adapted with permission)

(Wang et al. 2007), and other adverse health effects (Hu et al. 2008)), they updated a previous mathematical model (Byczkowski et al. 1996) to describe the kinetics and dose response induced by TCE *in vivo*. This model had several unknown parameters that were determined experimentally using an *in vitro* system in which precision-cut mouse liver slices were exposed to TCE vapors and the lipid peroxidation was quantified using an assay for thiobarbituric acid reactive substances. Through a series of simulations and comparisons to available data, these researchers concluded that their BBPD model adequately described both the PK and PD of TCE-induced lipid peroxidation. When fully validated, we anticipate that models such as this one will have the potential to provide researchers with tools to evaluate and quantify the effect of TCE dose on oxidative stress in the liver.

Unlike the non-cancer endpoint study just described, the work of Chen (2000) focused on developing a dose-response model for liver tumors induced by TCE. This biologically-based dose-response (BBDR) model was constructed using the general approach of Cohen and Ellwein (1990) and the stochastic models of Chen and Farland (1991) and Tan and Chen (1995). According to the study author, the utility of such a model could be to quantitatively describe TCE, DCA, and TCA bioassay results, clarify the role of these compounds on tumor induction, and evaluate how interactions among these chemicals could potentially impact low-dose extrapolation. By comparing model simulations and literature data on tumor incidence, Chen demonstrated that DCA could be responsible for most of the tumor response found in TCE and TCA bioassays. Aside from this important result, the author clarified the importance of biological assumptions on low-dose risk estimates, and emphasized the need for more flexible BBDR models and further laboratory studies to clarify the biological processes underlying dose-response relationships for TCE.

11.5 Quantitative Structure Activity Relationships

Quantitative structure activity relationships (QSARs) are mathematical models that link the structural characteristics of a chemical with its chemical or biological activity (Hansch and Leo 1995). When QSARs are used for property predictions, they are often called quantitative structure property relationships (QSPRs). The structural characteristics, or descriptors, are generally electronic, geometrical, topological, or constitutional properties of the molecule, while the biological activity is typically a physicochemical property of the molecule or some appropriate toxicological/pharmacological endpoint. For example, suppose that a researcher is interested in determining a measure of acute toxicity (LD50) for a large family of chemical congeners. One approach would be to experimentally determine this value for each chemical. This could be quite laborious, and each new chemical of interest would have to be tested. Another approach would be to (i) determine the value of LD50 for only certain of the congeners, (ii) identify easily-calculated chemical properties of the congeners that are good predictors (descriptors) of LD50 [say lipophilicity (logP), molar refractivity (MR), H-bonding acceptor ability (HBA), and H-bonding donor ability (HBD)], and (iii) create a mathematical correlation to predict LD50 based on these descriptors, e.g., $LD50=\alpha^*\log P+\beta^*MR+\chi^*HBA+\delta^*HBD$, where α, β, χ, δ would be determined from the experimental data obtained for the limited set of congeners. If the chosen descriptors and correlation form were appropriate, this equation could be used to predict the unknown values of LD50 for all of the remaining congeners.

QSAR models have been developed in several areas related to TCE pharmacokinetics and pharmacodynamics. In particular, for pharmacokinetic applications QSARs have been used to estimate a number of essential physicochemical parameters, such as the partition coefficient (PC). The PC, which depends on the properties of both the chemical and the tissue, is the ratio of the equilibrium concentration of the chemical in the tissue to that in the blood or plasma. Payne and Kenny (2002) examined a number of QSAR equations for calculating blood-air, tissue-air, or tissue-blood partition coefficients of TCE and other volatile organic chemicals in human and rat tissues. By comparing the predictions from several published empirical, non-empirical (tissue composition-based), and semi-empirical equations for tissue-air and tissue-blood PCs in humans and rats, they concluded that (i) some of the model equations could be used to estimate human blood–air PCs, but that predictions for the rat (for which chemical binding with blood proteins was significant) were not well predicted by any of the equations, (ii) tissue–blood PCs were most accurately estimated for most chemicals by empirical equations, and (iii) no single choice of model equation was best under all circumstances and that the appropriate choice will depend on the chemical, tissue, and species of interest.

Another study involving the application of QSARs to PK analyses was conducted by Price and Krishnan (2011), who developed an integrated QSAR-PBPK modeling approach to predict the inhalation toxicokinetics of chemicals in TCE-containing mixtures. One of the major aims of the study was to use QSARs to estimate many of the model parameters for which experimental studies were usually required. In particular, the authors determined PCs and kinetic parameters for metabolism (V_{max} and K_m) based solely on chemical structure using a group contribution approach. They then used these estimated parameters within an interaction-based PBPK models to predict the ADME of chemicals in mixtures of up to ten components. Despite some apparent inaccuracies in the parameter estimates, the study authors concluded that their integrated modeling methodology was useful for initial assessments of the pharmacokinetics of components within chemical mixtures.

A limited number of studies have employed QSAR modeling to analyze and characterize the pharmacodynamics associated with TCE exposure. One such study was conducted by Niederlehner et al. (1998) who developed QSAR models to predict the response of the daphnid *Ceriodaphnia dubia* to six widely used industrial chemicals, including TCE. In particular, these investigators developed QSARs to relate relevant endpoints [lethal concentration (LC50) at 2 days and reproductive impairment (reproductive IC50)] with the octanol-water partition coefficient (Kow) for the chemical or chemical mixture of interest. The authors also constructed a QSAR to predict the toxicity of the applied chemical mixture as a function of the mixture composition. Based on the study results, they determined that the QSARs developed seemed consistent with those created by other investigators for other species of daphnid, and that while a predictive dose-additive relationship overestimated toxicity for the chemical mixtures, the fitted (QSAR) models were more consistent with the observed results.

Aside from predictions of endpoints related to acute toxicity, QSARs have been constructed and applied for the prediction of the genotoxicity of many chemicals (Worth et al. 2013). For example, Parry et al. (1996) utilized QSAR modeling to analyze the chromosome malsegregation in *Aspergillus nidulans* by a structurally-related series of halogenated hydrocarbons, including TCE. To develop the QSARs, these researchers correlated three endpoints of interest [induction of chromosome malsegregation leading to aneuploidy (LEC), arrest of mitosis (ARR), and lethality (D37)] to two chemical descriptors [the molar refractivity (MR) and the difference between the highest occupied molecular orbital and the lowest unoccupied molecular orbital (DIFF)] using a set of training compounds. Following this parameterization, they used these QSARs to predict the activities of an unrelated test set of congeneric chemicals. Based on these and other validations, the study authors concluded that the models developed were highly effective in their ability to predict the activity of previously untested chemicals, but also noted that the potential to use this QSAR-based approach to predict the activity of aneugenic chemicals in higher organisms is presently unknown.

11.6 Future Directions

There are a number of scientific, economic, and societal factors motivating a transformation in chemical risk assessment from one that relies heavily on data generated through the dosing of experimental animal, to one in which virtually all routine toxicity testing would be conducted *in vitro* by evaluating the response of human cells or cell lines in a series of high-throughput, toxicity pathway assays (National Research Council 2007). A key element in enabling such a transformation will be the development and use of computational modeling tools in the fields of "omics" and systems biology to help organize, analyze, integrate, and augment these assay data (Raunio 2011).

11.6.1 "Omics" Models

Omics refers to the scientific disciplines and collective technologies involved in analyzing the roles and actions of molecules within various cellular "omes", such as the genome, proteome, and metabolome (Mayer 2011). Computational models in this field seek to organize experimental omics data, simulate interactions within and between components of the system, help to decipher relevant biology, and predict outcomes of perturbations to the system.

Genomic modeling focusses on developing and using computational tools and methods to understand and interpret genome sequences, including such diverse techniques as phylogenetic analysis, biosequence analysis, and gene expression data analysis (Koonin 2001; Luscombe et al. 2001). These complex approaches are data intensive and can benefit from the structure provided by modeling. For example, biosequence analysis examines the structure or function of DNA, RNA or peptide sequences in order to answer questions about sequence homology, regulatory elements, single nucleotide polymorphisms (SNPs) and other features. The methods used for sequence analysis are quite diverse, but all generate large amounts of data. A genome-wide association study to examine SNPs can generate one billion genotypes. Modeling can help researchers to organize, synthesize, analyze, and interpret this vast array of diverse data.

Although few genomic models have been developed in the context of TCE exposure, one such study was conducted by Kim et al. (2011), who conducted a large-scale gene expression analysis on animals exposed to TCE and two other VOCs (dichloromethane and ethylbenzene). A principal aim of this study was to determine if characteristic molecular signatures could be derived for each toxicant from gene expression profiles. Through the use of gene expression analysis, the study investigators were able to find such molecular signatures and identify many genes that could be used to discriminate between VOC-exposed animals and healthy controls. The authors concluded that such expression signatures could be used as surrogate markers for detecting and characterizing biological responses to VOC exposure in the environment.

Metabolomic modeling centers on describing and quantifying the metabolic pathways and spatially- and temporally-varying inventory of metabolites in cells, tissues, organs, or organisms, and linking this information to specific disease states or toxic insults. To date, the only model of this type for TCE is that of Mayeno et al. (2005), who developed a computer-based simulation tool, BioTRaNS (biochemical tool for reaction network simulation), that predicts metabolites from exposure to multiple chemicals and interconnects their metabolic pathways. In this study, the investigators used TCE and three other common drinking water pollutants (PERC, methyl chloroform, and chloroform) as test cases, and through a combination of simulations, discovered new interconnected metabolic pathways and previously-unreported metabolites, predicted reactive intermediates, such as epoxides and acid chlorides, and uncovered points in the metabolic pathways where typical endogenous compounds, such as glutathione or carbon dioxide, were consumed or generated. Example predicted metabolic pathways for a mixture of the four test chemicals using a simplified set of reaction rules are shown in Fig. 11.3; for results using more complete reaction rules, see the original publication (Mayeno et al. 2005). Aside from the results obtained in this particular study, the study authors suggested that BioTRaNS has the potential to aid in risk assessment and provide new and important insights into metabolites and the interrelationship between diverse chemicals that may remained unnoticed through experimentation alone.

11.6.2 Systems Biology Models

In contrast to omics, systems biology centers on an *integration* of data from multiple levels of complexity *across* "omes" into a "systems view" of biological and pathological processes. In the field of toxicology, systems biology frequently

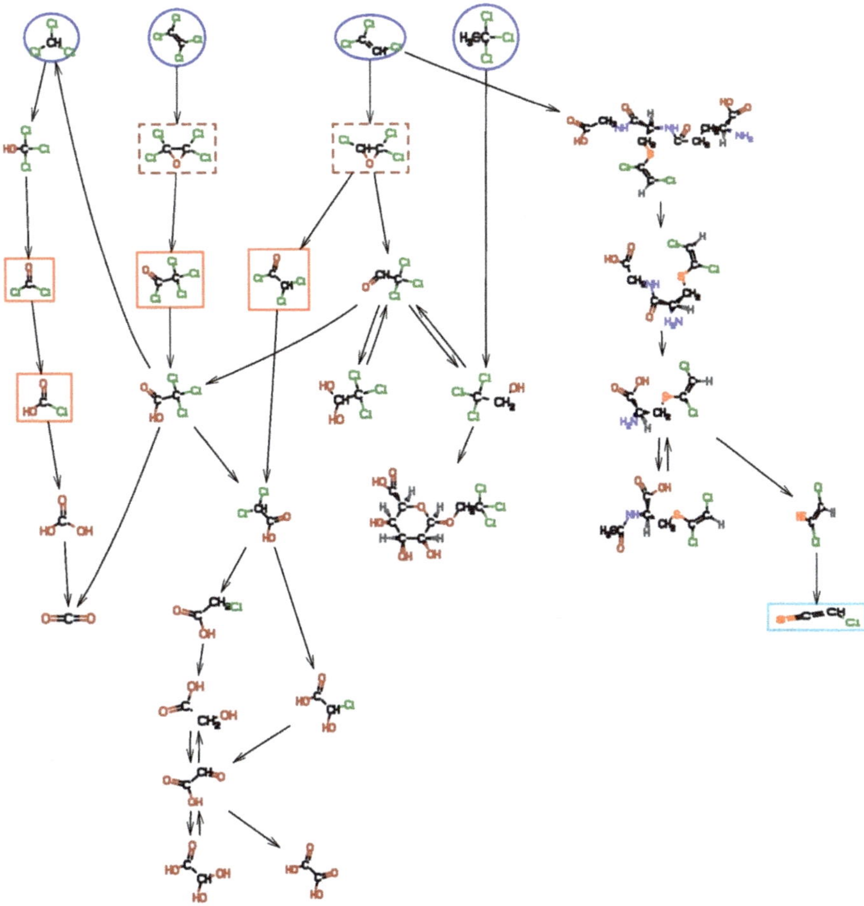

Fig. 11.3 BioTRaNS-generated biotransformation pathways for a mixture of trichloroethylene, PCE, methyl chloroform, and chloroform. Reactive metabolites are highlighted as follows: epoxides (*brown*, *box*, *dashed*); acid chlorides (*orange*, *box*, *solid*); thioketene (*turquoise*, *box*, *solid*); starting chemicals (*blue*, *ellipse*, *solid*)

involves an analysis of how xenobiotic-induced perturbations in gene and protein expression are linked to toxicological outcomes. The goal of systems biology modeling is to create holistic computational models of the functioning of the cell, multicellular systems, and ultimately the organism. These *in silico* models have the potential to elucidate linkages within and across the exposure-dose-effect continuum and may provide virtual test systems for evaluating the toxic responses of cells, tissues, and organisms.

Systems biology models and approaches for TCE are uncommon, though one such study was undertaken by Pleil (2009), who used a holistic approach and conceptual pathway model to begin to characterize and quantify the relationship between environmental exposures and human disease. By analyzing data obtained in several exposure studies focused on TCE and methyl tertiary butyl ether, he determined the relative roles of contaminant concentration level, biological media (breath or blood), and the contaminant type on the variability of biomarker measurements. As a result of these analyses, Pleil found that the observed variance in biomarkers depended more on the variability in exposures than on interindividual differences in internal biological parameters, and suggested that in the longer term, such a systems biology approach has the potential to inform the assessment of susceptibility ranges along many relevant toxicological pathways.

11.7 Example: Modeling of TCE-induced Autoimmunity

Exposure to TCE has been found to trigger or exacerbate autoimmune responses and/or autoimmune diseases. Chemically-induced autoimmunity (Bigazzi 1988) is a complex process, involving, *inter alia*, exposure to the chemical, its absorption, distribution, metabolism, and elimination, interactions of the parent chemical and/or its metabolites with biological targets, epigenetic and other cellular alterations, and an immune response. Each of these elements, in itself, is an intricate process.

Although the role of TCE in inducing autoimmune disease has been qualitatively investigated and described in a number of references (Cooney and Gilbert 2012; Gilbert 2010; Gilbert et al. 2009; Cooper et al. 2009) including sections of this book, mathematical models describing the pathogenesis of TCE-induced autoimmunity are lacking. Mathematical modeling can be beneficial in a number of ways: for example, in testing hypotheses and gaining insights into the mechanism of the disease process, such as critical events leading to the disease, the time course of molecular and cellular processes during disease progression, the relative importance of processes and cell types involved. Moreover, once a model has been validated, its application may facilitate (a) reduced number of animals required in testing and more efficient experimental designs, (b) improved and personalized treatment regimens, as well as disease prevention, and (c) better prediction of the sequelae and/or prognosis of a disease. An excellent introduction to mathematical modeling of biological systems is presented by de Pillis and Radunskaya (2012).

Here, we illustrate an example of how to approach a first-generation model for TCE-induced autoimmunity (TAI). Before proceeding to develop a model, we should first identify the goals of the model, i.e., what questions do we want answered? For instance, suppose we wish to know if the magnitude of TCE-induced autoimmune hepatitis (AIH) could be estimated based on a quantitative measure of a biomarker, in a biological fluid such as blood or urine. In this case, the model should be focused on the liver (target organ), hepatitis (endpoint), and biomarkers in blood and urine (predictor variables). A literature search revealed no previous models of TCE-AIH, although other autoimmune diseases and processes have been modeled. Often, if the exact system of interest has not been modeled, models describing analogous or related systems should be examined to consider whether the methodologies and approaches used for those systems can be adapted or serve as starting points.

Next, to develop this predictive model, an understanding of postulated and known mode-of-action of TCE-induced hepatitis would be necessary, as well as related aspects such as the ADME of TCE. If we are examining hepatitis of the liver, why is ADME important? It is because of what we wish the model to predict: here, the model must link biomarkers in blood and urine to pathologic features in the liver. Moreover, knowledge of the proposed pathogenetic mechanism of the disease indicates the importance of ADME: specifically, (a) TCE is transported via blood to the liver, metabolized, and eliminated, in a time-dependent fashion; (b) metabolites of TCE are believed to trigger or contribute to the disease onset, and (c) the biomarkers of interest are those in blood and urine.

The most straight-forward biomarkers would be TCE itself and its metabolites, in blood and urine. However, as hepatic concentrations are likely to more relevant than the concentrations in the blood, an important modeling aspect would be to relate blood or urine concentrations to those in the liver. This can be accomplished through the use of PBPK modeling (described above). Metabolism of TCE to reactive intermediates is mediated by enzymes, such as the cytochrome P450 2E1 enzyme, which show inter-individual variability and are chemically inducible. Adducts formed between reactive metabolites and biological targets, as well as perturbations to cellular processes (e.g., oxidative stress and consequent products like aldehydes) may contribute to the initiation, progression, and behavior of the autoimmune response, a process mediated by specific immune cell types and cell-signaling proteins such as cytokines. Further, TCE exposure has been associated with epigenetic alterations which may modulate the immune response. All of these processes involve temporal and spatial aspects. Thus, monitoring the appearance and disappearance of these specific liver events over time and relating them to biomarkers through mathematical models may lead to the discovery of biomarkers that can be used to better understand and predict disease progression. Further, a time-dependent evaluation of specific events within specific immune cell populations may also provide further insights. Thus, the experimental data should include "longitudinal" time points, collected over the course of the study.

A first-generation conceptual model for TCE-induced AIH is shown in Fig. 11.4. Note that the conceptual model does not include all known or proposed processes related to TCE-AIH pathogenesis; instead, only certain key steps are selected at this stage, in accordance to the principle of parsimony. To simulate the tissue distribution of TCE and other biomarkers, a PBPK model is coupled to the AIH model [step (0)]. Often, a PBPK model is linked with PD model to simulate the experimentally observed dynamic processes. Given a conceptual model, a mathematical model can be readily derived by writing an equation for each step. Specifically, the mathematical representation, corresponding to the beginning steps of the conceptual model in Fig. 11.4, is as follows:

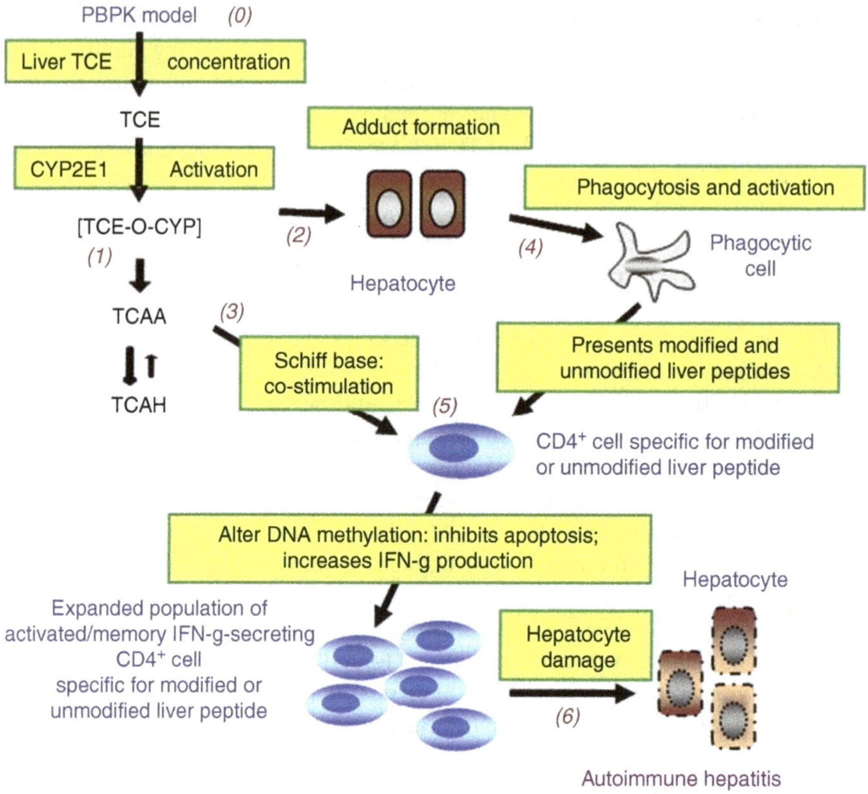

Fig. 11.4 Preliminary conceptual model TCE-induced autoimmune hepatitis

Step (0): PBPK Model: organ/tissue specific [TCE] over time

Step (1): $R_{\text{met},i} = \dfrac{\mathrm{d}M_i}{\mathrm{d}t} = f_{i,\text{met}} \times \dfrac{V_{\max,\text{met}}[\text{TCE}]}{K_{\text{m}} + [\text{TCE}]}$

Step (2): $\dfrac{\mathrm{d}A_{\text{add}}}{\mathrm{d}t} = k_{\text{add}} R_{\text{met},a} - \dfrac{v_{\max,a} A_{\text{add}}}{K_{\text{rep},a} + A_{\text{add}}} + f_2$

where

$R_{\text{met},i}$ = rate of TCE metabolism, where i = adducts or TCAH;

M_i = metabolite i;

t = time;

$f_{i,\text{met}}$ = fraction of metabolites that are of type i;

$V_{\max,\text{met}}$ = maximum rate of formation of metabolites;

K_m = Michaelis constant for metabolite formation;

[TCE] = TCE concentration at the site of metabolism;

A_{add} = amount of adduct formed;

k_{add} = rate constant for adduct formation;

$v_{\max,a}$ = maximum repair capacity due to damage by adduct formation;

$K_{\text{rep},a}$ = "half-saturation constant" for repair capacity; and

f_2 = addition term to be added later during model refinement.

Although TCE and its metabolites are the most straight-forward biomarkers, other biomarkers may better correlate TCE-AIH disease pathology. The selection of the biomarkers to be examined should be based on knowledge of the pathogenetic mechanism. The experimental work required to identify predictive biomarkers could be a major effort.

Finally, once a first generation model has been developed, the model can be further enhanced to allow for the investigation of a variety of other relevant research questions, such as the details of the mechanism of disease induction, the contribution of different immune cell types, and genetic predisposition to TCE-AIH, just to list a few potential applications.

Acknowledgements The authors thank Dr. Arthur N. Mayeno for his contribution to the materials contained in this chapter.

References

Abbas R, Fisher JW (1997) A physiologically based pharmacokinetic model for trichloroethylene and its metabolites, chloral hydrate, trichloroacetate, dichloroacetate, trichloroethanol, and trichloroethanol glucuronide in B6C3F1 mice. Toxicol Appl Pharmacol 147:15–30

Albanese RA, Banks HT, Evans MV, Potter LK (2002) Physiologically based pharmacokinetic models for the transport of trichloroethylene in adipose tissue. Bull Math Biol 64:97–131

Allen BC, Fisher JW (1993) Pharmacokinetic modeling of trichloroethylene and trichloroacetic acid in humans. Risk Anal 13:71–86

Andersen ME, Gargas ML, Clewell HJ, Severyn KM (1987) Quantitative evaluation of the metabolic interactions between trichloroethylene and 1,1-dichloroethylene in vivo using gas uptake methods. Toxicol Appl Pharmacol 89:149–157

Asher WE, Luo W, Campo KW, Bender DA, Robinson KW, Zogorski JS, Pankow JF (2007) Application of a source apportionment model in consideration of volatile organic compounds in an urban stream. Environ Toxicol Chem 26:1606–1613

Atteia O, Höhener P (2010) Semianalytical model predicting transfer of volatile pollutants from groundwater to the soil surface. Environ Sci Technol 44:6228–6232

Barton HA, Clewell HJ (2000) Evaluating noncancer effects of trichloroethylene: dosimetry, mode of action, and risk assessment. Environ Health Perspect 108(Suppl):323–334

Barton HA, Creech JR, Godin CS, Randall GM, Seckel CS (1995) Chloroethylene mixtures: pharmacokinetic modeling and in vitro metabolism of vinyl chloride, trichloroethylene, and trans-1,2-dichloroethylene in rat. Toxicol Appl Pharmacol 130:237–247

Béliveau M, Krishnan K (2005) A spreadsheet program for modeling quantitative structure-pharmacokinetic relationships for inhaled volatile organics in humans. SAR QSAR Environ Res 16:63–77

Bigazzi PE (1988) Autoimmunity induced by chemicals. J Toxicol Clin Toxicol 26:125–156
Bogen KT, Gold LS (1997) Trichloroethylene cancer risk: simplified calculation of PBPK-based MCLs for cytotoxic end points. Regul Toxicol Pharmacol 25:26–42
Bois FY (2000) Statistical analysis of Clewell et al. PBPK model of trichloroethylene kinetics. Environ Health Perspect 108(Suppl):307–316
Bois F, Gelman A, Jiang J, Maszle D, Zeise LG (1996) Population toxicokinetics of tetrachloroethylene. Arch Toxicol 70:347–355
Boyes WK, Evans MV, Eklund C, Janssen P, Simmons JE (2005) Duration adjustment of acute exposure guideline level values for trichloroethylene using a physiologically-based pharmacokinetic model. Risk Anal 25:677–686
Bruckner JV, Keys DA, Fisher JW (2004) The Acute Exposure Guideline Level (AEGL) program: applications of physiologically based pharmacokinetic modeling. J Toxicol Environ Health A 67:621–634
Brusseau ML, Nelson NT, Zhang Z, Blue JE, Rohrer J, Allen T (2007) Source-zone characterization of a chlorinated-solvent contaminated Superfund site in Tucson, AZ. J Contam Hydrol 90: 21–40
Brusseau ML, Russo AE, Schnaar G (2012) Nonideal transport of contaminants in heterogeneous porous media: 9 – impact of contact time on desorption and elution tailing. Chemosphere 89:287–292
Bushnell PJ, Shafer TJ, Bale AS, Boyes WK, Simmons JE, Eklund C, Jackson TL (2005) Developing an exposure-dose-response model for the acute neurotoxicity of organic solvents: overview and progress on in vitro models and dosimetry. Environ Toxicol Pharmacol 19: 607–614
Byczkowski JZ, Channel SR, Pravecek TL, Miller CR (1996) Mathematical model for chemically induced lipid peroxidation in precision-cut liver slices: computer simulation and experimental calibration. Comput Methods Programs Biomed 50:73–84
Byczkowski JZ, Channel SR, Miller CR (1999) A biologically based pharmacodynamic model for lipid peroxidation stimulated by trichloroethylene in vitro. J Biochem Mol Toxicol 13: 205–214
Caldwell JC, Evans MV, Krishnan K (2012) Cutting edge PBPK models and analyses: providing the basis for future modeling efforts and bridges to emerging toxicology paradigms. J Toxicol 2012:852384
Chambon JC, Broholm MM, Binning PJ, Bjerg PL (2010) Modeling multi-component transport and enhanced anaerobic dechlorination processes in a single fracture-clay matrix system. J Contam Hydrol 112:77–90
Chambon JC, Binning PJ, Jørgensen PR, Bjerg PL (2011) A risk assessment tool for contaminated sites in low-permeability fractured media. J Contam Hydrol 124:82–98
Chen CW (2000) Biologically based dose-response model for liver tumors induced by trichloroethylene. Environ Health Perspect 108(Suppl):335–342
Chen C, Farland W (1991) Incorporating cell proliferation in quantitative cancer risk assessment approaches, issues, and uncertainties. In: Butterworth B, Slaga T, Farland W, McClain RM (eds) Chemically induced cell proliferation: implication for risk assessment. John Wiley & Sons, New York
Chen C-C, Shih M-C, Wu K-Y (2010) Exposure estimation using repeated blood concentration measurements. Stochastic Environ Res Risk Assess 24:445–454
Chen C-C, Shih M-C, Wu K-Y (2012) Exposure reconstruction using a physiologically based toxicokinetic model with cumulative amount of metabolite in urine: a case study of trichloroethylene inhalation. Stochastic Environ Res Risk Assess 26:21–31
Chiu WA (2011) Trichloroacetic acid: updated estimates of its bioavailability and its contribution to trichloroethylene-induced mouse hepatomegaly. Toxicology 285:114–125
Chiu WA, Okino MS, Evans MV (2009) Characterizing uncertainty and population variability in the toxicokinetics of trichloroethylene and metabolites in mice, rats, and humans using an updated database, physiologically based pharmacokinetic (PBPK) model, and Bayesian approach. Toxicol Appl Pharmacol 241:36–60

Clement TP, Sun Y, Hooker BS, Petersen JN (1998) Modeling multispecies reactive transport in ground water. Ground Water Monit Remediation 18:79–92

Clewell HJ, Andersen ME (1994) Physiologically-based pharmacokinetic modeling and bioactivation of xenobiotics. Toxicol Ind Health 10:1–24

Clewell HJ, Andersen ME (2004) Applying mode-of-action and pharmacokinetic considerations in contemporary cancer risk assessments: an example with trichloroethylene. Crit Rev Toxicol 34:385–445

Clewell HJ, Gentry PR, Gearhart JM, Allen BC, Andersen ME (1995) Considering pharmacokinetic and mechanistic information in cancer risk assessments for environmental contaminants: examples with vinyl chloride and trichloroethylene. Chemosphere 31:2561–2578

Clewell HJ, Gentry PR, Gearhart JM (1997) Investigation of the potential impact of benchmark dose and pharmacokinetic modeling in noncancer risk assessment. J Toxicol Environ Health 52:475–515

Clewell HJ, Gentry PR, Covington TR, Gearhart JM (2000) Development of a physiologically based pharmacokinetic model of trichloroethylene and its metabolites for use in risk assessment. Environ Health Perspect 108(Suppl):283–305

Cohen SM, Ellwein LB (1990) Proliferative and genotoxic cellular effects in 2-acetylaminofluorene bladder and liver carcinogenesis: biological modeling of the ED01 study. Toxicol Appl Pharmacol 104:79–93

Cohen Y, Ryan PA (1985) Multimedia modeling of environmental transport: trichloroethylene test case. Environ Sci Technol 19:412–417

Cojocel C, Beuter W, Müller W, Mayer D (1989) Lipid peroxidation: a possible mechanism of trichloroethylene-induced nephrotoxicity. Toxicology 55:131–141

Cooney C, Gilbert KM (2012) Toxicology, epigenetics, and autoimmunity. Toxicology and epigenetics. Wiley, Cooney and Gilbert (2012): West Sussex, United Kingdom, p 688

Cooper GS, Makris SL, Nietert PJ, Jinot J (2009) Evidence of autoimmune-related effects of trichloroethylene exposure from studies in mice and humans. Environ Health Perspect 117:696–702

Cronin WJ, Oswald EJ, Shelley ML, Fisher JW, Flemming CD (1995) A trichloroethylene risk assessment using a Monte Carlo analysis of parameter uncertainty in conjunction with physiologically-based pharmacokinetic modeling. Risk Anal 15:555–565

Csanády GA, Göen T, Klein D, Drexler H, Filser JG (2010) Trichloroacetic acid in urine as biological exposure equivalent for low exposure concentrations of trichloroethene. Arch Toxicol 84:897–902

De Pillis LG, Radunskaya AE (2012) Best practices in mathematical modeling. Methods Mol Biol 929:51–74

Dobrev ID, Andersen ME, Yang RSH (2002) In silico toxicology: simulating interaction thresholds for human exposure to mixtures of trichloroethylene, tetrachloroethylene, and 1,1,1-trichloroethane. Environ Health Perspect 110:1031–1039

Droz PO, Fernández JG (1978) Trichloroethylene exposure. Biological monitoring by breath and urine analyses. Br J Ind Med 35:35–42

Easterling M, Evans M, Kenyon E (2000) Comparative analysis of software for physiologically based pharmacokinetic modeling: simulation, optimization, and sensitivity analysis. Toxicol Mech Methods 10:203–229

el-Masri HA, Constan AA, Ramsdell HS, Yang RS (1996a) Physiologically based pharmacodynamic modeling of an interaction threshold between trichloroethylene and 1,1-dichloroethylene in Fischer 344 rats. Toxicol Appl Pharmacol 141:124–132

El-Masri HA, Tessari JD, Yang RS (1996b) Exploration of an interaction threshold for the joint toxicity of trichloroethylene and 1,1-dichloroethylene: utilization of a PBPK model. Arch Toxicol 70:527–539

U.S. Environmental Protection Agency, (2001) Health assessment document for trichloroethylene synthesis and characterization. Washington, DC

U.S. Environmental Protection Agency (2011) IRIS toxicological review of trichloroethylene. Washington, DC

Evans MV, Chiu WA, Okino MS, Caldwell JC (2009) Development of an updated PBPK model for trichloroethylene and metabolites in mice, and its application to discern the role of oxidative metabolism in TCE-induced hepatomegaly. Toxicol Appl Pharmacol 236:329–340

Fernández JG, Droz PO, Humbert BE, Caperos JR (1977) Trichloroethylene exposure. Simulation of uptake, excretion, and metabolism using a mathematical model. Br J Ind Med 34:43–55

Fisher JW, Allen BC (1993) Evaluating the risk of liver cancer in humans exposed to trichloroethylene using physiological models. Risk Anal 13:87–95

Fisher JW, Whittaker TA, Taylor DH, Clewell HJ, Andersen ME (1990) Physiologically based pharmacokinetic modeling of the lactating rat and nursing pup: a multiroute exposure model for trichloroethylene and its metabolite, trichloroacetic acid. Toxicol Appl Pharmacol 102:497–513

Fisher JW, Mahle D, Abbas R (1998) A human physiologically based pharmacokinetic model for trichloroethylene and its metabolites, trichloroacetic acid and free trichloroethanol. Toxicol Appl Pharmacol 152:339–359

Gilbert KM (2010) Xenobiotic exposure and autoimmune hepatitis. Hepat Res Treat 2010(248157)

Gilbert KM, Przybyla B, Pumford NR, Han T, Fuscoe J, Schnackenberg LK, Holland RD, Doss JC, Macmillan-Crow LA, Blossom SJ (2009) Delineating liver events in trichloroethylene-induced autoimmune hepatitis. Chem Res Toxicol 22:626–632

Greenberg MS, Burton GA, Fisher JW (1999) Physiologically based pharmacokinetic modeling of inhaled trichloroethylene and its oxidative metabolites in B6C3F1 mice. Toxicol Appl Pharmacol 154:264–278

Hack CE, Chiu WA, Jay Zhao Q, Clewell HJ (2006) Bayesian population analysis of a harmonized physiologically based pharmacokinetic model of trichloroethylene and its metabolites. Regul Toxicol Pharmacol 46:63–83

Haddad S, Tardif G-C, Tardif R (2006) Development of physiologically based toxicokinetic models for improving the human indoor exposure assessment to water contaminants: trichloroethylene and trihalomethanes. J Toxicol Environ Health A 69:2095–2136

Hansch C, Leo A (1995) Exploring QSAR, vol 1, Fundamentals and applications in chemistry and biology. American Chemical Society, Washington, DC

Hissink EM, Bogaards JJP, Freidig AP, Commandeur JNM, Vermeulen NPE, Van Bladeren PJ (2002) The use of in vitro metabolic parameters and physiologically based pharmacokinetic (PBPK) modeling to explore the risk assessment of trichloroethylene. Environ Toxicol Pharmacol 11:259–271

Hu C, Jiang L, Geng C, Zhang X, Cao J, Zhong L (2008) Possible involvement of oxidative stress in trichloroethylene-induced genotoxicity in human HepG2 cells. Mutat Res 652:88–94

Isaacs KK, Evans MV, Harris TR (2004) Visualization-based analysis for a mixed-inhibition binary PBPK model: determination of inhibition mechanism. J Pharmacokinet Pharmacodyn 31:215–242

Johnson PC, Ettinger RA (1991) Heuristic model for predicting the intrusion rate of contaminant vapors into buildings. Environ Sci Technol 25:1445–1452

Johnson GR, Gupta K, Putz DK, Hu Q, Brusseau ML (2003) The effect of local-scale physical heterogeneity and nonlinear, rate-limited sorption/desorption on contaminant transport in porous media. J Contam Hydrol 64:35–58

Johnston JE, Gibson JM (2011) Probabilistic approach to estimating indoor air concentrations of chlorinated volatile organic compounds from contaminated groundwater: a case study in San Antonio, Texas. Environ Sci Technol 45:1007–1013

Keys DA, Bruckner JV, Muralidhara S, Fisher JW (2003) Tissue dosimetry expansion and cross-validation of rat and mouse physiologically based pharmacokinetic models for trichloroethylene. Toxicol Sci 76:35–50

Kim S, Kim D, Pollack GM, Collins LB, Rusyn I (2009) Pharmacokinetic analysis of trichloroethylene metabolism in male B6C3F1 mice: formation and disposition of trichloroacetic acid, dichloroacetic acid, S-(1,2-dichlorovinyl)glutathione and S-(1,2-dichlorovinyl)-L-cysteine. Toxicol Appl Pharmacol 238:90–99

Kim JK, Jung KH, Noh JH, Eun JW, Bae HJ, Xie HJ, Jang J-J, Ryu JC, Park WS, Lee JY, Nam SW (2011) Identification of characteristic molecular signature for volatile organic compounds in peripheral blood of rat. Toxicol Appl Pharmacol 250:162–169

Koizumi A (1989) Potential of physiologically based pharmacokinetics to amalgamate kinetic data of trichloroethylene and tetrachloroethylene obtained in rats and man. Br J Ind Med 46:239–249

Koonin EV (2001) Computational genomics. Curr Biol 11:R155–R158

Li T, Schultz I, Keys DA, Campbell JL, Fisher JW (2008) Quantitative evaluation of dichloroacetic acid kinetics in human–a physiologically based pharmacokinetic modeling investigation. Toxicology 245:35–48

Liao KH, Tan Y-M, Clewell HJ (2007) Development of a screening approach to interpret human biomonitoring data on volatile organic compounds: reverse dosimetry on biomonitoring data for trichloroethylene. Risk Anal 27:1223–1236

Lipscomb JC, Fisher JW, Confer PD, Byczkowski JZ (1998) In vitro to in vivo extrapolation for trichloroethylene metabolism in humans. Toxicol Appl Pharmacol 152:376–387

Luscombe NM, Greenbaum D, Gerstein M (2001) What is bioinformatics? A proposed definition and overview of the field. Methods Inf Med 40:346–358

Lyons MA, Yang RSH, Mayeno AN, Reisfeld B (2008) Computational toxicology of chloroform: reverse dosimetry using Bayesian inference, Markov chain Monte Carlo simulation, and human biomonitoring data. Environ Health Perspect 116:1040–1046

Maslia ML, Aral MM, Williams RC, Williams-Fleetwood S, Hayes LC, Wilder LC (1996) Use of computational models to reconstruct and predict trichloroethylene exposure. Toxicol Ind Health 12:139–152

Mayeno AN, Yang RSH, Reisfeld B (2005) Biochemical reaction network modeling: predicting metabolism of organic chemical mixtures. Environ Sci Technol 39:5363–5371

Mayer B (2011) Bioinformatics for Omics data. Humana Press, Clifton

McKone TE (1996) Alternative modeling approaches for contaminant fate in soils: uncertainty, variability, and reliability. Reliability Eng Syst Saf 54:165–181

Mumtaz MM, Ray M, Crowell SR, Keys D, Fisher J, Ruiz P (2012) Translational research to develop a human PBPK models tool kit-volatile organic compounds (VOCs). J Toxicol Environ Health A 75:6–24

National Center for Environmental Assessment (2011) Exposure factors handbook. Washington, DC

National Research Council (2007) Toxicity testing in the 21st century: a vision and a strategy. National Academies Press, Washington, DC

Niederlehner BR, Cairns J, Smith EP (1998) Modeling acute and chronic toxicity of nonpolar narcotic chemicals and mixtures to Ceriodaphnia dubia. Ecotoxicol Environ Saf 39:136–146

Office of Solid Waste and Emergency Response (2012) EPA's vapor intrusion database: evaluation and characterization of attenuation factors for chlorinated volatile organic compounds and residential buildings. U.S. Environmental Protection Agency, Office of Solid Waste and Emergency Response, Washington, DC

Oostrom M, Hofstee C, Walker RC, Dane JH (1999) Movement and remediation of trichloroethylene in a saturated heterogeneous porous medium. J Contam Hydrol 37:159–178

Parry JM, Parry EM, Bourner R, Doherty A, Ellard S, O'Donovan J, Hoebee B, De Stoppelaar JM, Mohn GR, Onfelt A, Renglin A, Schultz N, Söderpalm-Berndes C, Jensen KG, Kirsch-Volders M, Elhajouji A, Van Hummelen P, Degrassi F, Antoccia A, Cimini D, Izzo M, Tanzarella C, Adler ID, Kliesch U, Hess P (1996) The detection and evaluation of aneugenic chemicals. Mutat Res 353:11–46

Payne MP, Kenny LC (2002) Comparison of models for the estimation of biological partition coefficients. J Toxicol Environ Health A 65:897–931

Pederson BM, Thibodeaux LJ, Valsaraj KT, Reible DD (2001) Testing a multimedia compartmental model with monitoring data. Environ Toxicol Chem 20:2114–2121

Pleil JD (2009) Influence of systems biology response and environmental exposure level on between-subject variability in breath and blood biomarkers. Biomarkers 14:560–571

Price K, Krishnan K (2011) An integrated QSAR-PBPK modelling approach for predicting the inhalation toxicokinetics of mixtures of volatile organic chemicals in the rat. SAR QSAR Environ Res 22:107–128

Raunio H (2011) In silico toxicology – non-testing methods. Front Pharmacol 2:33
Reddy M, Yang RS, Andersen ME III, Clewel HJ (2005) Physiologically based pharmacokinetic modeling: science and applications. Wiley-Interscience, Reddy et al. (2005): Hoboken, New Jersey
Reisfeld B, Mayeno A (2012a) Computational toxicology, vol II, Methods in molecular biology. Humana Press, Clifton
Reisfeld B, Mayeno A (2012b) Computational toxicology, vol I, Methods in molecular biology. Humana Press, Clifton
Reisfeld B, Mayeno AN, Lyons M, Yang RSH (2007) Physiologically-based pharmacokinetic/pharmacodynamic modeling. In: Ekins S (ed) Computational toxicology: risk assessment for pharmaceutical and environmental chemicals. John Wiley & Sons, Hoboken
Reisfeld B, Ivy JH, Lyons M, Wright J, Rogers J, Mayeno AN (2013) DoseSim: a tool for pharmacokinetic/pharmacodynamic analysis and dose reconstruction. Bioinformatics 29:400–401
Reynolds DA, Kueper BH (2001) Multiphase flow and transport in fractured clay/sand sequences. J Contam Hydrol 51:41–62
Rodriguez CE, Mahle DA, Gearhart JM, Mattie DR, Lipscomb JC, Cook RS, Barton HA (2007) Predicting age-appropriate pharmacokinetics of six volatile organic compounds in the rat utilizing physiologically based pharmacokinetic modeling. Toxicol Sci 98:43–56
Sato A, Endoh K, Kaneko T, Johanson G (1991) A simulation study of physiological factors affecting pharmacokinetic behaviour of organic solvent vapours. Br J Ind Med 48:342–347
Sexton K, Mongin SJ, Adgate JL, Pratt GC, Ramachandran G, Stock TH, Morandi MT (2007) Estimating volatile organic compound concentrations in selected microenvironments using time-activity and personal exposure data. J Toxicol Environ Health A 70:465–476
Shen D (2007) Toxicokinetics. In: Klaassen CD (ed) Casarett & Doull's toxicology: the basic science of poisons. McGraw-Hill Professional, Shen (2007): NY, New York, p 1280
Simmons JE, Boyes WK, Bushnell PJ, Raymer JH, Limsakun T, McDonald A, Sey YM, Evans MV (2002) A physiologically based pharmacokinetic model for trichloroethylene in the male long-evans rat. Toxicol Sci 69:3–15
Simmons JE, Evans MV, Boyes WK (2005) Moving from external exposure concentration to internal dose: duration extrapolation based on physiologically based pharmacokinetic derived estimates of internal dose. J Toxicol Environ Health A 68:927–950
Simon TW (1997) Combining physiologically based pharmacokinetic modeling with Monte Carlo simulation to derive an acute inhalation guidance value for trichloroethylene. Regul Toxicol Pharmacol 26:257–270
Sohn MD, McKone TE, Blancato JN (2004) Reconstructing population exposures from dose biomarkers: inhalation of trichloroethylene (TCE) as a case study. J Expo Anal Environ Epidemiol 14:204–213
Staats DA, Fisher JW, Connolly RB (1991) Gastrointestinal absorption of xenobiotics in physiologically based pharmacokinetic models. A two-compartment description. Drug Metab Dispos 19:144–148
Stenner RD, Merdink JL, Fisher JW, Bull RJ (1998) Physiologically-based pharmacokinetic model for trichloroethylene considering enterohepatic recirculation of major metabolites. Risk Anal 18:261–269
Thomas RS, Bigelow PL, Keefe TJ, Yang RS (1996) Variability in biological exposure indices using physiologically based pharmacokinetic modeling and Monte Carlo simulation. Am Ind Hyg Assoc J 57:23–32
USAF-EPA TCE PBPK workgroup (2004) Development of a physiologically-based pharmacokinetic model of trichloroethylene and its metabolites for use in risk assessment. Report number AFRL-HE-WP-TR-2006-0049, Air Force Research Laboratory, Wright-Patterson AFB, Ohio
Valcke M, Krishnan K (2011) Evaluation of the impact of the exposure route on the human kinetic adjustment factor. Regul Toxicol Pharmacol 59:258–269
Wang G, Cai P, Ansari GAS, Khan MF (2007) Oxidative and nitrosative stress in trichloroethene-mediated autoimmune response. Toxicology 229:186–193

Wang X, Unger AJA, Parker BL (2012) Simulating an exclusion zone for vapour intrusion of TCE from groundwater into indoor air. J Contam Hydrol 140–141:124–138

Worth AP, Lapenna S, Serafimova R (2013) QSAR and metabolic assessment tools in the assessment of genotoxicity. Methods Mol Biol (Clifton, N.J.) 930:125–162

Yokley KA, Evans MV (2007) An example of model structure differences using sensitivity analyses in physiologically based pharmacokinetic models of trichloroethylene in humans. Bull Math Biol 69:2591–2625

Yu S, Unger AJA, Parker B (2009a) Simulating the fate and transport of TCE from groundwater to indoor air. J Contam Hydrol 107:140–161

Yu S, Freitas JG, Unger AJA, Barker JF, Chatzis J (2009b) Simulating the evolution of an ethanol and gasoline source zone within the capillary fringe. J Contam Hydrol 105:1–17

Tan W, Chen C (1995) A nonhomogeneous stochastic model of carcinogenesis for assessing risk of environmental agents. In: Axelrod D, Kimmel M, Arino O (eds) Mathematical population dynamics. Wuerz Publishing, Winnipeg

About the Editors

Kathleen M. Gilbert, Ph.D., is Professor of Microbiology and Immunology at the University of Arkansas for Medical Sciences, Little Rock, AR. She is an immunotoxicologist who is funded by the National Institute of Environmental Health Sciences to study environmental pollutants such as trichloroethylene that alter the immune system in a manner that promotes autoimmune disease. Dr. Gilbert has directed the Arkansas Center for Environmental Exposure Research (ACEER) since its inception in 2002. ACEER uses a variety of cutting-edge tools (i.e. transcriptomics, metabolomics, mathematical modeling) to assess toxicant-induced autoimmunity at both the whole-organism and molecular level. She recently served on a panel to review the EPA TSCA Workplan Chemical Risk Assessment for Trichloroethylene.

Sarah J. Blossom is Associate Professor of Pediatrics at the University of Arkansas for Medical Sciences, Little Rock, AR, whose research focus is to study the impact of trichloroethylene-induced maternal and early life oxidative stress/inflammatory dysfunction as it relates to neurodevelopmental disorders, autoimmunity, and congenital heart defects. Dr. Blossom's work is funded by the National Institutes of Health, the Center for Disease Control, and the Arkansas Biosciences Institute. Blossom is currently Associate Center Director for the Arkansas Center for Environmental Exposure Research (ACEER), where she works with Dr. Gilbert to foster key collaborations across the state and nation with the common goal of studying both biological pathways of toxicants (mechanisms, health effects, treatment strategies) as well as non-biomedical pathways (remediation, mathematical modeling, and novel detection devices) in an effort to diminish the impact of pollutants on human health and the environment.

K.M. Gilbert, S.J. Blossom (eds.), *Trichloroethylene: Toxicity and Health Risks*, Molecular and Integrative Toxicology, DOI 10.1007/978-1-4471-6311-4,

Index

K.M. Gilbert, S.J. Blossom (eds.), *Trichloroethylene: Toxicity and Health Risks*, Molecular and Integrative Toxicology, DOI 10.1007/978-1-4471-6311-4,

GPSR Compliance
The European Union's (EU) General Product Safety Regulation (GPSR) is a set of rules that requires consumer products to be safe and our obligations to ensure this.

If you have any concerns about our products, you can contact us on

ProductSafety@springernature.com

In case Publisher is established outside the EU, the EU authorized representative is:

Springer Nature Customer Service Center GmbH
Europaplatz 3
69115 Heidelberg, Germany

www.ingramcontent.com/pod-product-compliance
Lightning Source LLC
Chambersburg PA
CBHW060247290126
38964CB00003B/15
* 9 7 8 1 4 4 7 1 6 3 1 0 7 *